Circular and Transformative Economy

The main aim of this book is to illustrate circular models for sustainable resource management. It highlights the benefits of transformative approaches in integrating, simplifying, and facilitating understanding of complex systems and transforming systems towards greater sustainability while achieving multiple social, economic, and environmental outcomes. It provides pathways towards strategic policy decisions on socio-economic transformation supported by case studies.

Features:

- Discusses exploration of a transitional path to the circular economy, explored from the point of view of waste and technology.
- Explains transformational change towards sustainable socio-ecological interactions.
- Reviews provision of pathways towards sustainability through scenario development.
- Provides assessment of progress towards Sustainable Development Goals.
- Presents cross-sectoral and multicentric approaches towards circularity.

This book is aimed at researchers and professionals in water and environmental engineering, circular economy, sustainability, and environmental studies.

Africa Circular Economy Series

CRC Press (Taylor and Francis Group) introduces the **Africa Circular Economy Series** (with a Science and Technology outlook but preferably more interdisciplinary). Under this initiative, we invite scholars, academicians, researchers, and professionals to contribute to this series. We are interested in discussing monographs, references, textbooks, short-form books and handbooks to add to our book programs at undergraduate, postgraduate, and doctoral levels. All the books published under this series will be part of our Global Publishing program.

Circular and Transformative Economy: Advances towards Sustainable Socio-economic Transformation

Edited by Luxon Nhamo, Sylvester Mpandeli, Stanley Liphadzi and Tafadzwanashe Mabhaudhi

Circular and Transformative Economy
Advances towards Sustainable Socio-economic Transformation

Edited by
Luxon Nhamo, Sylvester Mpandeli,
Stanley Liphadzi and Tafadzwanashe Mabhaudhi

CRC Press
Taylor & Francis Group
Boca Raton London New York

CRC Press is an imprint of the
Taylor & Francis Group, an **informa** business

Designed cover image: Anja Van Der Merwe

First edition published 2024
by CRC Press
2385 NW Executive Center Drive, Suite 320, Boca Raton FL 33431

and by CRC Press
4 Park Square, Milton Park, Abingdon, Oxon, OX14 4RN

CRC Press is an imprint of Taylor & Francis Group, LLC

© 2024 selection and editorial matter, Luxon Nhamo, Sylvester Mpandeli, Stanley Liphadzi and Tafadzwanashe Mabhaudhi; individual chapters, the contributors

Funded by University of KwaZulu-Natal, Durban, South Africa.

Library of Congress Cataloging-in-Publication Data
Names: Nhamo, Luxon, editor. | Mpandeli, Sylvester, editor. | Liphadzi, Stanley, editor.
Title: Circular and transformative economy : advances towards sustainable socio-economic transformation / edited by Luxon Nhamo, Sylvester Mpandeli, Stanley Liphadzi and Tafadzwanashe Mabhaudhi.
Description: 1 edition. | Boca Raton, FL : CRC Press, 2024. | Series: Africa circular economy series | Includes bibliographical references and index.
Identifiers: LCCN 2023050302 (print) | LCCN 2023050303 (ebook) | ISBN 9781032356013 (hardback) | ISBN 9781032356037 (paperback) | ISBN 9781003327615 (ebook)
Subjects: LCSH: Circular economy. | Sustainable development. | Environmental engineering. | Water-supply—Environmental aspects.
Classification: LCC HC79.E5 C57 2024 (print) | LCC HC79.E5 (ebook) |
DDC 338.9/27—dc23/eng/20231120
LC record available at https://lccn.loc.gov/2023050302
LC ebook record available at https://lccn.loc.gov/2023050303

ISBN: 9781032356013 (hbk)
ISBN: 9781032356037 (pbk)
ISBN: 9781003327615 (ebk)

DOI: 10.1201/9781003327615

Typeset in Times
by codeMantra

Contents

Going Circular: A Foreword

The grand challenges our growing populations face, including climate change, environmental degradation, resource depletion and scarcity, and worsening poverty and inequality, among others, are intricately linked to our current linear economic model, which is based on a 'take-make-consume-waste' system that aggravates resource depletion as much as environmental degradation. It exacerbates climate change as it does not consider a more energy-conscious life-cycle maximisation but builds on continuous production and consumption, making landfills a key contributor to global methane emissions. The linear economy will lead to severe resource insecurity and potential global conflicts if allowed to continue, threatening sustainable development, particularly in the nexus of soaring population growth, diminishing resources and environmental damage as it is eminent in many low- and middle-income countries.

Today, humans use as many ecological resources as if we live on 1.75 Earths, thus exceeding our planetary boundaries (Global Footprint Network, 2023). The fundamental question that needs to be addressed is how the globe can meet the growing demand for resources. A **transformation towards a more circular economy** has emerged as a potential solution applying the principles of resource recovery for recycling and reuse. The circular economy model is envisioned as a systems approach that benefits businesses, society, and the environment as it unlocks new values in a resource-constrained world (Ellen MacArthur Foundation, n.d.).

A recent review by the Water Research Commission of South Africa (WRC) showed that the circular economy has gained momentum since 2018. However, much of this progression has been driven by the Global North (Naidoo et al. 2021). Research on the circular economy from the Global South, especially in Africa, is still in its infancy. This extends to related regulations, incentives, and other required policy responses to facilitate private sector support for resource recovery (Lazurko et al. 2018). The WRC review also showed that despite the circular economy being an integrated approach, energy received the most attention. Embedding the circular economy within a more polycentric and transformative approach, such as the water-energy-food nexus, might help counter this bias (Naidoo et al. 2021).

Indeed, with Africa's urban population set to double in the next 30 years, African cities will require greater volumes of water and food and, in turn, generate more waste. Ensuring a secure and healthy food supply to African cities and managing its waste flows in a safe way that preserves the continent's food cultures, environment, and rich biodiversity will be one of the major challenges in the decades ahead. The good news is that positive examples of circular innovations are also emerging in Africa's food and agricultural sector (Ellen MacArthur Foundation, 2021).

Multi-disciplinary research is required to support this movement and identify feasible opportunities for integrating economic, social, and sustainable natural resource management outcomes to benefit overall system resilience and the Sustainable Development Goals (Otoo et al., 2016).

This book will contribute towards this target. It has come at an opportune time to present case studies and lessons learnt for a more environmentally friendly circular economy model with a particular focus on Africa. This book highlights the benefits of transformative approaches in integrating, simplifying, and facilitating the understanding of complex systems and promoting the transformation towards greater sustainability amidst the grand challenges humanity faces.

Pay Drechsel
Leader Circular Economy of the CGIAR Initiative on Resilient Cities and Senior Fellow/Advisor – Research Quality Assurance at the International Water Management Institute (IWMI), Colombo, Sri Lanka

FURTHER READING

Ellen MacArthur Foundation, 2021. Circular economy in Africa: examples and opportunities (Food and Agriculture). https://www.ellenmacarthurfoundation.org/circular-economy-in-africa/overview (accessed 22/2/2023).

Ellen MacArthur Foundation, n.d. https://archive.ellenmacarthurfoundation.org/explore/the-circular-economy-in-detail (accessed 22/2/2023).

Global Footprint Network, 2023. https://www.footprintnetwork.org/ (accessed 22/2/2023).

Lazurko, A., Drechsel, P., Hanjra, M.A., 2018. Financing resource recovery and reuse in developing and emerging economies: enabling environment, financing sources and cost recovery. Colombo, Sri Lanka: IWMI, 39p. (Resource Recovery and Reuse Series 11). doi: 10.5337/2018.220.

Naidoo, D., Nhamo, L., Lottering, S., Mpandeli, S., Liphadzi, S., Modi, A.T., Trois, C., Mabhaudhi, T., 2021. Transitional pathways towards achieving a circular economy in the water, energy, and food sectors. *Sustainability* 13, 9978. doi: 10.3390/su13179978.

Otoo, M., Drechsel, P., Danso, G., Gebrezgabher, S., Rao, K., Madurangi, G., 2016. Testing the implementation potential of resource recovery and reuse business models: from baseline surveys to feasibility studies and business plans. Colombo, Sri Lanka: IWMI, 59p. (Resource Recovery and Reuse Series 10). doi: 10.5337/2016.206.

Preface

Circular and transformative approaches have emerged as an alternative to the existing linear system, which has now reached its limits when addressing the current complex and interlinked grand challenges facing humankind. Although linear models have been beneficial for centuries, they have reached their threshold. This is because linear approaches often over-emphasise a limited set of system attributes, notably efficiency, at the expense of other aspects. Linear and sector-centric approaches inadvertently compromise resilience-building initiatives, allowing trade-offs to cascade from one sector to another. For example, the COVID-19 pandemic exposed the fragility of linear models when addressing today's interconnected challenges. The pandemic showed that focusing on one sector during a shock only aggravates the stresses in other sectors as decision-makers often view the world from a linear and silo perspective. However, the COVID-19 lockdowns resulted in job losses, increasing debt burden, company closures, and economic recessions.

The increasing complexities associated with today's interlinked grand challenges, including resource insecurity, inequality, poverty, population growth, and environmental degradation and climate change, require a paradigm shift by adopting circular and transformative approaches that are capable of addressing trade-offs and enhancing synergies, facilitating humankind to remain operating within planetary boundaries.

Although the fourth Industrial Revolution has brought considerable advances and opportunities for development, its reliance on complex, cross-cutting, and interconnected systems when delivering goods and services exposed the systems to severe disruptions and shocks of severe magnitude. This is evidenced by the disruptions being caused by climate change and pandemics in global supply chains. Sector-based or system-specific resilience initiatives are often associated with systemic risks, which emanate from strategies that lead to suboptimal efficiencies in one sector at the expense of other sectors. Transformative approaches, catalysed by the circular economy, provide the pathways to transition from linearity to circularity, providing the roadmap towards cross-sectoral sustainability and socio-economic resilience.

The chapters in this book highlight the benefits of transformative approaches in integrating, facilitating, and simplifying an understanding of complex systems and transforming them towards greater sustainability while achieving multiple social, economic, and environmental outcomes. This forms the basis for sustainable development and sound human-environmental health outcomes, enhances resilience initiatives, and informs coherent and strategic policy decisions. The premise is to provide policy- and decision-makers with a practical handbook on circular models for sustainable resource management. It provides pathways towards strategic policy decisions on socio-economic transformation. Through evidence and case studies, the overarching goal is to showcase how circular and transformative approaches (circular economy, nexus planning, just transition, one health, horizon scanning, scenario planning, and sustainable food systems) can enhance sustainable socio-economic transformation.

Luxon Nhamo, Sylvester Mpandeli, Stanley Liphadzi,
and Tafadzwanashe Mabhaudhi

About the Editors

Luxon Nhamo is a Research Manager at the Water Research Commission of South Africa (WRC) and an Honorary Research Fellow at the University of KwaZulu-Natal (UKZN), South Africa. He has over 20 years of progressive research experience in agricultural water management, environmental Geographic Information Systems (GIS) and remote sensing, water-energy-food nexus, climate change adaptation, and early warning systems.

Sylvester Mpandeli is an Executive Manager at the Water Research Commission of South Africa and an Adjunct Professor at the University of Venda. He is Vice President of the International Commission on Irrigation and Drainage (ICID) and the South African National Committee on Irrigation and Drainage (SANCID) Chairman. He is a member of the Gauteng Province Premier's Advisory Team and a South African Weather Services board member.

Stanley Liphadzi is a Group Executive Manager at the Water Research Commission (WRC) and an Adjunct Professor at the University of Venda. He leads the Research & Development Branch in the WRC in the production of new knowledge and innovation in water and sanitation. Stanley's research interest is in systems thinking for sustainable development.

Tafadzwanashe Mabhaudhi is a Professor of Climate Change, Food Systems and Health at the London School of Hygiene and Tropical Medicine. Previously, he was the Research Group Leader: Sustainable and Resilient Food Systems at the International Water Management Institute (IWMI). He holds Honorary Professor appointments at the University of KwaZulu-Natal (UKZN) and the University of Nottingham, Malaysia. He has more than ten years of research experience, translating it into policy outcomes. He has published more than 200 papers and received several awards.

Contributors

Taruvinga Badza
Water, Sanitation & Hygiene Research
& Development Centre, University
of KwaZulu-Natal (UKZN)
Pietermaritzburg, South Africa

Mcloud Kayira Chirwa
Alliance for a Green Revolution
in Africa (AGRA)
Nairobi, Kenya

Tinashe Lindel Dirwai
International Water Management
Institute (IWMI)
Pretoria, South Africa

Nosipho Dlamini
School of Engineering, University
of KwaZulu-Natal (UKZN)
Pietermaritzburg, South Africa

Protase Echessah
Alliance for a Green Revolution
in Africa (AGRA)
Nairobi, Kenya

Webster Gumindoga
Construction and Civil Engineering
Department, University of Zimbabwe
Harare, Zimbabwe

Samkelisiwe Hlophe-Ginindza
Water Research Commission (WRC)
Pretoria, South Africa

Nonhlanhla Kalebaila
Water Research Commission (WRC)
Pretoria, South Africa

Mpho Kapari
Water Research Commission (WRC)
Pretoria, South Africa

Buyisile Kholisa
Water Research Commission (WRC)
Pretoria, South Africa

Erna Kruger
Mahlathini Development Foundation
(MDF)
Pietermaritzburg, South Africa

Edward Kurwakumire
Geomatics Department, Tshwane
University of Technology
Pretoria, South Africa

Betsie le Roux
Food and Water Research (FAWR)
Pretoria, South Africa

Stanley Liphadzi
Water Research Commission (WRC)
Pretoria, South Africa

Tafadzwanashe Mabhaudhi
International Water Management
Institute (IWMI), Pretoria,
South Africa and the University
of KwaZulu-Natal (UKZN),
Pietermaritzburg, South Africa

James Magidi
Geomatics Department, Tshwane
University of Technology
Pretoria, South Africa

Betty Maimela
Mahlathini Development Foundation
  (MDF)
Pietermaritzburg, South Africa

Anabela Manhica
Alliance for the Green Revolution
  in Africa (AGRA)
Nairobi, Kenya

Everisto Mapedza
International Water Management
  Institute (IWMI)
Pretoria, South Africa

Nomvuselelo Mgwenya
TruSense Consulting Services
Pretoria, South Africa

Jennifer Molwantwa
Water Research Commission (WRC)
Pretoria, South Africa

Sylvester Mpandeli
Water Research Commission (WRC)
Pretoria, South Africa

Lindiwe Carol Mthethwa
Faculty of Education, University
  of Zululand
Richards Bay, South Africa

William Musazura
University of KwaZulu-Natal
Pietermaritzburg, South Africa

Eustina Musvoto
TruSense Consulting Services
Pretoria, South Africa

Assan Ng'ombe
Alliance for the Green Revolution
  in Africa (AGRA)
Nairobi, Kenya

Luxon Nhamo
Water Research Commission (WRC)
Pretoria, South Africa

Charles Nhemachena
Alliance for a Green Revolution
  in Africa (AGRA)
Nairobi, Kenya

Daniel Njiwa
Alliance for a Green Revolution
  in Africa (AGRA)
Nairobi, Kenya

Alfred Oduor Odindo
Water, Sanitation & Hygiene Research
  & Development Centre, University
  of KwaZulu-Natal (UKZN)
Pietermaritzburg, South Africa

Dalia Saad
School of Chemistry, Molecular
  Sciences Institute, Wits University
Johannesburg, South Africa

Aidan Senzanje
School of Engineering, University
  of KwaZulu-Natal (UKZN)
Pietermaritzburg, South Africa

Mendy Zibuyile Shozi
Water, Sanitation & Hygiene Research
  & Development Centre, University
  of KwaZulu-Natal (UKZN)
Pietermaritzburg, South Africa

Nafiisa Sobratee-Fajurally
International Water Management
  Institute (IWMI)
Pretoria, South Africa

Kristina Söderholm
Luleå University of Technology
Luleå, Sweden

Patrik Söderholm
Luleå University of Technology
Luleå, Sweden

Cuthbert Taguta
University of KwaZulu-Natal (UKZN)
Pietermaritzburg, South Africa

Attie van Niekerk
Nova Institute, and Centre for Faith and
      Community, Faculty of Theology
      and Religion, University of Pretoria
Pretoria, South Africa

John Ngoni Zvimba
Water Research Commission (WRC)
Pretoria, South Africa

Acknowledgements

The editors are most grateful to the chapter contributors, as this book would not have been possible without their commitment. It has been a delight working with these pleasant and humble subject experts. We would also like to appreciate the constructive comments of the anonymous reviewers, whose comments we used to enhance the book's quality. Last but not least, this book project would not have come to fruition without the support of the Water Research Commission of South Africa (WRC), the International Water Management Institute (IWMI), Nexus Gains Initiative of the CGIAR, the Centre on Climate Change and Planetary Health at the London School of Hygiene and Tropical Medicine, and the Centre for Transformative Agricultural and Food Systems (CTAFS), University of KwaZulu-Natal (UKZN). We are indebted to Lani van Vuuren and Mpho Kapari, who dissected every word we scribbled, and Anja Van Der Merwe, who worked on graphics.

1 Understanding circularity and transformative approaches and their role in achieving sustainability

Luxon Nhamo, Sylvester Mpandeli, Stanley Liphadzi, and Tafadzwanashe Mabhaudhi

1.1 INTRODUCTION

The increasing complexities with today's interlinked challenges related to resource insecurities, the emergence of novel infectious diseases, socio-economic decline and environmental degradation require systemic approaches that address trade-offs, enhance synergies, minimise resource depletion, and promote waste reduction while operating within the planetary boundaries (Kimani-Murage et al., 2021; Menton et al., 2020; Naidoo et al., 2021a). Today's age, which is dubbed the 4th Industrial Revolution, depends on sophisticated, cross-cutting, cross-sectoral, and interconnected systems to conveniently deliver goods and services (Nhamo and Ndlela, 2021). Although this globalisation has come with considerable technological advances and opportunities for development, it has also exposed the globe and its systems to severe disruptions and shocks, as demonstrated by climate change and pandemics which often cause disruptions in global supply chains (Magableh, 2021; Shang et al., 2021).

As in any complex system, tensions always manifest between efficiency and resilience, the ability to anticipate, absorb, recover, and adapt to unexpected disruptions (Nhamo and Ndlela, 2021). These tensions indicate the connectedness between the attributes of a system, and therefore, addressing the tensions individually is bound to exacerbate existing challenges. Therefore, sector-based or system-specific resilience interventions are often accompanied by systemic risks resulting from initiatives that lead to suboptimal efficiencies in one sector at the expense of others (Nhamo and Ndlela, 2021). Cross-sectoral challenges require cross-sectoral interventions to realise integrated and multi-centric solutions (Naidoo et al., 2021b). Therefore, transformative approaches are cross-sectoral and polycentric decision support tools capable of systematically and holistically addressing cross-sectoral challenges. This is enhanced by promoting the reuse and recycling of resources, ensuring that resources stay in use for longer periods, thus mitigating resource depletion and reducing environmental waste (Mastos et al., 2021). Therefore, transformative approaches promote circularity and contribute towards achieving Sustainable Development Goals

DOI: 10.1201/9781003327615-1

1

(SDGs), particularly Goals 11 (sustainable cities and communities) and 12 (responsible consumption and production) and the other interlinked goals.

Transformative approaches, including the circular economy, nexus planning, just transition, one health, scenario planning, strategic foresight, horizon scanning, and sustainable food systems, emphasise cross-sectoral interventions and enhance socio-economic resilience against current challenges and future shocks (Batisha, 2022). They are considered pathways towards sustainable development and are envisioned to guide the transformational change agenda by promoting resource use efficiency and addressing current cross-sectoral challenges in an integrated manner (Nhamo et al., 2021). Thus, transformative approaches are polycentric as they comprise multiple decision-making centres where each centre has substantive autonomy and is also located at varying levels (Nhamo et al., 2020; Thiel, 2017). Polycentric denotes many centres of decision-making which are formally independent of each other yet rely on each other and hence are circular and cross-sectoral (Patala et al., 2022; Thiel, 2017). Therefore, the existing interlinked, cross-sectoral, and interconnected challenges suggest an urgent need to transition from the current linear system to polycentric, cross-sectoral, integrated, and circular systems, as global and local systems now resemble stress and are over-stretched.

A key attribute of transformative approaches is that they focus on the positive interlinkages that envision creating synergies and aim to transform the socio-ecological and economic system instead of addressing single issues. This book, therefore, highlights the benefits of transformative approaches in integrating, simplifying, and facilitating the understanding of complex systems and promoting the transformation towards greater sustainability while achieving multiple social, economic, and environmental outcomes. This forms the basis for sustainable development, and sound human and environmental health outcomes, as circularity promotes resilience and informs policy on proactive interventions during a crisis or a shock. The premise is to provide policy- and decision-makers with a practical handbook on circular models for sustainable resource management. This book provides pathways towards strategic policy decisions on socio-economic transformation. Through evidence and case studies, the overarching goal is to highlight how circular and transformative approaches can enhance sustainable socio-economic transformation.

1.2   DEFINING TRANSFORMATIVE APPROACHES

Current linear approaches have reached their limits as they cannot address the interlinked grand challenges associated with contemporary global systems of globalisation and liberalisation. The traditional linear economic models are inadequate when addressing current "wicked problems" that cascade from one sector to another (Naidoo et al., 2021a). Therefore, transformative approaches are directly linked to sustainable development, and their operationalisation is envisioned to address economic and environmental challenges. Thus, they are important in promoting integrated resource management that leads to circularity and a green economy (D'Amato, 2021; Hysa et al., 2020). Transformative approaches embrace the interconnectedness and interlinkages between sustainability dimensions and address sustainable development holistically (D'Amato, 2021; Naidoo et al., 2021b), especially when global population, consumption, and pollution continue to increase. The intricate interlinkages between socio-economic and environmental issues and how they interact with humankind create trade-offs that obstruct sustainable development progress (Dörgő et al., 2018). Yet, transformative approaches enhance positive interactions and interlinkages that generate synergies, focusing on changing the socio-ecological system instead of addressing single issues.

As already alluded to, there is a need for humankind to shift from linearity to circularity. A transformative and circular approach is an open-ended and cross-sectoral process of producing, structuring, and applying solutions-oriented knowledge to provide integrated strategic policies and solutions towards sustainable development (Cornell et al., 2013; David Tàbara et al., 2019). The key characteristics of transformative approaches are integrative, inclusive, adaptive, iterative, and pluralistic (Visseren-Hamakers et al., 2021), enabling any production system's efficiency.

These attributes are increasingly making transformative approaches significant in socio-economic and environmental research and policy-making, particularly after the COVID-19 pandemic when economies started experiencing massive strains compounded by climate change–related impacts (Shang et al., 2021). Therefore, transitioning towards sustainable development should be built around transformative approaches which provide integrated strategic policies formulated around integrated solutions from a cross-sectoral perspective (Nhamo and Ndlela, 2021; Wittman et al., 2017). Therefore, transformative approaches are evolutionary, open-ended, non-linear, and based on searching, learning, and experimentation (Geels et al., 2016).

1.2.1 UNDERSTANDING LINEARITY AND CIRCULARITY

A linear model is based on the 'take-make-consume-waste' system (Figure 1.1), which often results in environmental degradation and exacerbates climate change (Hysa et al., 2020). It is an inefficient model that does not consider resource conservation in the product value chain but is characterised by discarding by-products before they are fully utilised. Therefore, the lifespan of resources is significantly reduced (Reike et al., 2018). The intensive resource consumption nature of the linear model contributes to climate change as the environment is its main victim.

On the other hand, the circular economy is a departure from the traditional linear model as it focuses on decoupling economic growth from finite resource consumption (Velenturf and Purnell, 2021). The main difference between the circular and linear economy is that the former is regenerative and restorative by design and maintains resources, products, and related materials at their highest value at all times (Jørgensen and Pedersen, 2018). Instead of the take-make-consume-waste process of the linear model, the mantra of the circular model is its 3R model, which represents the reduce-reuse-recycle process (Figure 1.1, Table 1.1). As already alluded, the circular model emphasises services rather than products for profits, unlike the case of the linear model (Reike et al., 2018).

Thus, the shift from a linear to a circular model is motivated by the need to promote sustainable economic systems that are environmentally friendly, more eco-effective, and less eco-efficient (Figure 1.2). This is particularly important as how humankind drives the economy determines the effectiveness of the initiatives implemented to

FIGURE 1.1 Difference between the linear and circular models.

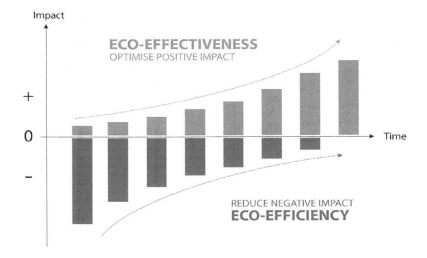

FIGURE 1.2 The difference between eco-effectiveness and eco-efficiency (Adapted from Toxopeus et al., 2015).

TABLE 1.1
Comparison between a linear and a circular economy

Attribute	Linear economy	Circular economy
Business model	Take-make-consume-waste	Reduce-reuse-recycle
Efficiency process	Eco-efficiency	Eco-effectivity
Product time frame	Short-term, focuses on profits	Long-term, multiple product values
Product life cycle	Downcycling quickly discards by-products into the environment	Upcycling, allowing products to stay long in use, regenerative and restorative
Vision	Focuses on products for profits	Focuses on the sustainability of product services

combat climate change (Velenturf and Purnell, 2021). The distinction between the circular and linear systems is in how value is created or maintained. The linear system focuses on profitability, irrespective of the product life cycle, whereas the circular system emphasises sustainability (Kirchherr et al., 2017).

Eco-effectiveness, as an attribute of the circular economy (Figure 1.2), is when residual flows are reused for a function that is the same (functional recycling) or even higher (upcycling) than the original function of the resource (Morseletto, 2020). As a result, the value is fully retained or even increased. Yet, the eco-efficient system of the linear model is characterised by downcycling, whereby a product, or part of it, is reused for a low-grade application that reduces the value of the material and makes it difficult to reuse the same material flow again (Bocken et al., 2016; Toxopeus et al., 2015).

1.3 SIGNIFICANCE OF CIRCULARITY AND TRANSFORMATIVE APPROACHES

The efforts to balance industrial development, environmental quality, social well-being, and economic growth are anticipated to achieve sustainable development by at least 2030 before the damage becomes irreversible (Mensah and Ricart Casadevall, 2019). This can only be achieved by formulating coherent and strategic policies that guide efficient resource use and low-carbon emission while at the same time promoting countries' overall economic growth. Contemporary grand challenges, including climate change, the emergence of novel infectious diseases, environmental degradation, resource depletion, increasing inequality, and poverty, among other challenges, are interlinked and cut across sectors (Naidoo et al., 2021b). The intricate interlinkages of the challenges and how they trigger crises in all sectors calls for integrated and transformative approaches (nexus planning, circular economy, just transition, one health, horizon scanning, sustainable food systems, and scenario planning) that apply cross-sectoral interventions and seek integrated solutions (Nhamo and Ndlela, 2021; Whitmee et al., 2015). An advantage of these transformative approaches is that they complement each other during implementation as they inform and enhance each other (Naidoo et al., 2021b). A good example is the circular economy that is increasingly gaining prominence among policy and decision-makers, academics, stakeholders, and environmentalists due to the need to achieve sustainability amid mounting challenges compounding resource insecurity and environmental degradation (Naidoo et al., 2021a). Transitioning from linearity to circularity catalyses the transformational change agenda as humankind thrives to achieve the SDGs by 2030.

Not only do they enhance the transformational agenda and environmental sustainability, but they also provide the pathways towards achieving national and regional targets like regional integration, employment creation, poverty alleviation, inclusive economic growth, climate action, and good health and well-being (Mabhaudhi et al., 2019; Nhamo et al., 2018). Transformative and circular models provide the tools to inform strategic policy decisions and priority areas for intervention and investment decisions (Adamides, 2020; Naidoo et al., 2021a). For example, smart systems and technologies that include product service systems (PSS) and performance models are a result of transformational thinking. They are guiding decisions that link the circular economy and the Internet of Things (IoT) and are accelerating transformational change towards a green economy (Ingemarsdotter et al., 2019; Naidoo et al., 2021a). In this digital era of globalisation, the circular economy model is catalysed by digital technologies that include the IoT, Big Data, and Data Analytics that facilitate the smooth tracking and flow of products, components, and materials, allowing the derived data to be used to improve resource management and inform decision-making across various phases of the production cycle (Kristoffersen et al., 2020).

1.4   CONCLUSIONS

The circular economy has become an important strategy to enhance resilience and build adaptation against climate change. The concept has gained much attention in recent years, particularly with policy and decision-makers and the major fields of engineering and natural resources that are fast adopting and implementing circular economy principles. In contrast to the linear production model based on the use and disposal principle, the circular economy approach creates and sustains a regenerative system that promotes resource use efficiency and minimises greenhouse gas emissions. Unlike the linear model, which has left a heritage of an unstable socio-ecological and economic environment, the circular model's regenerative values enhance the elimination of environmental damage and produce invaluable social benefits. Transitioning to a circular economy requires the efforts of all sectors through cross-sectoral systemic transition through radical changes in societal values, norms, and behaviours. Thus, transitioning to the circular economy involves overcoming tensions that might bring transformative pressures.

REFERENCES

Adamides, G. (2020) A review of climate-smart agriculture applications in Cyprus. *Atmosphere* 11, 898.
Batisha, A. (2022) Horizon scanning process to foresight emerging issues in Arabsphere's water vision. *Scientific Reports* 12, 12709.
Bocken, N.M., De Pauw, I., Bakker, C., Van Der Grinten, B. (2016) Product design and business model strategies for a circular economy. *Journal of Industrial and Production Engineering* 33, 308–320.
Cornell, S., Berkhout, F., Tuinstra, W., Tàbara, J.D., Jäger, J., Chabay, I., de Wit, B., Langlais, R., Mills, D., Moll, P. (2013) Opening up knowledge systems for better responses to global environmental change. *Environmental Science & Policy* 28, 60–70.
D'Amato, D. (2021) Sustainability narratives as transformative solution pathways: Zooming in on the circular economy. *Circular Economy and Sustainability*, 1, 231–242.
David Tàbara, J., Jäger, J., Mangalagiu, D., Grasso, M. (2019) Defining transformative climate science to address high-end climate change. *Regional Environmental Change* 19, 807–818.
Dörgő, G., Sebestyén, V., Abonyi, J. (2018) Evaluating the interconnectedness of the sustainable development goals based on the causality analysis of sustainability indicators. *Sustainability* 10, 3766.

Geels, F.W., Kern, F., Fuchs, G., Hinderer, N., Kungl, G., Mylan, J., Neukirch, M., Wassermann, S. (2016) The enactment of socio-technical transition pathways: A reformulated typology and a comparative multi-level analysis of the German and UK low-carbon electricity transitions (1990–2014). *Research Policy* 45, 896–913.

Hysa, E., Kruja, A., Rehman, N.U., Laurenti, R. (2020) Circular economy innovation and environmental sustainability impact on economic growth: An integrated model for sustainable development. *Sustainability* 12, 4831.

Ingemarsdotter, E., Jamsin, E., Kortuem, G., Balkenende, R. (2019) Circular strategies enabled by the internet of things—A framework and analysis of current practice. *Sustainability* 11, 5689.

Jørgensen, S., Pedersen, L.J.T. (2018) The circular rather than the linear economy, in: Shrivastava, P., Zsolnai, L. (Eds.), *RESTART sustainable business model innovation*. Springer, Geneva, Switzerland, pp. 103–120.

Kimani-Murage, E., Gaupp, F., Lal, R., Hansson, H., Tang, T., Chaudhary, A., Nhamo, L., Mpandeli, S., Mabhaudhi, T., Headey, D.D. (2021) An optimal diet for planet and people. *One Earth* 4, 1189–1192.

Kirchherr, J., Reike, D., Hekkert, M. (2017) Conceptualizing the circular economy: An analysis of 114 definitions. *Resources, Conservation and Recycling* 127, 221–232.

Kristoffersen, E., Blomsma, F., Mikalef, P., Li, J. (2020) The smart circular economy: A digital-enabled circular strategies framework for manufacturing companies. *Journal of Business Research* 120, 241–261.

Mabhaudhi, T., Nhamo, L., Mpandeli, S., Nhemachena, C., Senzanje, A., Sobratee, N., Chivenge, P.P., Slotow, R., Naidoo, D., Liphadzi, S. (2019) The water–energy–food nexus as a tool to transform rural livelihoods and well-being in Southern Africa. *International Journal of Environmental Research and Public Health* 16, 2970.

Magableh, G.M. (2021) Supply chains and the COVID-19 pandemic: A comprehensive framework. *European Management Review* 18, 363–382.

Mastos, T.D., Nizamis, A., Terzi, S., Gkortzis, D., Papadopoulos, A., Tsagkalidis, N., Ioannidis, D., Votis, K., Tzovaras, D. (2021) Introducing an application of an industry 4.0 solution for circular supply chain management. *Journal of Cleaner Production* 300, 126886.

Mensah, J., Ricart Casadevall, S. (2019) Sustainable development: Meaning, history, principles, pillars, and implications for human action: Literature review. *Cogent Social Sciences* 5, 1653531.

Menton, M., Larrea, C., Latorre, S., Martinez-Alier, J., Peck, M., Temper, L., Walter, M. (2020) Environmental justice and the SDGs: From synergies to gaps and contradictions. *Sustainability Science* 15, 1621–1636.

Morseletto, P. (2020) Targets for a circular economy. *Resources, Conservation and Recycling* 153, 104553.

Naidoo, D., Nhamo, L., Lottering, S., Mpandeli, S., Liphadzi, S., Modi, A.T., Trois, C., Mabhaudhi, T. (2021a) Transitional pathways towards achieving a circular economy in the water, energy, and food sectors. *Sustainability* 13, 9978.

Naidoo, D., Nhamo, L., Mpandeli, S., Sobratee, N., Senzanje, A., Liphadzi, S., Slotow, R., Jacobson, M., Modi, A., Mabhaudhi, T. (2021b) Operationalising the water-energy-food nexus through the theory of change. *Renewable and Sustainable Energy Reviews* 149, 10.

Nhamo, L., Mabhaudhi, T., Mpandeli, S., Dickens, C., Nhemachena, C., Senzanje, A., Naidoo, D., Liphadzi, S., Modi, A.T. (2020) An integrative analytical model for the water-energy-food nexus: South Africa case study. *Environmental Science and Policy* 109, 15–24.

Nhamo, L., Mpandeli, S., Senzanje, A., Liphadzi, S., Naidoo, D., Modi, A.T., Mabhaudhi, T. (2021) Transitioning toward sustainable development through the water–energy–food nexus, in: Ting, D., Carriveau, R. (Eds.), *Sustaining tomorrow via innovative engineering*. World Scientific, Singapore, pp. 311–332.

Nhamo, L., Ndlela, B. (2021) Nexus planning as a pathway towards sustainable environmental and human health post Covid-19. *Environment Research* 192, 110376.

Nhamo, L., Ndlela, B., Nhemachena, C., Mabhaudhi, T., Mpandeli, S., Matchaya, G. (2018) The water-energy-food nexus: Climate risks and opportunities in southern Africa. *Water* 10, 567.

Patala, S., Albareda, L., Halme, M. (2022) Polycentric governance of privately owned resources in circular economy systems. *Journal of Management Studies* 59, 1563–1596.

Reike, D., Vermeulen, W.J., Witjes, S. (2018) The circular economy: New or refurbished as CE 3.0?—Exploring controversies in the conceptualization of the circular economy through a focus on history and resource value retention options. *Resources, Conservation and Recycling* 135, 246–264.

Shang, Y., Li, H., Zhang, R. (2021) Effects of pandemic outbreak on economies: Evidence from business history context. *Frontiers in Public Health* 9, 632043.

Thiel, A. (2017) The scope of polycentric governance analysis and resulting challenges. *Journal of Self-Governance and Management Economics* 5, 32.

Toxopeus, M.E., De Koeijer, B., Meij, A. (2015) Cradle to cradle: Effective vision vs. efficient practice? *Procedia CIRP* 29, 384–389.

Velenturf, A.P., Purnell, P. (2021) Principles for a sustainable circular economy. *Sustainable Production and Consumption* 27, 1437–1457.

Visseren-Hamakers, I.J., Razzaque, J., McElwee, P., Turnhout, E., Kelemen, E., Rusch, G.M., Fernandez-Llamazares, A., Chan, I., Lim, M., Islar, M. (2021) Transformative governance of biodiversity: Insights for sustainable development. *Current Opinion in Environmental Sustainability* 53, 20–28.

Whitmee, S., Haines, A., Beyrer, C., Boltz, F., Capon, A.G., de Souza Dias, B.F., Ezeh, A., Frumkin, H., Gong, P., Head, P. (2015) Safeguarding human health in the Anthropocene epoch: Report of The Rockefeller Foundation–Lancet Commission on planetary health. *The Lancet* 386, 1973–2028.

Wittman, H., Chappell, M.J., Abson, D.J., Kerr, R.B., Blesh, J., Hanspach, J., Perfecto, I., Fischer, J. (2017) A social–ecological perspective on harmonizing food security and biodiversity conservation. *Regional Environmental Change* 17, 1291–1301.

2 Voluntary agreements and systemic lock-in in the circular economy
The certification of sewage sludge in Sweden

Patrik Söderholm and Kristina Söderholm

2.1 INTRODUCTION

2.1.1 BACKGROUND AND MOTIVATION

The European Union (EU) promotes a transition to a circular economy in which the values of products, materials and resources are maintained (European Commission, 2015). Through waste prevention and the reuse or recycling of generated waste, avoiding the often-significant environmental costs associated with the extraction of virgin natural resources is possible. There are also concerns about the future availability of some virgin resources; their long-run supply could be threatened due to depletion and/or restricted to relatively few countries in politically unstable regions. At the same time, however, the generated waste may contain high contamination levels. This implies that reusing and recycling resources and materials involve difficult trade-offs. Specifically, it is important to identify sustainable management practices that can address the often-conflicting goals of increased circulation of waste on the one hand and decreased exposure to toxic elements on the other (Brunner, 2010; Johansson et al., 2020).

This chapter departs from this dual objective of reusing waste. At the same time, mitigating pollution addresses the opportunities and challenges of managing this through voluntary agreements between stakeholders (e.g., suppliers, end users, and public agencies). Introducing circular economy policies and regulations has proved difficult (e.g., Bengtsson and Tillman, 2004; Söderholm, 2020). One important reason for this is that various stakeholders and actors, including scientists from different disciplines, often have conflicting views regarding the extent to which resource recycling can be promoted without jeopardizing pollution control.

In this context, it is interesting to observe how voluntary agreements between stakeholders have emerged to address barriers in the circular economy. These barriers include, for instance, cases in which one firm manufactures a product in a way that increases the cost of recycling for the downstream processor. In such a

DOI: 10.1201/9781003327615-2

case, a voluntary agreement between the manufacturer and the recycler can internalize this cost and encourage the manufacturer to change the product design to enable downstream recycling (e.g., Nicolli et al., 2012). Many voluntary agreements involve efforts to internalize related barriers in the markets for environmental virtue, not least information problems between firms and their stakeholders (Potoski and Prakash, 2013). In the circular economy, information about the presence of trace elements and pollutants in various materials and waste fractions is an apt example (Johansson, 2018). Voluntary agreements – or green clubs – can help alleviate such information problems, e.g., by investing in and requiring in-depth analyses of waste streams and building trust for waste recovery among stakeholders.

In the chapter, we address the challenges of sewage sludge management in Sweden, with a particular emphasis on the lessons that can be drawn following the introduction of a voluntary certification scheme aiming to improve sludge quality, thereby facilitating its use in the agricultural sector.

2.1.2 THE CASE OF SEWAGE SLUDGE

The water used by households and industries will typically mix with surface water run-off and be transported to a WWTP. At the plant, the wastewater is treated mechanically, biologically, and chemically to remove micro-organisms and other substances that may be harmful to people and/or the natural environment before it re-enters the water cycle. Sewage sludge (biosolids) is the matter, i.e., the solid residues, resulting from this treatment. Following anaerobic digestion, sewage sludge can be managed in different ways. In the EU, the main reuse route is the application on agricultural soil (48%), particularly in countries such as Denmark, France, Ireland, Portugal, Slovakia, Spain, and Sweden (e.g., EurEau, 2021). Other significant sewage sludge destinations in the EU Member States include incineration, landfill, and land reclamation. Clearly, sewage sludge management is also a topic of significant interest in the Global South (e.g., LeBlanc et al., 2008; Tesfamariam et al., 2015).

Using sewage sludge in the agricultural sector is a relevant empirical illustration of the trade-offs in addressing both circular economy and non-toxic environment concerns. Sewage sludge contains valuable resources – not least phosphorus and nitrogen. Applying sludge to arable land provides an opportunity to make use of the nutrients in the sludge and reduce the production and use of mineral fertilizers, which contribute to significant greenhouse gas emissions.[1] However, the sludge also acts as a sink for various pollutants, i.e., toxic elements, organic contaminants, pathogens, pharmaceutical residues, and microplastic. Thus, applying sewage sludge on agricultural soil will diffuse these pollutants, and the content of heavy metals (e.g., cadmium, mercury, lead and zinc) remains several times higher in sludge compared to mineral fertilizers (Swedish Environmental Protection Agency, 2011).

The levels of many other categories of substances – such as pathogens, pharmaceutical residues, and microplastics – have increased over time but are generally not at all detected in mineral fertilizers. These substances could cause harm to human health and the natural environment. This, combined with the uncertainties regarding the specific characteristics and impacts of undesirable pollutants, could turn sewage sludge in agriculture into a relatively risky and complex practice from both a health and business point of view (Bowler, 1999; Ekane et al., 2021).[2] The application of sludge on agricultural soil, especially for food crop production, has faced a lot of resistance from key stakeholders such as farming and consumer organizations (Hultman et al., 2000).

Sewage sludge management also represents a field in which voluntary agreements between key stakeholders have been launched, e.g., in Germany and Sweden (Johansson, 2018). In Sweden, the so-called REVAQ scheme involves the voluntary certification of WWTPs. It was launched in 2008 to allay the concerns about sludge applications on arable land. In brief, this scheme sets limits on key contaminants in the sludge (not least certain metals) and demands continuous reduction of these in the wastewater reaching plants. These requirements have helped build trust for sludge reuse in Sweden among the key stakeholders, including farmers and consumer organizations. In this chapter, we focus on the experiences of this voluntary sludge management agreement and the lessons that can be drawn from it.

2.1.3 OBJECTIVE AND RESEARCH CONTRIBUTION

The chapter aims to investigate and discuss the emergence, outcomes, and future challenges of the Swedish voluntary certification scheme REVAQ. By doing this, we contribute to existing research by addressing the tension between system optimization and system change in the context of voluntary environmental agreements (see also below). The chapter also sheds new empirical light on the challenges of sewage sludge management for agricultural purposes.

Previous literature on sewage sludge management is extensive (see Krogmann et al. (1997) for an early review). Social science research has addressed the conflicts surrounding sludge recycling. Past studies have been concerned with the nature and the causes of these conflicts, e.g., shedding new light on the role of media (e.g., Goodman and Brett, 2006), public education (e.g., LeBlanc et al., 2008), the risk perceptions of important stakeholders (e.g., Ekane et al., 2021), the management challenges in the presence of scientific uncertainty (e.g., Bengtsson and Tillman, 2004; Öberg and Mason-Renton, 2018), and the societal challenges in terms of difficulties in establishing a common knowledge base (e.g., Ekman Burgman, 2022; Ekman Burgman and Wallsten, 2021).

Related research has also investigated ways to solve these conflicts, including the involvement of the public in different decision-making processes (e.g., Mason-Renton and Luginaah, 2018; Pollans, 2017) and the adoption of sanitary norms in the infrastructure (e.g., Gerling, 2019). There exists, of course, plenty of previous research on alternative technological solutions that enable the recovery of nutrients from the sewage sludge (e.g., Jedelhauser and Binder, 2018), including the novel sanitation solutions that are more diverse in terms of, for instance, source separation and decentralization (e.g., urine diversion) (for a review, see Hoffmann et al., 2020). In this context, studies also address the barriers to socio-technical change in the sewage sludge management field (e.g., Barquet et al., 2020; Bugge et al., 2019; McConville et al., 2017a, 2017b; Söderholm et al., 2022).

In line with the latter strand of research, this chapter also builds on the sustainability transitions literature. This means that we depart from the notion that existing water and wastewater systems can be conceptualized as large socio-technical systems consisting of networks of actors and institutions (i.e., regulations, standards, codes of conduct, etc.) as well as material artefacts and knowledge (Geels, 2002; Kemp et al., 1998). A key feature of such systems is path dependency, i.e., where water and wastewater systems tend to be locked in into a few technological pathways. These pathways tend to be particularly self-reinforcing since the investments are characterized by high upfront costs and increasing returns from adoption (such as scale, learning and network economies). Existing institutions – e.g., laws and codes of conduct – could also contribute to path dependence; these often favour the incumbent actors and technologies (see Section 2.2 for a more in-depth discussion).

Unlike previous research, though, we devote particular attention to how voluntary agreements among incumbent actors in the socio-technical system will influence the choice between *system optimization*, such as improving the existing system in terms of reduced production costs and improved environmental performance, and *system change*, i.e., seeking to innovate beyond the existing system, and infrastructure (see Bugge et al., 2019). The latter will typically require the emergence of novel value chains, actor networks, and institutional change. The chapter highlights challenges that are of particular concern for the establishment of a circular economy. Specifically, while voluntary agreements can help internalize the external costs associated with upstream production (in this way facilitating recycling) and address information failures among stakeholders, such agreements risk favouring the incumbent actors that often prefer to prioritize system optimization over system change.

2.1.4 OUTLINE

Section 2.2 outlines some simple theoretical points of departure for the analysis. Section 2.3 outlines the development of Swedish sewage sludge management over time, including the roles of stakeholder perceptions, government regulations, and actor collaborations. The emergence, the outcomes, and the challenges of the Swedish voluntary certification scheme REVAQ are investigated in Section 2.4, while Section 2.5 ends the chapter with a concluding discussion.

2.2 THEORETICAL POINTS OF DEPARTURE

The water and wastewater sector can be conceptualized as a socio-technical system consisting of networks of actors and institutions – i.e., regulations, standards, and codes of conduct – as well as material artefacts and knowledge (Geels, 2002). This sector is also characterized by large-scale infrastructure with a long-term investment horizon, creating path dependency. As a result, the system will tend to be locked in into a certain pathway of economic, technological, and institutional development (Klitkou et al., 2019).

Several mechanisms often contribute to such systemic lock-in (Blanken et al., 2019; Eijlander and Mulder, 2019). First and perhaps foremost, the incumbent actors, not least the WWTPs, are specialists in existing technologies and are, by definition, the established actors who dominate the existing regime. Moreover, these incumbent actors possess substantial power and resources to influence the technological trajectories that will dominate the future. Second, the policies and institutions that have emerged over time reflect the interests and perspectives of the incumbent actors that comprise the system. These institutions include both legal rules but also informal norms and practices. For instance, in the wastewater sector, lock-in tends to be based on a paradigm that portrays centralized systems as more efficient than small-scale and decentralized systems (Barquet et al., 2020; Söderholm et al., 2022).

Sustainability-oriented research has devoted much attention to the long-term, multidimensional transformation processes that shift the established socio-technical system into more sustainable modes of production and consumption (e.g., Markard, 2011). This literature emphasizes the initial protection of path-breaking innovations, which will otherwise fail to compete with the incumbent socio-technical systems. Hence, so-called niches play a key role, i.e., breeding places for evolving new technological solutions, regulatory structures, user practices, and so forth (Kemp et al., 1998). These niches thus protect against the established technologies and create possibilities for innovation, e.g., learning-by-doing processes that help lower costs and improve environmental performance.

The transition to more sustainable production and consumption patterns tends to take place through a gradual configuration and reconfiguration based on what is happening within the system, e.g., in different competing niches, but also on events in what is often referred to as the landscape level (e.g., Geels, 2014). At the landscape level, comprehensive ecological, cultural, geopolitical, and macroeconomic changes could occur, typically affecting all socio-technical systems. In the sludge management context, important landscape-level changes could involve consumer preferences towards food, increased awareness of climate change, and technological trends (such as digitalization).

The above implies that the sustainable transition of the water and wastewater systems involves a tension between what can be achieved: (a) within the existing socio-technical system, i.e., through system optimization in terms of continuous incremental improvements, or (b) through nurturing and developing novel technological trajectories, i.e., innovation beyond the existing system. Either of these pathways requires coordination and communication across the actors in the value chain and the mobilization of support for what these actors – and their stakeholders – consider to be the most sustainable options (Bowler, 1999).

One important example of an actor-network collaboration, which tends to be closely associated with the system optimization pathway, is voluntary environmental agreements – or green clubs (van't Veld and Kotchen, 2010). In these agreements, actors in the system agree to comply with certain environmental standards and/or activities. The club aspect here refers to the fact that the agreement provides non-rival – yet excludable – reputation benefits to the participating actors, while green implies that this agreement generates environmental public goods. It should be clear that the Swedish REVAQ certification scheme meets this definition of a green club (see further Section 2.4).

Specifically, many voluntary agreements attempt to address information problems between the various actors in the socio-technical system and their stakeholders (Potoski and Prakash, 2013). Generating environmental public goods requires shared knowledge and collaboration among actors, e.g., information about trace elements in existing waste streams. By establishing a benchmark of best environmental practices, the actors that form part of the agreement will reap mutual reputation benefits. The benefits are made excludable, e.g., through a certification scheme exclusively for club members (Sandler and Tschirhart, 1980). In the case of REVAQ, the certification of the WWTPs that have joined the club – and thus have committed themselves to invest in (upstream) environmental improvements – signals that the quality of the generated sewage sludge is good enough for agricultural use.

Potoski and Prakash (2013) identify and discuss four collective challenges facing green clubs of this kind. These are (a) programme establishment, thus securing that the relevant actors invest resources to create the agreement despite the incentives to free ride on the efforts of others; (b) recruiting, i.e., offer the joining actors (excludable) benefits from joining the club; (c) monitoring, thus making sure that the joining actors adhere to the club requirements; and (d) marketing, in the sense that stakeholders (e.g., consumers) need to be made aware of the environmental public goods jointly provided by the club members.

By combining these challenges with the socio-technical system perspective introduced above, it is useful to make three remarks. *First*, addressing the above challenges facing green clubs, except for perhaps (a), involves *continuous* efforts on the part of the club members. Changes at the landscape level could lead to altered priorities and increased efforts. For instance, changes in consumer preferences could imply that the scope of the environmental activities needs to be broadened (e.g., reducing previously unattended trace elements in the waste streams), and any progress made informed to stakeholders. Failures to adapt to such changing circumstances may destabilize the collaboration, and the signals communicating the club members' environmental credentials (e.g., the certification of plants, processes, or products) could start to be questioned.

Second, it is reasonable to hypothesize that in the absence of technological niches challenging the existing socio-technical system, green clubs – their objectives and structure – will typically be shaped by a group of incumbent actors. As noted above, the institutions that have emerged over time tend to reflect the interests and perspectives of these actors. Therefore, they also have the resources and power to determine the nature of the green club activities. Another reason is that voluntary agreements are easier to establish if transaction costs, i.e., the costs of identifying potential partners and reaching an agreement, are low. This is typically the case in the existing socio-technical system, not least those systems building on large-scale infrastructure involving relatively few and easily identified actors. The establishment of voluntary agreements will also be facilitated by the fact that the existing institutions tend to favour the incumbents.

Third, and finally, there will naturally be important consequences of this strong position for the incumbent actors. Positive feedback effects in technology systems reinforce technology choices, e.g., firms often choose to build on accumulated technology-specific knowledge when developing novel and better-performing products and processes. This leads to path-dependent behaviour, and the costs of exploring alternative technology pathways increase. For instance, establishing new actor networks around the novel technology may be cumbersome due to coordination failures and uncertainties about which actors should take on which roles in the technological development (Story et al., 2011). In other words, establishing green clubs will not necessarily promote novel technological niches and risks reducing the scope for establishing new value chains and actor collaborations. Efforts to generate environmental public goods will focus on system optimization rather than system change.

In the remainder of this chapter, we discuss the issue of socio-technical change in the presence of green clubs and the challenges facing such agreements. The specific case of sewage sludge management in Sweden, including the voluntary certification scheme REVAQ, is studied based on a set of secondary sources. Specifically, the analysis relies on previous research work, articles in sector magazines (e.g., *VAV-nytt*) and debate articles in Swedish national newspapers. This material is rich, not least given the conflicts surrounding sewage sludge management in Sweden, and provides a good opportunity to grasp the arguments made by various system actors and the priorities these have made over time.

2.3 THE MANAGEMENT OF SEWAGE SLUDGE IN SWEDEN

During the second half of the 20th century, the volumes of sewage sludge soared, not least due to the growing number of households connected to the sewage system. Since the turn of the century, the total production of sewage sludge from WWTPs in Sweden has exceeded 200,000 metric tons (dry solids), with a modest decrease from 222,000 tons in 2000 to 211,000 tons in 2018 (Statistics Sweden, annual).

Figure 2.1 illustrates the use of this sludge in terms of the percentage shares applied to agriculture, landfills, and landfill covers over the period of 1988–2018. Another significant use (not displayed in the figure) includes other land applications, such as

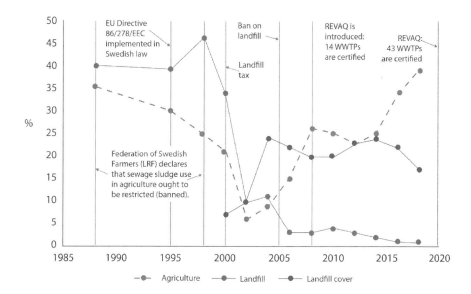

FIGURE 2.1   The use of sewage sludge in Sweden, 1988–2018 (percentage shares).

Source: Statistics Sweden (annual). Reports on discharges to water and sludge production (MI 22).

in the form of forest fertilizers, application in green areas (following composting), and topsoil production. Figure 2.1 also highlights a few key regulatory changes and events in the sewage sludge management field over the period.

Since the 1960s, the agricultural use of sewage sludge has typically been perceived as a low-cost solution to the disposal problem, which also benefits farmers.[3] Only in the 1970s were regulations put in place to mitigate the risks from pathogens in the sludge (Dagerskog and Olsson, 2020). In 1973, Sweden introduced limit values regarding the maximum allowed concentrations of heavy metals for applying sewage sludge to arable land. These limits have become more stringent over time (Hultman et al., 2000). During the late 1990s – following the implementation of the EU Directive (86/278EEC) regulating the use of sewage sludge in the agricultural sector in Swedish legislation – the government also introduced requirements on the maximum amount of sludge that can be applied on arable land in terms of limit values expressed in grams of various metals per hectare and year.[4] The Swedish requirements have overall been more stringent than those stipulated in the EU Directive.

Despite this regulatory progress, the 1970s and 1980s also witnessed the advent of an intensive debate on the health and environmental risks related to pollutants, both heavy metals and organic substances. There was hesitance from the public and farmers about using human waste as fertilizer. In the mid-1980s, the debate was particularly fuelled by concerns about the presence of organic micropollutants (e.g., dioxin) in the sludge.[5] Questions were also raised about other ways of managing the sludge, and researchers noted that incineration of digested sludge was one interesting alternative worthy of further evaluation (Hultman et al., 2000). Until now, though, incineration of sewage sludge (with or without phosphorus recovery) has been low in Sweden, representing around 1%–2% of total use over the period of 2010–2018 (and zero during earlier periods).

In 1988, the Federation of Swedish Farmers (LRF) claimed a ban on sewage sludge application on arable land (Balmer and Frost, 1990). This ban was, however, lifted following negotiations between LRF, the Swedish Water and Wastewater Association (SWWA), and the Swedish Environmental Protection Agency (SEPA). This ultimately led to forming of a national consultation group that aimed to stimulate the application of high-quality sludge on agricultural soil and agree upon various precautionary measures (Hultman et al., 2000). The agreement involved additional requirements on the metal content of the sludge. In 1999, the SWWA also introduced a certification scheme – i.e., essentially a forerunner to REVAQ – and gained support from the food industry (Johansson, 1999).

However, this agreement did not last long following a new recommendation of LRF to ban the application of sewage sludge on agricultural land. This time, it was primarily due to concerns about the presence of brominated flame retardants in the sludge and their potential negative effects on soils and organisms. The Swedish Chemicals Agency had also raised concerns about silver, cadmium, and polychlorinated biphenyls (PCB), and the ban was influenced by reports on hygienic risks related to wastewater from hospitals (Bengtsson and Tillman, 2004). Consequently, LRF argued that the existing voluntary agreement had not reached its objectives (Eksvärd, 1999), while the SWWA maintained that large enough security margins related to the contents of heavy metals and toxic organic materials were already applied (Hellström, 2000).

Figure 2.1 shows that the concerns about the content of the sewage sludge generated were followed by a significant decline in the agricultural use of sludge as a share of total sludge production. Following the ban in 1999, SEPA was set to evaluate the health and environmental aspects of sludge use. The agency was overall positive towards sludge application on arable land and noted that the present situation requires that "over a transition period, society will have to accept a balance between increased recovery and reduced pollution and risk for the spread of disease" (SEPA, 2002, p. 71). This standpoint remained in follow-up evaluations. One central conclusion of SEPA was that "sewage sludge can be applied to arable land in the short- as well as the long run with acceptable risks concerning the added metals and organic substance as well as infection control" (SEPA, 2010, p. 12).

The difficulties for the WWTPs in identifying suitable applications for their generated sludge intensified with the introduction of policy instruments aiming at abandoning landfills as a waste management option. Figure 2.1 displays that before the turn of the century, the share of sewage sludge destined for landfills was significant and typically above 40%. However, in 2002, the government introduced a tax on landfill disposal, and since 2005, there is also a ban on the landfill of organic wastes, including sludges from WWTPs. One consequence of these policies has been that no new landfills are created in Sweden, putting a cap on the demand for sludge for landfill cover purposes.

This situation put much pressure on WWTPs and SWWA to identify ways to make sludge application on agricultural soil more accepted. One important step was the introduction of the joint certification scheme REVAQ in 2008, which started as a smaller development project in 2002 (l'Ons et al., 2012). REVAQ is a voluntary agreement initially managed by SWWA, LRF, the Swedish Food Federation, and the Swedish Food Retailer's Federation in cooperation with SEPA. The current REVAQ system is owned and administered solely by the SWWA. The agreement's objective has been to avoid an unacceptable long-term accumulation of metals and undesired organic substances on agricultural land. A WWTP can be certified through REVAQ, and the plant owners then commit to, not least, upstream work in the form of removing the sources of metals and other contaminants before these reach the plant (Persson et al., 2015). Included in the REVAQ system are also requirements that the sludge must be thoroughly cleaned by one of a set of defined methods to prevent the distribution of pathogens and viruses to arable land. LRF and the SWWA recommend that solely sludge from REVAQ-certified WTTPs that comply with the above requirements – slightly below 50% of total sludge production in Sweden – should be used on agricultural soil. Overall, REVAQ has stricter regulations (e.g., standards) than are legislated.

Figure 2.1 shows that the share of agricultural use of sewage sludge has increased following the introduction of the REVAQ scheme. In this way, the scheme has been successful (see further Section 2.4). However, the launch of REVAQ has not settled the controversies regarding the use of sludge on arable land. For instance, Swedish flour mills do not accept grain fertilized with sewage sludge, primarily for fear of consumer backlash. The debate on microplastics in sewage sludge has led several farmers to refuse to accept sludge applications on their land (Johansson, 2018). Overall, Swedish farmers are largely against the spreading of sludge on their land,

and this resistance is often echoed in the food industry (Wallenberg and Eksvärd, 2018). Occasionally, the debate has been quite intensive in media also after the introduction of REVAQ, for instance with Wilson (2010) as well as Fagerberg et al. (2010) arguing against sludge use on arable land, and with Mattson and Davidsson (2010) as well as Eksvärd (2009) taking the opposite stance.

In 2020, a government inquiry delivered its report evaluating different options for managing the sewage sludge generated in Sweden (SOU, 2020:3). The evaluators proposed two options: a complete ban on sludge use and one option in which sludge use on arable land is allowed, albeit under yet stricter quality requirements. Both options included new quantitative targets for the recycling of phosphorous. The Swedish government will now need to consider the inquiry's suggestions and how to implement any subsequent new policy. Clearly, the outcome of this policy process could have major impacts on the current management system, where the second option is the one most consistent with a continued reliance on REVAQ (e.g., Rasmussen et al., 2020). So far, though, the government has not presented any proposal on the future management of sewage sludge in Sweden.

2.4 THE OUTCOMES AND CHALLENGES OF REVAQ

In this chapter, we describe and reflect on the emergence, outcomes, and challenges of the Swedish voluntary certification scheme REVAQ. The chapter makes three important points: (a) the establishment of REVAQ was in many ways a natural response of the incumbent actors to the uncertain policy environment and the need to gain trust among stakeholders for continued application of sewage sludge on arable land; (b) the preventive environmental work pursued as a result of the certification scheme has largely been successful, thus resulting in decreased flows of hazardous substances to soil; but (c) REVAQ faces challenges, in part related to an inability to promote system change, something which raises questions about its long-run stability.

2.4.1 The rationale behind the emergence of REVAQ

Establishing effective voluntary agreements targeting environmental outcomes (e.g., pollution prevention) can be viewed as a response to address an unfulfilled stakeholder demand for environmental protection and actors' expectations for rewards for their investments as part of such agreements (Potoski and Prakash, 2013). This is illustrated by the establishment of REVAQ – as a development project in 2002 and its formal launch in 2008. The WWTPs in Sweden are owned by the municipalities, and all municipalities are members of the SWWA (today known as Swedish Water). The SWWA is an important body for collaboration and dissemination of knowledge and has long promoted sewage sludge use in the agricultural sector. By teaming up with key stakeholders downstream in the value chain – i.e., LRF, the Swedish Food Federation, and the Swedish Food Retailer's Federation (and SEPA) – while at the same time emphasizing source separation upstream, it has been possible to build confidence and trust for this waste management option (Ekane et al., 2021).

As noted above, the certification scheme implies requirements for those WWTPs that choose to join the scheme regarding treatment processes and performance standards for metal and microbial contents in the sewage sludge. In 2008, the first 14 WWTPs were certified through REVAQ, and by 2020, this number had increased to 43 (REVAQ, 2021); the sludge produced in these plants represents around 50% of the total amount of sludge generated in Sweden. An important mission for REVAQ is also to ensure that information regarding the continuous improvements in sludge quality and traceability regarding the end use of the sludge is readily and easily available for all stakeholders.

An important driver behind the establishment of REVAQ has been the uncertainty about future regulations. Land application of sewage sludge in Sweden is governed through the EU Directive 86/278/EEC. Member States are allowed to impose even stricter regulations, and most countries, including Sweden, have done so. Still, the existing regulations date back to 1994 (SNFS, 1994:2) despite continuous work to revise them (Öberg and Mason-Renton, 2018), including the most recent public inquiry (SOU, 2020:3). The REVAQ requirements are overall more stringent and encompassing than the official government regulation (see further Section 4.3), and this has been key for building trust among stakeholders. Still, in the case of organic substances, there are no limit values. The key reason is that it is hard to monitor the content of specific organic substances, and it is, therefore, more practical to map the chemicals used and produced upstream (REVAQ, 2011).

One aspect of the often-uncertain Swedish policy instrument is that the attitudes towards applying sewage sludge on arable land differ across Swedish authorities. For instance, the Swedish Chemicals Agency, the Swedish Medical Products Agency, and the Swedish Board of Agriculture have been rather negative – or at least hesitant – towards such use, while the SEPA has been quite positive overall.[6] In practice, this implies that the individual municipalities that own and operate WWTPs have gained a lot of discretion to choose their own path. In Sweden, most municipalities are positive towards sewage sludge as a fertilizer (Johansson, 2018).

2.4.2 OUTCOMES

Prior evaluations of the REVAQ scheme, including the initial development project, typically conclude that the scheme has been quite successful in achieving safe and sustainable recycling of the nutrients in sewage sludge to agriculture (Cassel, 2012; Kärrman et al., 2007; Malmqvist et al., 2006; Rasmussen et al., 2020). Cassel (2012) argues that, at present, REVAQ constitutes the best available approach to reducing the concentration of various pollutants in sewage sludge. Regular sampling and analyses of the wastewater and the sludge are combined with tracing investigations made upstream from the WWTP. This is partly done by structured measurements in the sewer network, working from the plant and upstream, and partly by sampling and analysing the contributions from selected sources.

Over the years, there has been a strong focus on eliminating cadmium at the source before it enters the sewage system, much due to pressure from consumer organizations (Bengtsson and Tillman, 2004). Examples of cadmium sources that have been identified and removed include wastewater from landfills, airports, and electric power plants (Persson et al., 2015). This has required not only a lot of analyses but also information efforts aimed at changing behaviour. One example concerns a specific type of paint used by artists. It contains cadmium, and through information campaigns, artists are encouraged to refrain from flushing paint residues down the sink and into the sewage system. The REVAQ organization has also lobbied the government to ban cadmium in artist paint (Cassel, 2012).

This kind of environmental work has had positive impacts. Over the period of 2010–2020, the share of certified WWTPs that have a cadmium-phosphorous ratio below 20 mg Cd/kg P has increased from around 10% to 50%. In turn, the share of plants with a corresponding ratio above 30 mg Cd/kg P has instead decreased from around 20% in 2010 to 0% in 2020 (REVAQ, 2021). Similar progress can be observed for other prioritized metals and trace elements, including gold, mercury, silver, and tin. There are also examples of successful efforts to remove various organic contaminants (e.g., from car washes). Such measures are, however, more difficult to evaluate (Kärrman et al., 2007), not least because the organic contaminants often emerge from the considerable number of products sold to households.

While various information campaigns represent the most frequently adopted upstream measure among certified WWTPs, other measures have also been important. As noted above, one measure is to influence amendments to laws and regulations controlling what substances are allowed to be used in various sectors of the economy and which will eventually end up in WWTPs. The municipalities that own plants also have the right to express their opinion during the environmental permitting processes of new activities. There are plenty of examples of direct cooperation with the industries causing pollution (REVAQ, 2021).[7]

In sum, the Swedish REVAQ certification scheme has shown that one can reduce contaminants while at the same time increasing the recycling of nutrients by implementing a systematic, goal-oriented, and trustful between key incumbent stakeholders.

2.4.3 CHALLENGES

The REVAQ has aimed to establish a long-term plan for sewage sludge management. The scheme has represented a response to a perceived lack of legislation regulating the application of sewage sludge in agriculture. In Sweden, the national legislation is from the mid-1990s and only regulates seven different trace elements. For actors such as LRF and the owners of WWTPs, it has been important to go beyond these legislated requirements to build up trust for the use of sludge. REVAQ has, therefore, imposed more stringent requirements and regulated 60 different elements. As illustrated above, this scheme has led to important improvements in sludge quality, e.g., in terms of reduced cadmium content. Nevertheless, REVAQ faces several challenges in the future, and these are discussed below.

One important challenge relates to the fact that previous measures to improve sludge quality – not least through upstream work – have tended to target the most easily accessible measures (i.e., the low-hanging fruits). Some stakeholders, including LRF, have expressed that it is hard to conceive of significant further increases in the application of sludge on arable land even in the presence of the REVAQ certification, not least given that all wastewater is mixed and the WWTPs cannot control the inflow completely (Johansson, 2018).

REVAQ implies a lot of work on imposing demands on upstream industries, but the potential will likely be limited. For instance, the most important flows of cadmium stem from burning fossil fuels (Cassel, 2012), and these often need to be regulated in some other way. Moreover, it is also believed to become much harder to put corresponding demands on households. This is, however, critical for limiting the inflow of, for instance, pharmaceutical residues. Concerns have also been raised about other substances, such as polyfluoroalkyl substances, which are extremely persistent in the environment but have stayed undetected until recently (Ekman Burgman, 2022). The current REVAQ requirements do not cover emerging pollutants such as pharmaceutical residues and microplastics. Already, Kärrman et al. (2007), an evaluation of the early REVAQ project, recognized the need to devote more attention to such substances.

At the same time, the long-term objectives of REVAQ require continuous improvements. For instance, the requirements for the cadmium content have been gradually reduced over time, but many municipalities often lack the resources and the possibilities to reduce this further. The long-term target for cadmium is to attain a cadmium-phosphorous ratio of 17 mg Cd/kg P by 2025. However, in 2020, only about 20% of the certified WWTPs reached the 2025 target, and more than 50% of the municipalities expressed concerns about reaching it (REVAQ, 2021). Kärrman et al. (2019) corroborate this conclusion and note how a gradually higher proportion of the sludge does not comply with the REVAQ requirements.

Dagerskog and Olsson (2020) remark that in light of the uncertainty about the prospects of complying with increasingly stringent requirements and identifying the farmers willing to accept the sewage sludge, many municipalities may opt for other sludge management options. Phosphorus recovery could be achieved by replacing chemical treatment with biological treatment for so-called struvite crystallization or constructing new waste incineration plants that burn the sludge into ashes before the phosphorus is extracted (Nättorp et al., 2017).[8] Both options require significant investments. However, these options represent end-of-pipe solutions in that they do not require substantial water and sewage infrastructure changes. A more radical – and potentially more sustainable – solution involving a new infrastructure is source separation, in which the wastewater leaves the houses in different pipes (McConnville et al., 2017b). For instance, different systems that handle urine and/or toilet waste separately have been studied. In this way, the sludge volumes from the central system can be reduced, and a major share of the nutrients could be recovered.[9]

Still, such socio-technical changes face several barriers. One barrier is related to the past ability of the REVAQ scheme to induce continuous improvements in sludge quality, in this way making possible the continued expansion of sludge application on arable land (Rasmussen et al., 2020). Thus, regarding system optimization, the REVAQ system has been a success. Still, various actors have expressed concerns that the scheme risks leading to path dependence and systemic lock-in, and where the responsible authorities thus could be reluctant to intervene and promote potentially more efficient – yet so far immature – solutions (Johansson, 2018). In other words, REVAQ represents a green club that incumbent actors have established, and it has benefitted these same actors, who are unlikely to prioritize system change.

A related barrier is the lack of clearer political goals (and standpoints) concerning the increased phosphorus recovery and novel technological development. Without such political commitments, there will be little interest in making the necessary investments (see also Section 2.5). As noted above, the most recent public inquiry (SOU, 2020:3) lays out two main pathways for the management of sewage sludge in Sweden: (a) a ban on the application of sludge on arable land and (b) continuous improvements in sludge quality through stricter quality requirements. A ban on sludge use would likely force Swedish WWTPs to invest heavily in modern technologies, such as incineration and extracting phosphorus from the ash. Moreover, a move towards incineration would imply the end of the systematic upstream anti-pollution work taking place under REVAQ (Dagerskog and Olsson, 2020). The above discussion implies that even if the Swedish government chooses the second option, this does not automatically lead to continued reliance on the REVAQ scheme in the long run. A continued emphasis on upstream pollution prevention will likely remain important, but this could require other types of policy instruments and/or different organizational solutions.

2.5   CONCLUDING DISCUSSION

Our analysis of the Swedish certification scheme REVAQ has illustrated how this green club has been heavily shaped by the incumbent actors in the existing water and wastewater system. Similar outcomes can be detected in other countries. For instance, Bowler (1999) shows how a group of stakeholders in the United Kingdom mobilized support for agricultural application as the most sustainable sludge management option. Still, with primary incumbents in the driver's seat, there will be a focus on system optimization, i.e., improving the environmental quality of the sewage sludge within the realms of the existing system. At the same time, there is less focus on change in terms of creating niches that permit green innovation beyond the existing system. In Sweden, a ban on the application of sewage sludge on arable land would certainly steer the focus towards system change, but so far, the government has been silent about how to proceed on this issue.

Moreover, the chapter has also shown signs of instability in the REVAQ scheme. Previously unattended trace elements in the waste, e.g., microplastics and pharmaceutical residues, could make it difficult for the club members (e.g., the SWWA and LRF) to convince the key stakeholders (farmers and consumers) about the quality of the environmental public good that the club provides. If they fail in this key task, new actors (e.g., the WWTPs) will be reluctant to join the system, and some existing members could exit the club.

In this chapter, we do not evaluate the various future pathways for Swedish sludge management. Instead, the chapter ends by briefly elaborating on what is needed to promote system change in the wastewater treatment sector, e.g., with a focus on incineration and phosphorous recycling or even more decentralized solutions. For this purpose, building on the sustainability transitions literature, and its emphasis on three processes that enable technological niches to evolve and develop is useful. These three processes are (a) the articulation of expectations and visions, (b) the building of social networks, and (c) learning (Kemp et al., 1998).

The *articulation of expectations and visions* reflects the need for shared positive expectations that legitimate the nurturing of a niche. This process is important for attracting attention, resources, and new actors. The present Swedish sewage sludge management policies lack *direction*. It is useful to draw a parallel to Germany. In 2002, the German government introduced a sludge certification scheme, which involved more stringent requirements than the existing legislation, like the Swedish scheme. However, this scheme only lasted until 2016, when the German government decided to partially ban the application of sewage sludge on arable land.[10] Instead, incineration and the extraction of phosphorus from ash should be prioritized. This decision led to a "long-awaited clarity for WWTPs as well as farmers in an uncertain issue" (Johansson, 2018, p. 25). Moreover, the decision was accompanied by a rather long transition period, thus providing actors with an opportunity to change practices, technological development, and innovation in phosphorus recovery (Barquet et al., 2020).[11] Sweden and Denmark have instead set more general targets for recycling nutrients in sewage sludge and other wastes, which implies fewer prospects for system change.

The *building of social networks* represents a process in which experimentation in technological niches can bring new types of actors together. The emerging networks need to mobilize the commitment and resources of the actors and ensure that the alignment within the network is facilitated through regular interactions between the actors. In the sludge management context, the emergence of a wastewater system based on sludge incineration requires a "high degree of coordination and communication across what is arguably a more complicated value chain that adds incineration, chemical pre-processing and fertilizer production as intermediate steps between the wastewater treatment plant and farm" (Rasmussen et al., 2020, p. 3). Policy could play a key role here in helping to strengthen the collaborative practices in such novel networks. Policymakers could need to adopt a management style that, in part, goes beyond the traditional role of the regulator. It must be based on process leadership to encourage network actors to participate in joint problem-solving (Bengtsson and Tillman, 2004).

The third process is *learning*. Technology development is rooted in various learning processes necessary for reducing risk. For instance, learning by doing refers to the learning that takes place as modern technologies are up-scaled, including the tacit knowledge acquired during manufacturing. Learning by using instead refers to the learning that occurs in connection with using the products or processes, i.e., as users provide feedback to suppliers and devise new ways to use and/or integrate modern technology in existing production processes. Already in 2000, Hultman et al. (2000) noted that there had been a lack of evaluation of different sludge handling technologies in Sweden; most of the development has been in the hands of private companies. However, there are risks associated with such development efforts, and the government could, therefore, assume a more active role and share these risks. In this context, Kärrman et al. (2019) call for an integrated and coordinated research and development agenda at the national level.[12] Niches in the form of test beds in various municipalities could be established.

Finally, regardless of whether Swedish policymakers ultimately choose to favour incremental improvements in sludge quality followed by agricultural application or instead opt for a more radical change in the system (e.g., through incineration and phosphorous recovery), building trust among all stakeholders, not least lay citizens, will be very important. Managing the often-conflicting objectives of increased circulation of waste on the one hand and decreased exposure to various trace elements on the other is difficult and has to acknowledge the risk perceptions of the citizens. If the chosen solution does not align well with these risk perceptions, it will not matter whether it is a solution that focuses on system optimization or system change.

ACKNOWLEDGEMENTS

Financial support from the Swedish Research Council Formas (Grant No. 2018-00194), within the national research programme Sustainable Spatial Planning, is gratefully acknowledged, as are comments from two anonymous reviewers. Any remaining errors reside solely with the authors.

NOTES

1 There have also been concerns about the future availability of phosphorus reserves, both geologically and since existing reserves are controlled by only a few countries (Cordell and White, 2011).
2 Stringent regulations for the treated wastewater from WWTPs have led to more efficient treatment processes but this has also implied that an increasing percentage of pollutants in the wastewater has instead been transferred to the sludge treatment phase.
3 In the 1950s, the sludge was dumped in waterways, and following opposition, in international waters. The latter approach was however also abandoned when deemed environmentally unacceptable (Ekman Burgman, 2022).
4 Bauer et al. (2020) present an updated review of the legislation relating to sewage sludge disposal in Sweden compared to a selection of other EU Member States.
5 The scientific basis for the claims made about sludge representing hazardous waste was occasionally claimed to be relatively weak (e.g., Palm et al., 1989).
6 Johansson (2018) notes that in Denmark, there is no single agency for chemicals control; instead, this issue is the responsibility of the Danish Environmental Protection Agency. This creates, it can be argued, a more consistent stance towards sewage sludge management in Denmark compared to Sweden.
7 Since 2008, over 5000 different facilities – e.g., industries, car washes, hospitals, etc. – have been approached concerning the presence of undesired organic substances that could end up in the sewage system (REVAQ, 2021).
8 The Swedish waste management corporation Ragn-Sells has recently patented a new technology for recovering phosphorus from sludge ash (Dagerskog and Olsson, 2020). Lipinska (2018) also reports about the development of fermentation technologies, which contribute to both the reduction of sludge and to the production of energy from biogas generated in the process of methane formation.
9 One example is the urine-drying technology (Prithvi, 2019). It can be plugged into existing toilets, diverting, and drying out the urine in a separate box, and thus retains most of the nutrients without major retrofitting of pipes.
10 A similar development has taken place in Switzerland where a ban on sludge application was introduced in 2006 (Kärrman et al., 2019).
11 The implementation of a German ban on sludge application in agriculture is projected to take 12 years. A similar move in Sweden could likely take even longer, this since incineration already is the most common form of sludge treatment in Germany (Rasmussen et al., 2020).

12 One important challenge for policy makers who attempt to promote sustainable technology development is whether to focus on a single technological pathway or instead adopt a portfolio approach, thus supporting several pathways in parallel. The German approach, with its focus on mono-incineration followed by phosphoric acid production, has been criticized for its narrow scope, and the fact that other new technologies could have a greater environmental potential (e.g., permitting the recovery of also nitrogen and carbon) (Barquet et al., 2020).

REFERENCES

Balmer, P., Frost, R.C. (1990) Managing change in an environmentally conscious society: a case study, Gothenburg (Sweden). *Water Science and Technology*, 12, 45–56.

Barquet, K., Järnberg, L., Rosemarin, A., Macura, B. (2020) Identifying barriers and opportunities for a circular phosphorous economy in the Baltic Sea region. *Water Research* 171, 115433.

Bauer, T., Ekman Burgman, L., Andreas, L., Lagerkvist, A. (2020) Effects of different implementation of legislation relating to sewage sludge disposal in the EU. *Detrius* 10, 92–99.

Bengtsson, M., Tillman, A.-M. (2004) Actors and interpretations in an environmental controversy: the Swedish debate on sewage sludge use in agriculture. *Resources, Conservation and Recycling* 42, 65–82.

Blanken, M., Verweij, C., Mulder, K.F. (2019) Why novel sanitary systems are hardly introduced? *Journal of Sustainable Development of Energy, Water and Environment Systems* 7, 13–27.

Bowler, I.R. (1999) Recycling urban waste on farmland: an actor-network interpretation. *Applied Geography* 19, 29–43.

Brunner, P.H. (2010) Clean cycles and safe final sinks. *Waste Management & Research* 28, 575–576.

Bugge, M.M., Fevolden, A.M., Klitkou, A. (2019) Governance for system optimization and system change: the case of urban waste. *Research Policy* 48, 1076–1090.

Cassel, M. (2012) Styrkor och svagheter hos gällande styrmedel för avloppsslam. Bachelor's Thesis in Environmental Science, Lund University, Sweden.

Cordell, D., White, S. (2011) Peak phosphorus: clarifying the key issues of a vigorous debate about long-term phosphorus security. *Sustainability* 3, 2027–2049.

Dagerskog, L., Olsson, O. (2020) Swedish sludge management at the crossroads. SEI Policy Brief, Stockholm Environment Institute, Sweden.

Eijlander, S., Mulder, K.F. (2019) Sanitary systems: challenges for innovation. *Journal of Sustainable Development of Energy, Water and Environment Systems* 7, 193–212.

Ekane, N., Barquet, K., Rosemarin, A. (2021) Resources and risks: perceptions on the application of sewage sludge on agricultural land in Sweden, a case study. *Frontiers in Sustainable Food Systems* 5, 647780.

Ekman Burgman, L. (2022) What sewage sludge is and conflicts in Swedish circular economy policymaking. *Environmental Sociology* 8, 292–301.

Ekman Burgman, L., Wallsten, B. (2021) Should the sludge hit the farm? – How chemo-social relations affect policy efforts to circulate phosphorus in Sweden. *Sustainable Production and Consumption* 27, 1488–1497.

Eksvärd, J. (1999) Går det att få förtroende för slammet? *VAV-nytt* 5, 38–39.

Eksvärd, J. (2009) LRF försvarar gödsling med avloppsslam, Sveriges Radio, 19 April.

EurEau (2021) Wastewater treatment – sludge management. Briefing note, The European Federation of National Associations of Water Services, Brussels.

European Commission (2015) Closing the Loop – an EU action plan for the circular economy. COM(2015) 614/2, Brussels.

Fagerberg, B., Hagström, B., Eckerman, I., Barregård, L. (2010) Medicinska skäl mot spridning av avloppsslam på åkermark. *Läkartidningen*, March.

Geels, F.W. (2002) Technological transitions evolutionary reconfiguration processes: a multi-level perspective and case study. *Research Policy* 31, 1257–1274.

Geels, F.W. (2014) Reconceptualizing the co-evolution of firms-in-industries and their environments: developing an inter-disciplinary triple embeddedness framework. *Research Policy* 43, 261–227.

Gerling, D.M. (2019) Excrementalisms: revaluing what we have only ever known as waste. *Food, Culture & Society* 22, 622–638.

Goodman, J.R., Brett, P.G. (2006) Beneficial or biohazard? How the media frame biosolids. *Public Understanding of Science* 15, 359–375.

Hellström, T. (2000) Fakta på bordet ska bryta slammets oroscirkel. *VAV-nytt* 1, 55–57.

Hoffmann, S., Feldmann, U., Bach, P.M., Binz, C., Farrelly, M., Frantzeskaki, N., Hiessl, H., Inauen, J., Larsen, T.A., Lienert, J., Londong, J., Lüthi, C., Maurer, M., Mitchell, C., Morgenroth, E., Nelson, K.L., Scholten, L., Truffer, B., Udert, K.M. (2020) A research agenda for the future of urban water management: exploring the potential of non-grid, small-grid, and hybrid solutions. *Environmental Science and Technology* 54, 5312–5322.

Hultman, B., Levlin, E., Stark, K. (2000) Swedish debate on sludge handling. Division of Water Resources Engineering, Royal Institute of Technology, Stockholm.

Jedelhauser, M., Binder, C.R. (2018) The spatial impact of socio-technical transitions – the case of phosphorus recycling as a pilot of the circular economy. *Journal of Cleaner Production* 197, 856–869.

Johansson, B. (1999) Göteborg fick första P-märket. *VAV-nytt* 4, 41–43.

Johansson, N. (2018) How can conflicts, complexities and uncertainties in a circular economy be handled? A cross European study of the institutional conditions for sewage sludge and bottom ash utilization. Royal Institute of Technology, Stockholm.

Johansson, N., Velis, C., Corvellec, H. (2020) Towards clean material cycles: is there a policy conflict between circular economy and non-toxic environment? *Waste Management & Research* 38, 705–707.

Kärrman, E., Andersson, C., von Bahr, B., Berg, J., Nilsson, J. (2019) Översikt över återvinning av fosfor och kväve från avlopp i nio utvalda länder. RISE Report 2019:119, Stockholm.

Kärrman, E., Malmqvist, P.-A., Rydhagen, B., Svensson, G. (2007) Utvärdering av ReVAQ-projektet. Report No. 2007–02, Svenskt Vatten Utveckling, Stockholm.

Kemp, R., Schot, J., Hoogma, R. (1998) Regime shifts to sustainability through processes of niche formation: the approach of strategic niche management. *Technology Analysis and Strategic Management* 10, 175–198.

Klitkou, A., Fevolden, A.M., Capasso, M. (Eds.) (2019) *From waste to value. Valorisation pathways for organic waste streams in circular bioeconomies.* Routledge, New York.

Krogmann, U., Boyles, L.S., Martel, C.J., McComas, K.A. (1997) Biosolids and sludge management. *Water Environment Research* 69, 534–550.

LeBlanc, R.J., Matthews, P., Richard, R.P. (2008) *Global atlas of excreta, wastewater sludge, and biosolids management: moving forward the sustainable and welcome uses of a global resource.* United Nations Human Settlements Programme, Nairobi.

Lipinska, D. (2018) The water-wastewater-sludge sector and the circular economy. *Comparative Economic Research* 21, 121–137.

l'Ons, D., Mattsson, A., Davidsson, F., Mattsson, J. (2012) REVAQ – the Swedish certification system for sludge application to land. Experiences at the Rya WWTP in Gothenburg and challenges for the future. Paper presented at the 17th European Biosoldis and Organic Resources Conference, Leeds.

Malmqvist, P.-A., Kärrman, E., Rydhagen, B. (2006) Evaluation of the REVAQ project to achieve safe use of wastewater sludge in agriculture. *Water Science & Technology* 54, 129–135.

Markard, J. (2011) Transformation of infrastructures: sector characteristics and implications for fundamental change. *Journal of Infrastructural Systems (ASCE)* 17, 107–117.

Mason-Renton, S.A., Luginaah, I. (2018) Conceptualizing waste as a resource: urban biosolids processing in the rural landscape. *Canadian Geographer* 62, 266–281.

Mattson, A., Davidsson, F. (2010) Nödvändigt att återföra fosfor via avloppsslam. Göteborgs-posten, 31 January.

McConville, J., Kvarnström, E., Jönsson, H., Kärrman, E., Johansson, M. (2017a) Source separation: challenges and opportunities for transition in the Swedish wastewater sector. *Resources, Conservation and Recycling* 120, 144–156.

McConville, J., Kvarnström, E., Jönsson, H., Kärrman, E., Johansson, M. (2017b) Is the Swedish wastewater sector ready for a transition to source separation? *Desalination and Water Treatment* 91, 320–328.

Nättorp, A., Remmen, K., Remy, C. (2017) Cost assessment of different routes for phosphorus recovery from wastewater using data from pilot and production plants. *Water Science and Technology* 76, 413–424.

Nicolli, F., Johnstone, N., Söderholm, P. (2012) Resolving failures in recycling markets: the role of technological innovation. *Environmental Economics and Policy Studies* 14, 261–288.

Öberg, G., Mason-Renton, S.A. (2018) On the limitation of evidence-based policy: regulatory narratives and land application of biosolids/sewage sludge in BC, Canada and Sweden. *Environmental Science and Policy* 84, 88–96.

Palm, O., Dahlberg, A.G., Holmström, H. (1989) Sludge quality from municipal wastewater treatment plants in Sweden – past and future trends. *Vatten* 45, 30–35.

Persson, T., Svensson, M., Finnson, A. (2015) REVAQ certified wastewater treatment plants in Sweden for improved quality of recycled digestate nutrients. A case story from the IEA Bioenergy Task 37, International Energy Agency, Paris.

Pollans, L.B. (2017) Trapped in trash: 'modes of governing' and barriers to transitioning to sustainable waste management. *Environment & Planning A* 49, 2300–2323.

Potoski, M., Prakash, A. (2013) Green clubs: collective action and voluntary environmental programs. *Annual Review of Political Science* 16, 399–419.

Prithvi, S. (2019) The urine drying pilot is operational. Kretsloppsteknik, 11 March.

Rasmussen, M., Olsson, O., Trimmer, C., Barquet, K., Rosemarin, A. (2020) Implications of new national policies on management of sewage sludge for a Swedish municipality. Policy Brief in the Bonus Return research program (www.bonusreturn.eu), Stockholm Environment Institute, Stockholm.

REVAQ (2011) Regler för certifieringssytemet REVAQ – Utgåva 2.1, Bromma.

REVAQ (2021) REVAQ årsrapport 2020, Bromma.

Sandler, T., Tschirhart, J. (1980) The economic theory of clubs: an evaluative survey. *Journal of Economic Literature* 18, 1481–1521.

SNFS 1994:2 Föreskrifter om skydd för miljön, särskilt marken, när avloppsslam används i jordbruket. Statens naturvårdsverks författningssamling, Stockholm.

Söderholm, K., Vidal, B., Hedström, A., Herrmann, I. (2022) Flexible and resource-recovery sanitation solutions: what hindered their implementation? A 40-year Swedish perspective. *Journal of Urban Technology* 30:1, 23–45.

Söderholm, P. (2020) The green economy transition: the challenges of technological change for sustainability. *Sustainable Earth* 3, 6.

SOU 2020:3 Hållbar slamhantering: betänkande från utredningen om en giftfri och cirkulär återföring av fosfor från avloppsslam. Government Offices of Sweden, Stockholm.

Story, V., O'Malley, L., Hart, S. (2011) Roles, role performance, and radical innovation competences. *Industrial Marketing Management* 40, 952–966.

Swedish Environmental Protection Agency (2002) Aktionsplan för återföring av fosfor ur avlopp. Report 5214, Stockholm.

Swedish Environmental Protection Agency (2010) Redovisning av regeringsuppdrag 21 – Uppdatering av aktionsplan för återföring av fosfor ur avlopp, Stockholm.

Swedish Environmental Protection Agency (2011) Halter av 61 spårelement i avloppsslam, stallgödsel, handelsgödsel, nederbörd samt i jord och gröda. Report 6580, Stockholm.

Tesfamariam, E.H., Annandale, J.G., de Jager, P.C., Ogbazghi, Z., Malobane, M.E., Mbetse, C.K.A. (2015) Quantifying the fertilizer value of wastewater sludges for agriculture. WRC Report No. 2131/1/15., Water Research Commission, Pretoria.

van't Veld, K., Kotchen, M.J. (2010) Green clubs. NBER Working Paper 16627, National Bureau of Economic Research (NBER), Cambridge.

Wallenberg, P., Eksvärd, J. (2018) Lantbrukets syn på kretslopp. MAt i Cirkulära Robusta System (Macro) system, Stockholm.

Wilson, L. (2010) Mot allt sunt förnuft att säga ja till rötslam. Göteborgs-posten, 1 June.

3 Global status of circular economy adaptation within wastewater services

Transition pathways and the role of innovation

*John Ngoni Zvimba, Eustina Musvoto,
Nomvuselelo Mgwenya, and Buyisile Kholisa*

3.1 INTRODUCTION

In recent years, the concept of a circular economy (CE) has received global prominence in politics, business, and research agendas. Research has identified numerous potential benefits that can be derived from transitioning from a traditional linear economy to a CE. These include economic, social, improved resource security, and reduced greenhouse gas (GHG) emissions (European Environment Agency, 2016). Despite these potential benefits, it is acknowledged that transforming the traditional linear economic model is a big challenge that entails transforming the current production and consumption patterns. In this regard, innovative transformational technologies such as digital and engineering technologies in combination with creative thinking have been identified as factors that can drive fundamental changes across entire value chains that are not restricted to specific sectors or materials (Accenture, 2014; Acsinte & Verbeek, 2015; Vanner et al., 2014). Such a major transformation would, in turn, significantly impact the economy, environment, and society. Understanding these impacts is crucial for researchers and policymakers in designing future policies in the field (European Commission, 2017; Rizos et al., 2017; Vanner et al., 2014).

Although the water sector has not yet fully transitioned to a CE, water utilities have been early adopters of technologies and business practices that support the CE (Jazbec et al., 2020). This has been in response to various threats and challenges the sector has faced in recent years (i.e., water scarcity, increasing energy prices, more stringent regulations, rapid urbanisation and climate change impacts). Impeding regulatory environments and opaque market conditions are the main obstacles to the water sector's transition to a CE (International Water Association, 2016). Thus,

DOI: 10.1201/9781003327615-3

29

to define a clear role for water utilities in transitioning to a CE, the IWA developed a framework targeted at decision-makers in water utilities and key stakeholders. The framework identified three key interrelated pathways (water, energy, and materials) to achieving CE principles in the water sector. In addition, consumers, industry, regulation, infrastructure, and urban and basin economies have been identified as the main factors that drive and enable the transition of the water sector to a CE (International Water Association, 2016). Water utilities must anticipate, respond to, and influence these factors to accelerate the pathways to achieving a CE. In transitioning to a CE, water utilities must also change their current operation and seek new management approaches, partnerships, and business opportunities.

The IWA framework further identified WWTPs as one of the key junctions in the three pathways to transitioning to a CE. This is mainly because, within the man-made water cycle, wastewater carries 50%–100% of waste resources lost, mostly in the form of unrecovered water, energy, and hydrochar materials (Musvoto & Mgwenya, 2022). The wastewater treatment sector is also responsible for approximately 3% of electricity consumption globally, accounting for about 56% of the operational carbon footprint of urban water systems (Batstone et al., 2015). Several researchers have studied WWTPs and their potential for recovering valuable resources (Swartz et al., 2013; Van Vuuren et al., 2014; Zvimba & Musvoto, 2020). These studies have shown that energy efficiency in WWTPs and more efficient utilisation of wastewater energy potential can lead to energy-positive WWTPs. Also, implementing energy conservation measures and using renewable energy sources significantly improve the WWTPs' energy efficiency. Furthermore, resource and materials recovery from wastewater, such as using carbon to produce high-value by-products (biopolymers and fine chemicals) and nutrients (phosphorus and nitrogen), which are useful in agriculture, reduces the global environmental impact of their industrial production.

To successfully achieve the above, an understanding of the status of CE adaptation in the global wastewater sector and appropriate frameworks and strategies that can be adapted for application by LMIC are required. Furthermore, the role of innovative technologies as accelerators for transitioning the water sector to a CE requires critical evaluation. Such technologies can support the utilisation of WWTPs, identified as a key junction in the IWA framework, as wastewater biorefinery platforms are at the centre of the transition for improved wastewater management and resource recovery.

3.2   CIRCULAR ECONOMY ADAPTATION PROGRESS

Regions that have made considerable progress in promoting the CE are the European Union (EU), China, Japan, South Korea, and parts of the USA.

3.2.1   THE EUROPEAN UNION

The CE concept emerged in Europe in the 1980s and 1990s and is reported to have been formally used in an economic model for the first time by Pearce and Turner (1991). However, before this, early policies of EU member states drawing on ideas that can be traced to the 1960s and 1970s had promoted elements of circularity in certain parts of the economy. For example, driven by a desire to divert waste from landfills, the Netherlands and Germany pioneered the concepts of waste prevention and reduction. The waste hierarchy was introduced to the Dutch Parliament in 1979 (McDowall et al., 2017). The concept has become increasingly prominent in the past decade and is now adopted as part of the EU economic policy and strategy.

Research has shown that numerous potential benefits are derived from transitioning from a linear economy to a CE, and the benefits of implementing CE principles within EU countries were found to include (European Environment Agency, 2016):

- improved resource security and decreased import dependency,
- reduced environmental impact, including a drastic reduction in GHG emissions,
- economic benefits that include new opportunities for growth and innovation, as well as savings related to improved resource efficiency, and
- social benefits ranging from new job creation across all skill levels to changes in consumer behaviour, leading to better health and safety outcomes.

Through transitioning to a CE, the EU predicts a doubling of economic and environmental benefits, 11% growth in average disposable incomes and a halving of carbon dioxide emissions by 2030 (Ellen MacArthur Foundation, 2015). Specific benefits to countries and sectors within the EU have further been highlighted in subsequent studies (Bačová et al., 2016; European Environment Agency, 2016). While the benefits of the CE are being increasingly acknowledged, there is still a range of barriers that need to be overcome. The barriers identified as major challenges for CE implementation are technological, policy and regulatory, financial and economic, consumer and social (European Commission, 2014; Galvão et al., 2018; Rizos et al., 2017).

The significance of these barriers differs for materials, products, and sectors. Several actions are required at the EU, national, regional, and local levels to drive transformation, depending on the nature of the barrier faced. Various drivers are often required in a sector or value chain to overcome these barriers and consider the multiple factors that often influence each other. Due to its complexity, the transition to a CE is a multi-level governance challenge, requiring actions in the public and private sectors and at an individual level. Thus, identification and detailed understanding of specific barriers are very important so that appropriate mitigation measures can be implemented.

Studies in the EU have shown that the transition to a CE requires systemic change and a more holistic, integrated approach that considers the multiple connections and influences within and between sectors, value chains and stakeholders (European Commission, 2014; Humphris-Bach et al., 2016). With this approach, key factors such as different incentives, distribution of economic rewards and impacts of specific measures along a value chain across different sectors and policy areas should be considered. Complementary tools and approaches that can easily be advanced by the private and public sectors and individuals at all levels, from local to the EU, are required. Policy intervention beyond private initiatives has been identified as a key driver in overcoming some barriers to transitioning to a CE. Identified potential policy actions include regulatory measures, economic incentives, targeted and increased funding efforts to engage and link actors along the value chain and initiatives to raise awareness of the benefits of the CE and available solutions.

In 2015, the European Commission adopted an action to help accelerate the EU's transition towards a CE, boost global competitiveness, promote sustainable economic growth, and generate new jobs (European Commission, 2015). The action plan sets out measures to 'close the loop' of product lifecycles, from production and consumption to waste management and the market for secondary raw materials. It also identifies five priority sectors to speed up the transition along their value chain, and these include (i) plastics, (ii) food waste, (iii) critical raw materials, (iv) construction & demolition, and (v) biomass & bio-based materials. In this regard, close cooperation with member states, regions and municipalities, businesses, research bodies, citizens and other stakeholders involved in the CE is promoted in the action plan.

3.2.2 OTHER REGIONS

Apart from the EU, other regions that have made significant progress in promoting a CE are China, Japan, South Korea, and part of the USA.

The concept of CE is not new in China, as it dates back to the 1990s, with origins in cleaner production, industrial ecology and ecological modernisation. The thinking was inspired by implementation examples in Europe, the United States and Japan (Geng et al., 2009). In 2003, the central government formally accepted the concept as a new development strategy that culminated in the 2009 CE Promotion Law, the natural framework for advancing CE. Subsequently, various action plans that provide further details for specific sectors, as well as clarity on the implementation of the provisions of the CE Promotion Law, have been put in place (McDowall et al., 2017). Since its implementation, the Promotion Law has evolved to include concern for eco-design, potential product regulations and restrictions on some classes of disposable goods, green consumption, and extended producer responsibility.

In addition, the Promotion Law requires establishing target responsibility systems in support of the CE and measuring and evaluating progress against indicators. To promote CE, the Chinese government has invested significantly in demonstration projects, deployed tax incentives and allowed reuse/recycling activities previously banned, such as selling relatively clean wastewater. It is estimated that extending such practices would save Chinese businesses and households 32 trillion yuan (US$4.6 trillion) in 2030, equivalent to 14% of its projected gross domestic product that year (Geng et al., 2019). Although the Chinese CE agenda is framed on the same principles as the EU (waste minimisation, raw materials, and resource efficiency), there are differences in policy focus areas. EU policies focus on consumption and product design more than China, focusing on specific manufacturing sectors (McDowall et al., 2017) and measures to increase efficiency and reduce waste pollution in manufacturing.

Japan and South Korea also have national strategies for enabling CE. Japan has legislated on eco-design and made producers responsible for the after-use of their products, thereby boosting markets for secondary materials. These CE initiatives have saved materials, waste, energy, and emissions. In Kawasaki, Japan, reusing industrial and municipal wastes in cement manufacturing has reduced GHG emissions by about 15% (41,300 tonnes per year) since 2009 and saved up to 272,000 tonnes of virgin materials annually. Like China, South Korea has operated industrial parks that use the principles of a CE to link companies' supply chains and reuse or recycle common materials.

The United States has hundreds of corporate recycling and a handful of regional programmes, such as the San Francisco, California Zero Waste scheme. However, beyond this, few broad federal initiatives have been comparable to those pursued by China and the EU (Klimentov, 2018). To develop new CE opportunities and realise their ambitions faster in the USA, the Ellen MacArthur Foundation launched a US chapter of its Circular Economy 100 (CE 100) programme in 2016. The CE 100 is a pre-competitive innovation programme that enables organisations to develop new opportunities and realise their CE ambitions faster. It brings together corporates, governments and cities, academic institutions, emerging innovators, and affiliates in a unique multi-stakeholder platform. Specially developed programme elements help members learn, build capacity, network, and collaborate with key organisations around the CE (Ellen MacArthur Foundation, 2015).

The launch followed a study by the US Chamber of Commerce Foundation that showed that the 5,589 largest publicly traded companies in the US sent 342 million metric tons of waste to landfills and incinerators in 2014 (Bowdish, 2016). Companies generate 7.81 metric tons of waste for every million dollars in revenue. Reducing paper waste by a mere 1% would save these companies nearly $1 billion. To date, the members of the CE 100 programme include large corporations like Walmart, Microsoft, Coca-Cola, Google, Nike, and other institutions.

3.2.3 SOUTH AFRICA

South Africa does not yet have a unified national policy and strategy for transitioning to a CE. However, lessons learnt from other regions and increased awareness of potential opportunities stimulate serious discussions and initiatives on a CE in the public and private sectors. Despite the lack of a national policy on CE, legislation like the National Environmental Management Act (Republic of South Africa, 2009) is driving progress in some areas of CE aspects, such as waste recycling and converting waste to energy. Efforts are also being made at the government and sector level to cooperate with other regions that have gained traction in transitioning to a CE. Examples of these efforts include the following:

- The CE Mission with the EU, whose main objectives are to increase cooperation between the EU and LMIC in the field of environmental policy, achieve a better understanding of the environmental challenges faced by LMIC and promote green solutions through business partnerships abroad (European Commission, 2018). The Terms of Reference for the Forum on Environment, Climate Change, Sustainable Development and Water between the EU and South Africa include an agreement to further cooperate in areas that include biodiversity, CE and water resources management issues, among others. The cooperation also involves private sector operators.
- Membership to the Platform for Accelerating the Circular Economy (PACE), a public-private collaboration platform and project accelerator. PACE aims to shape global public-private leadership and accelerate action towards the CE. Project focus areas include plastics, electronics, food & bioeconomy, a business model, and market transformation across China, ASEAN, Europe, and Africa.

In addition to policy and industry initiatives, research is required into specific aspects of applying the CE to South Africa. Some questions that need addressing at a policy and strategic level include identifying drivers for CE in South Africa that would benefit the country most. The drivers need to be relevant to South Africa as an LMIC and resource-rich country so that opportunities in the economy can be identified and applied in shaping the CE agenda in the country. In addition, risks to South Africa from other countries adopting CE also need to be understood, for example, the EU (de Jong et al., 2016).

South Africa currently exports €6.1 billion worth of critical raw materials (72.1% of GDP) to the EU, which would be substantially reduced if the EU moved fully to a CE. In addition, €8.4 billion in mineral exports (22.7% of GDP) could be threatened.

3.3 CIRCULAR ECONOMY SOLUTIONS IN THE WATER SECTOR

3.3.1 OVERVIEW

For the water and wastewater (collectively water) sector, transitioning to a CE aligns with the United Nations Sustainable Development Goals (SDGs). Water has a dedicated goal in SDG6 (ensure availability and sustainable management of water and sanitation for all), and its attainment will be reliant upon contributing to and benefiting from the attainment of other SDGs, most notably in the context of the CE, SDG12 (ensure sustainable consumption and production patterns). This interdependence across goals manifests at a national level, highlighting the need for greater cooperation amongst sectors, incentivising innovation and enabling meaningful engagement with citizens (International Water Association, 2016).

To define a clear role for water utilities in transitioning to a CE, the IWA developed a framework targeted at decision-makers in water utilities and key stakeholders. The framework identified three key interrelated pathways to achieving CE principles in the water sector: water, materials, and energy. Graphical illustrations of these pathways are given in Figure 3.1.

3.3.2 WATER PATHWAY

To reduce the inefficiency in existing water systems that worsen the gap between supply and demand, the IWA framework recommends that the water pathway be developed as a closed loop. Three factors to achieving this are diversified resource options, efficient conveyance systems and optimal reuse. Options to be considered include upstream investment to ensure optimal conservation measures and pollution control to minimise treatment costs, rainwater harvesting, greywater recycling, wastewater reuse, reduction of water loss/leakage in potable water distribution systems and reduction in water consumption.

3.3.3 MATERIALS PATHWAY

In the materials, pathway resource recovery from wastewater operations must compete with other products in the market for successful incorporation into the CE. Key issues to be considered include resource recovery efficiency, production scale, pricing,

Water Pathway

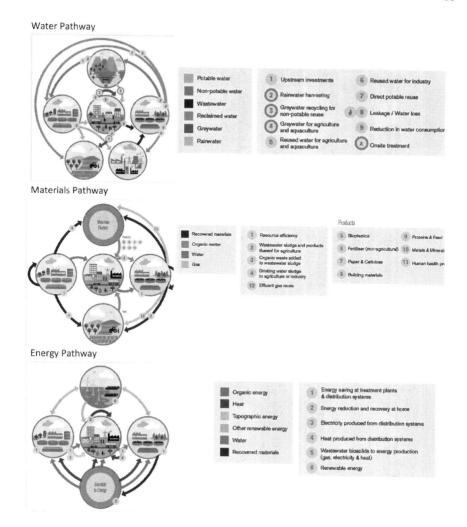

FIGURE 3.1 An illustration of the IWA framework pathways to a circular economy in the water sector (Adapted from IWA 2016).

quality, and consumer acceptance. Therefore, Water Utilities must collaborate with the industry to understand and address these issues. Options that can be considered for successful materials recovery include resource efficiency, drinking water, sludge reuse in agriculture and/or industry, wastewater sludge (WWS), products reuse for agriculture, co-processing of external biomass (e.g. municipal solid waste, agricultural waste, woody biomass, industrial waste) with WWS and recovery of high-value niche products from wastewater operations (e.g. bioplastics, non-agricultural fertiliser, paper and cellulose, building materials).

3.3.4 ENERGY PATHWAY

Water and wastewater operations consume much energy, while certain treatment processes contribute to GHG emissions. The IWA framework recommends that the objective of the energy pathway should be to reduce carbon-based energy consumption, increase renewable energy production and consumption and contribute to the zero-carbon emissions initiative. Options to be considered in the energy pathway include energy saving at treatment plants and in conveyance systems, energy reduction and recovery in the home, electricity production from water conveyance systems, heat production from wastewater conveyance systems, energy generation from WWS and use of renewable energy for water and wastewater operations.

Throughout the pathways, there are critical junctions where water, energy, or materials intersect, and opportunities arise to transition to a CE. By analysing these junctions, utilities can gain insights and take action to create partnerships for transitioning to the CE. These junctions include (i) water-wise communities, (ii) industry, (iii) WWTPs, (iv) drinking water treatment plants, (v) agriculture, (vi) natural environment and (vii) energy generation (IWA, 2016).

Overall, the main factors that drive and enable the transition of the water sector to a CE are consumers, industry, regulation, infrastructure, and urban and basin economies. Water utilities must anticipate, respond to, and influence these factors to accelerate the pathways to achieving a CE. The challenge for utilities is to shift these factors from traditionally enabling a conventional linear economy model to a CE model. In transitioning to a CE, water utilities must also change their current operation and seek new management approaches, partnerships, and business opportunities (IWA, 2016).

3.3.5 CHALLENGES AND BARRIERS IN THE WATER SECTOR

Similar to other sectors, the benefits of transitioning to a CE in the water sector have been shown through theoretical models and practical experience in areas such as energy generation and wastewater effluent reuse, where partial circularity has been achieved. However, the full transition still faces significant challenges and barriers, particularly in applying WWTPs as biorefineries at the centre of the transition and subsequent recovery and reuse of associated by-products. The most significant barriers that have been identified include the following:

- **Regulation**: Lack of laws and regulations to facilitate the transition to a CE, such as setting appropriate environmental standards for the use of recycled products, specifying health regulations related to the reuse and recycling of products, regulation on recovered product categorisation (as 'waste' or a 'resource') and certification, limiting disposal of wastewater solids to landfills and encouraging investment and innovation in the reuse and recycling industry. The absence of integrated policies and existing legislative barriers have been identified as significant barriers to developing wastewater biorefineries.
- **Economics**: The cost of wastewater reuse is not economically competitive due to the water pricing policy (Greyson, 2007; Hislop & Hill, 2011). In most jurisdictions, water is priced very cheaply for political reasons to

induce sustainable use in the long run (European Commission, 2014). The water market price or value should reflect not only internal costs but also external costs, including those of an economic, social, or environmental nature (Abu-Ghunmi et al., 2016). Without supportive policies and if prices do not reflect the true economic costs of products, barriers to implementing a CE will persist (European Commission, 2014). Thus, it is widely recommended that with the emphasis on the importance of investment by the private sector, the wastewater treatment sector also needs to adopt a full-cost recovery model that charges users of the reclaimed water a price that covers the full cost incurred in wastewater treatment.

- **Public Perception:** Public perception regarding resources recovered from wastewater and wastewater reuse is a significant barrier that needs to be thoroughly investigated and understood to enhance the market value of recovered water and materials.
- **Technology:** The full-scale implementation of innovative recovery technologies is still limited. The impacts of emerging technologies on most wastewater product recovery have not yet been completely assessed in terms of sustainability and economics, and in many cases, the technology readiness level (TRL) is still below 5 (Puyol et al., 2017).

Research and studies indicate that to overcome these barriers, widespread full-scale implementation of circular solutions for wastewater requires a standardised approach to evaluate fit-for-purpose developing technologies addressing environmental, cost, social, market, and political aspects (e.g., the policy favouring GHG reduction over resource recovery) and legislative barriers. Financial instruments, incentives and adequate regulatory mechanisms are also required to support public and private engagement in CE pathways.

3.4 TECHNICAL EVALUATION OF THE EHTP PROCESS AS AN EMERGING TECHNOLOGY

3.4.1 Technology overview

To support the successful transitioning of the water sector to a CE, the role of innovation at an appropriate TRL is quite critical. Emerging biochemical and thermochemical technologies that recover traditional resources like energy, nutrients, and other by-products across the wastewater treatment cycle play a significant role. These technologies can accelerate the adoption of CE principles by converting WWTPs into resource recovery facilities at the centre of that transition and need to be assessed in terms of sustainability and economics. One such technology is the emerging enhanced hydrothermal polymerisation (EHTP) process (Figure 3.2). The technology processes a wide range of waste biomass in an anaerobic chemical environment at temperatures around 180°C–240°C, autogenous pressure of <3.5 MPa and retention time of less than two hours. Under these conditions, most organics remain as they are or are converted to liquid. The amount of gas produced is relatively small

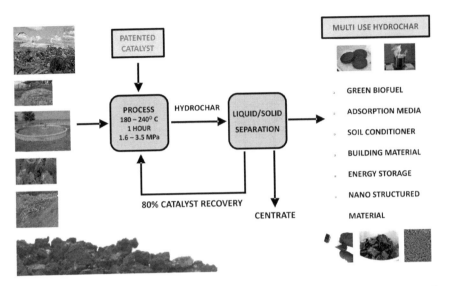

FIGURE 3.2 Schematic representation of the EHTP process (Adapted from Musvoto et al., 2018).

(~5% of solid feedstock) and low in carbon dioxide (CO_2), with no methane (CH_4) generated. Thus, the process has minimal GHG effects. Therefore, the process is quite similar to hydrothermal carbonisation (HTC), except that catalysts are selected to reduce decarboxylation reactions and CO_2 evolution.

Previous studies have demonstrated the potential feasibility of the EHTP technology, at a pilot scale, to co-process WWS and other external biomass into a multiuse hydrochar (Musvoto et al., 2018, 2019). In this regard, pilot scale studies have been successfully applied to process the following:

- WWS on its own includes primary sludge (PS), waste-activated sludge (WAS), anaerobically digested sludge (DS), and combined sludge.
- Faecal sludge (FS) from ventilated improved pit (VIP) toilets.
- WWS and FS, in combination with other waste biomass from the community.

3.4.2 Co-processing of Sludge and Other Biomass
Using an Emerging Technology

Data from evaluation studies of the EHTP process have shown that the process treated both WWS and FS to produce a completely sterile hydrochar with no microbial life (Musvoto et al., 2018, 2019). The hydrochar had a higher calorific value than the original sludge feedstock except for pre-processed feedstock such as DS, as illustrated in (Figure 3.3).

Furthermore, the EHTP process was observed to destroy selected endocrine-disrupting compounds (EDCs) and pharmaceuticals present in WWS. Analysis of the hydrochar has also shown that it has multiple potential uses, such as:

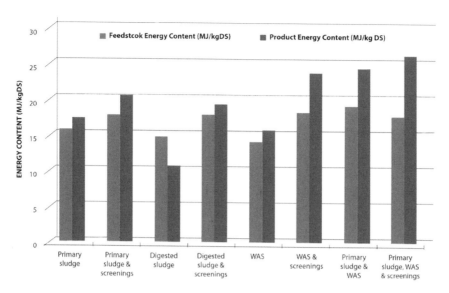

FIGURE 3.3 Energy content values for different feedstock combinations processed by the EHTP process and generated products (hydrochar).

- Biofuel that can be used for combined heat and power (CHP) generation at WWTPs, co-combustion with coal or other green biofuels in power stations, as a substitute for coal in pulverised coal injection processes and domestic use as a replacement for polluting fuels like firewood, coal, and kerosene.
- In agriculture, as a fertiliser/soil conditioner.
- Building materials in cement and brick making.
- Adsorption media for tertiary water/wastewater effluent treatment instead of conventional coal-derived granular-activated carbon (GAC).

3.4.3 Technology performance

The performance of the EHTP process has been evaluated based on compliance with regulations and potential disposal and beneficial use routes for wastewater solids in South Africa.

3.4.3.1 Compliance with South African sludge management regulations

The original sludge feedstock and generated hydrochar were classified according to the classification given in the Department of Water and Sanitation (DWS) Guidelines for the Utilisation and Disposal of Wastewater Sludge (Herselman & Moodley, 2009; Snyman & Herselman, 2006). The sludge classification system is based on three classes: microbiological content, stability and level of pollution caused by organic and inorganic contaminants. Table 3.1 illustrates the sludge classification system.

The poorest sludge quality is classified as Class C3c biosolids, while Class A1a biosolids represent a high-quality sludge with high beneficial value and possible commercial application.

TABLE 3.1
South African sludge classification system

Microbial class	A	B	C
Stability class	1	2	3
Pollution class	a	b	c

TABLE 3.2
Microbiological content of feedstock and EHTP hydrochar

	Escherichia coli (colonies/g)		Helminth Ova (count/dry gram)	
Parameter	Feedstock	Hydrochar	Feedstock	Hydrochar
	Sludge from WWTP			
PS & WAS	5×10^7	0	60	0
DS	5.1×10^5	0	5	0
	Faecal sludge from VIP latrines			
Area (A) FS[a]	6.2×10^4	0	0	0
Area B FS[b]	1.5×10^4	10	151	0

[a] Samples from pit latrines frequently emptied (once a week or less).
[b] Samples from stockpiled FS that have undergone significant biological degradation.

3.4.3.1.1 Microbiological class

A summary of the microbiological content of WWS and FS feedstock and the hydrochar produced from the EHTP process is given in Table 3.2. Comparing the microbiological content in the feedstock and produced hydrochar with the limits in the DWS Guidelines shows that both the WWS and FS feedstock, including anaerobically DS, fall into Class C (Herselman & Moodley, 2009; Snyman & Herselman, 2006). The EHTP process removed all microbial life and produced a Class A hydrochar.

3.4.3.1.2 Stability class

Since the EHTP is a thermal process, it is designed to produce hydrochar that satisfies the stability Class 1 of the DWS Guideline (Herselman & Moodley, 2009; Snyman & Herselman, 2006).

3.4.3.1.3 Pollutant class

Ultimate analysis was carried out on feedstock and hydrochar following EHTP processing to determine the concentration of metals stipulated in the DWS Guidelines, as shown in Table 3.3. The results generally showed an increase in the content of heavy metals in the hydrochar for all feedstock samples. This indicates that heavy metals are retained in the solid product in the EHTP reactor and not transferred into the liquid during the EHTP processing of sludge. In this regard, the classification of the hydrochar in terms of the DWS Guidelines depends on the metal content of

TABLE 3.3

Concentration of regulated metals in sludge from a typical WWTP (Musvoto et al., 2018)

	Primary sludge			WAS			Digested sludge		
	Feed	Product	% Increase	Feed	Product	% Increase	Feed	Product	% Increase
Compulsory metals (mg/kg)									
Arsenic (As)	12	11	-6.7	0	20	100.0	20	0	-100.0
Cadmium (Cd)	0	0		0	0		0	0	
Chromium (Cr)	202	289	43.1	152	371	143.9	277	290	4.6
Copper (Cu)	266	427	60.5	184	495	169.2	326	398	22.1
Lead (Pb)	82	152	83.9	143	384	168.6	301	245	-18.8
Mercury (Hg)	0	0		0	0		0	0	
Nickel (Ni)	48	87	81.8	0	0		73	0	-100.0
Zinc (Zn)	2,053	2,886	40.6	1,324	3,262	146.4	2,318	3,039	31.1
Some of the recommended benchmark metals (mg/kg)									
Manganese (Mn)	541	384	-29.1	898	1,445	61.0	1,069	1,225	14.6
Molybdenum (Mo)	16	23	45.6	7	17	131.0	10	19	84.8
Selenium (Se)	19	22	15.1	9	14	53.7	20	27	31.4
Strontium (Sr)	103	104	1.2	90	142	57.3	123	153	24.0
Thallium (Ti)	2,254	3,780	67.7	1,384	3,679	165.9	2,489	3,632	45.9
Vanadium (V)	84	151	80.8	44	113	154.6	87	97	12.5

the original feedstock (Herselman & Moodley, 2009; Snyman & Herselman, 2006). Although the EHTP process increased the heavy metal content of the hydrochar, the metal content is still low enough for the hydrochar to be classified as Class A. Similar results were obtained for FS, where the heavy metal content is very low, and the increase through the EHTP process does not change the hydrochar pollutant class.

3.4.3.1.4 Other micropollutants

The efficiency of the EHTP process in removing contaminants of emerging concern (CECs) was also evaluated at both laboratory and pilot scales (Musvoto et al., 2019). WWS feedstock was processed in the EHTP reactor, and both the feedstock and produced hydrochar and process supernatant were analysed for selected pharmaceuticals, oestrogens and per–polyfluoroalkyl substances. The results showed significant destruction of the selected CECs following the processing of WWS using the EHTP process.

3.4.3.1.5 Process water

About 10%–20% of the initial solids content was converted to liquid in the EHTP process. The process, therefore, produces an exceptionally low volume of process water consisting of the initial water content and water generated from liquified solids. Like the hydrochar given in Table 3.2, process water analysis has shown that it is completely sterile (no microbial life). However, the process water contains a high TCOD, TKN, and P concentration and is characterised by low pH. At centralised WWTPs, the process water can be returned to the inlet works after pH adjustment and co-treated with the incoming wastewater.

3.4.4 Beneficial uses

3.4.4.1 Biofuel

The EHTP process produced hydrochar with a higher calorific value than the feedstock except in cases where the feedstock has been previously pre-processed (e.g., DS, old FS). Combining pre-processed sludge with untreated sludge and/or other waste biomass (e.g., inlet works screenings, waste biomass from the community) increases the calorific value of the hydrochar. The calculated characteristics necessary to describe the energy content of both the feedstock and hydrochar are higher heating value (HHV), fuel ratio, hydrochar yield (Hy), energy densification (Ed) and energy yield (Ey). These characteristics for selected feedstock and produced hydrochar are summarised in Table 3.4.

The sludge that was not pre-processed (PS and WAS) as combined sludge feedstock produced hydrochar with higher calorific values and energy densification above 1, showing that the EHTP process improves energy densification in the feedstock. The fuel ratio (Fixed Carbon/Volatile Content) and the ash content for hydrochar were also higher than the feedstock. WWTPs processing WWS could produce better quality biofuel than the plants processing FS due to lower ash content on the feed and hydrochar.

Table 3.5 summarises the elemental composition and calculated O/C and H/C ratios of the feedstock and hydrochar. These ratios decreased during the EHTP process due to dehydration and decarboxylation reactions. The O/C and H/C ratios were plotted on a Van Krevelen diagram (Figure 3.4), a widely accepted method for comparing the fuel properties of coals and other biofuels (Peters et al., 2016).

TABLE 3.4

Proximate analysis results and biofuel characteristics (processing temp. 190°C–200°C)

	Volatile (%)	Ash (%)	Fixed C (%)	HHV (MJ/kgDS)	Fuel ratio	H_y (%)	Ed	E_y (%)
Sludge feedstock								
PS/WAS Feedstock	68.5	17	11.6	20.3	0.17			
PS/WAS Hydrochar	68.1	19.9	14.8	25.4	0.22	62.7	1.25	78.4
PS/WAS + Screenings Feedstock	73.0	6.7	13.1	22.3	0.18			
PS/WAS + Screenings Hydrochar	78.9	14	14.5	27.6	0.18	47.9	1.24	59.3
DS Feedstock	60.7	29.6	9.7	18.6	0.16			
DS Hydrochar	44.4	44.1	11.7	16.4	0.26	64.7	0.88	57.0
DS/Screenings Feedstock	75.4	11.1	13.5	22.0	0.18			
DS/Screenings Hydrochar	70.2	13.4	16.5	25.0	0.24	60.1	1.14	68.3
Faecal sludge feedstock								
Area B Coarse Screened FS Feedstock	49.0	43.5	7.3	12.6	0.15			
Area B Coarse Screened FS Hydrochar	35.6	54.2	10.0	10.6	0.28	45.4	0.84	38.3
Area B Fine Screened FS Feedstock	46.9	46.1	6.7	10.8	0.14			
Area B Fine Screened FS Feedstock Hydrochar	30.3	59.9	9.8	9.2	0.32	53.9	0.86	46.2
Area B Coarse Screened FS/PS&WAS	51.8	39.1	8.2	12.4	0.16			
Area B Coarse Screened FS/PS&WAS Hydrochar	37.2	52.6	10.1	13.2	0.27	63.1	1.07	67.3
Area B Fine Screened FS/PS&WAS	50.6	40.1	9.0	11.2	0.18			
Area B Fine Screened FS/PS&WAS Hydrochar	35.3	54.1	10.5	12.4	0.30	51.8	1.11	57.5
Area (A) FS	64.2	25.3	10.3	17.6	0.16			
Area (A) FS Hydrochar	49.5	40.8	9.8	13.5	0.20	72.0	0.77	55.5
Area (A) FS/PS &WAS	66.7	18.8	14.4	20.4	0.22			
Area (A) FS/PS &WAS Hydrochar	64.0	20.8	15.2	23.4	0.24	60.0	1.15	68.9

TABLE 3.5

Elemental analysis and H/C and O/C ratios

Sample	Elemental analysis (% DS)					H/C	O/C
	C	N	H	S	O		
Sludge and faecal sludge							
Primary Sludge Hydrochar	36.9	2.0	5.1	1.3	12.0	0.14	0.33
Primary Sludge + Screenings Hydrochar	36.2	1.7	6.6	0.7	20.9	0.18	0.58
WAS Feedstock	31.0	12.8	3.0	1.3	40.7	0.10	1.31
WAS Hydrochar	41.9	13.0	2.8	0.9	34.0	0.07	0.81
Digested Sludge Feedstock	28.0	3.6	4.6	1.3	14.7	0.16	0.52
Digested Sludge Hydrochar	26.8	2.3	4.0	0.9	11.6	0.15	0.43
Composted Sludge feedstock	24.4	14.4	3.3	1.3	50.3	0.14	2.06
Composted Sludge Hydrochar	34.2	16.4	2.6	0.9	49.4	0.08	1.44
Area B Coarse Screened FS Feedstock	27.3	2.0	3.7	0.9	17.9	0.13	0.66
Area B Coarse Screened FS Hydrochar	24.9	1.7	3.0	0.7	10.3	0.12	0.41
Area B Fine Screened FS Feedstock	25.7	2.1	3.8	0.9	21.5	0.15	0.84
Area B Fine Screened FS Hydrochar	15.4	1.3	1.9	0.5	21.0	0.12	1.36
Area B Fine Screened FS + PS & WAS Feedstock	30.0	2.3	4.6	0.9	23.1	0.15	0.77
Area B Fine Screened FS + PS & WAS Hydrochar	32.4	2.0	3.9	0.7	8.3	0.12	0.26
Area B Coarse Screened FS PS & WAS Feedstock	19.2	1.6	2.9	0.6	35.6	0.15	1.85
Area B Coarse Screened FS+P & WAS Hydrochar	22.5	1.7	3.0	0.5	18.2	0.13	0.81
Area (A) FS Feedstock	39.9	3.5	5.7	0.8	18.0	0.15	0.45
Area (A) FS Hydrochar	39.6	2.3	5.6	0.5	9.7	0.14	0.24
Area (A) FS + PS & WAS Feedstock	38.5	4.9	6.0	0.7	31.1	0.16	0.81
Area (A) FS + PS & WAS Hydrochar	52.3	2.5	7.1	0.5	16.8	0.14	0.32

Other fuels

Wood	50.0		6.0		44	0.12	0.88
Peat	54.8	0.9	5.4	0.1	35.8	0.10	0.65
Lignite	70.0	25.0	5.0		25	0.07	0.36
Coal (Pittsburgh Seam)	75.5	1.2	5.0	3.1	4.9	0.07	0.06
Bituminous Coal	83.0	2.0	5.0		11	0.06	0.13
Anthracite	83.0	2.0	3.5		2	0.04	0.02

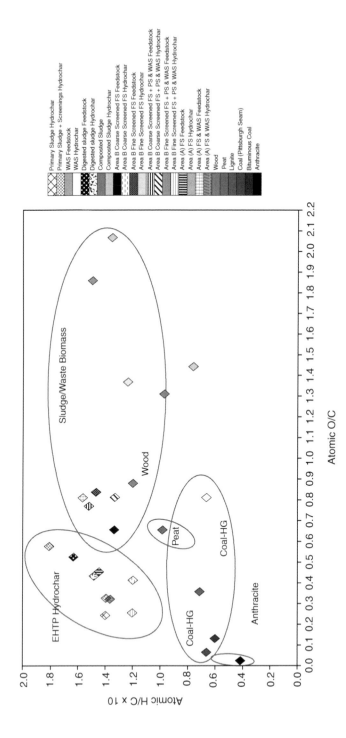

FIGURE 3.4 Van Krevelen diagram for sludge feedstocks, hydrochars from the EHTP process, coals and other fuels.

The highest-ranked coals have the lowest O/C and H/C ratios, and are normally in the bottom left corner of the diagram.

The EHTP process enhances the fuel properties of biomass by removing hydrogen and oxygen, resulting in carbon densification in the hydrochar. The sludge feedstocks and combined sludge and other biomass had oxygen and hydrogen content higher than low-grade brown coal. After EHTP processing, hydrochar oxygen and hydrogen contents are reduced, and the hydrochar O/C ratio of values between low bituminous coal and brown coal is achieved while the H/C ratios are higher than coal.

CO_2 emissions from fuels depend primarily on their carbon content and their hydrogen–carbon ratio. Over the years, fossil fuel usage trends have tended toward a higher hydrogen-to-carbon (H/C) ratio. The higher the H/C ratio, the higher the energy efficiency of the fuel and the lower the CO_2 emissions from its combustion. Primitive fuel, such as wood, had twice the carbon content compared to its successor, coal. However, coal, with a lower H/C ratio, is twice as energy efficient than wood. Later, coal was succeeded by oil, which had a much higher H/C ratio and thus benefited over wood and coal in having higher energy efficiency and lower CO_2 emissions. Natural gas has an even lower carbon content compared to oil. However, the ratio of hydrogen to carbon is still lower in biofuels. In fact, biofuels such as hydrogen have zero H/C ratios.

The EHTP process improves the fuel characteristics of sludge and other waste biomass by producing a hydrochar with lower H/C and O/C ratios and higher calorific value. Hydrochar also has a higher H/C ratio than traditional fuels such as coal and will thus have a lesser carbon emission when combusted as a biofuel.

However, it must be noted that the ash content of the hydrochar is higher than that of high-grade coal and will, therefore, impact the combustion efficiency of the hydrochar.

3.4.4.2   Agriculture

The hydrochar produced from the EHTP process has higher concentrations of nutrients and carbon than the feedstock. Thus, the hydrochar can be used as a soil conditioner/fertiliser provided that the heavy metal concentrations do not exceed the limits in the Sludge Guidelines (Herselman & Moodley, 2009; Snyman & Herselman, 2006). In this regard, a detailed investigation of the application of hydrochar generated from sludge for agricultural purposes is required. This should use the already developed Sludge Application Rate Advisor (Tesfamariam et al., 2015), a useful tool for sludge classification, rate application, and metal accumulation prediction.

3.4.4.3   Adsorption media

Preliminary laboratory tests have shown that hydrochar produced from processing woody biomass in the EHTP process can be applied as adsorption media and has characteristics like some commercial-grade activated carbon. Studies are being undertaken to investigate the efficacy of hydrochar from processing sludge as an adsorption media. The use of hydrochar from processing sludge as an adsorption media can be useful as a polishing step to treat final effluent from WWTPs as part of wastewater reclamation and recycling in support of the water pathway within a CE.

3.4.4.4 Other applications

Hydrochar also has the potential to be used as a building material (cement and brick making) and cathode in microbial fuel cells (MFCs), as well as an energy storage device due to the presence of nitrogen functional groups. Further investigations on these applications need to be undertaken to ensure diversified applications are feasible.

3.4.5 Applications for the EHTP process

The field tests have indicated that the EHTP process can be applied to process sludge independently and in combination with other waste biomass to produce a sterile hydrochar with various potential uses.

Based on the results from the field testing, the EHTP process can be applied for wastewater solids and other community waste biomass management within a CE as follows:

- process untreated WWS or further treat pre-DS at centralised WWTPs in combination with other waste biomass from the community. FS from low-cost sanitation systems can also be co-processed.
- FS from low-cost sanitation systems at a centralised facility or a facility for a few households. Application for individual households at a small scale is also feasible.

These applications are graphically illustrated in Figures 3.5 and 3.6, as previously reported by Zvimba et al. (2021). Figure 3.5 illustrates the incorporation of the EHTP process into WWTPs infrastructure. Generally, the incorporation of the EHTP process into the current WWTP infrastructure demonstrates the application of the technology for the treatment of WWS in combination with other biomass to a quality higher than

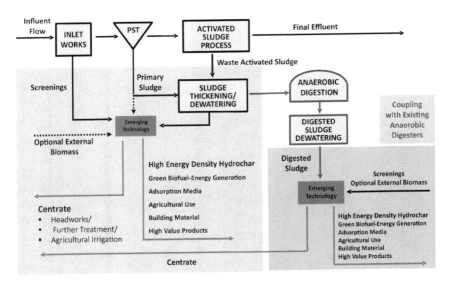

FIGURE 3.5 Schematic layout of EHTP process retrofit into existing WWTP infrastructure.

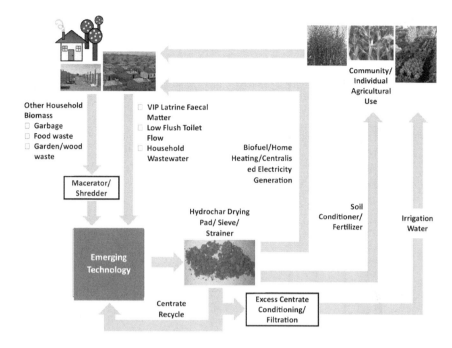

FIGURE 3.6 Detailed schematic illustration for application of the EHTP processing faecal sludge from low-cost sanitation systems at a centralised facility in combination with another household biomass.

generally achieved by commonly applied biochemical conversion processes utilised by the wastewater services sector, including further reduction of sludge quantity.

Thus, the EHTP process provides significant flexibility based on its ability to process different sludge combinations, including incorporating screenings and other external biomass generating high energy content hydrochar. This further indicates the need for coupling the emerging technology with current WWTPs, thereby avoiding redundancy of existing infrastructure and advancing the vision of establishing resource efficiency within wastewater management to support transitioning to CE.

Besides retrofitting the emerging technology into existing infrastructure, as outlined in Figure 3.5, the EHTP process can be applied as a standalone technology for greenfield applications. Figure 3.6 shows the closed-loop CE concept for applying the EHTP process for processing FS from a low-cost sanitation system at a centralised facility, as previously reported by Zvimba et al. (2021).

Figure 3.6 demonstrates the possible integration of waste management as a wide range of biomass generated within communities can be potentially processed using the EHTP process to generate materials useful for meeting community resource requirements in support of energy and food security.

As illustrated in Figure 3.6, adopting the EHTP technology for processing different waste streams facilitates the transition to a CE with possibilities of creating new business models and jobs, developing new skills and investments within communities, and reducing the carbon footprint as key social, economic, and environmental benefits. Therefore, the wastewater services sector needs to rethink its sludge management strategy, envisaging maximum benefits from resource recovery in support of the CE implementation.

3.5 APPROPRIATE TECHNOLOGIES FOR COUPLING WITH THE EHTP PROCESS TO PROMOTE A CIRCULAR ECONOMY

3.5.1 OVERVIEW

To fully implement the CE framework with WWTPs as resource recovery centres, coupling various technologies with the EHTP process is required to exploit opportunities within the three interrelated pathways (water, energy, and materials). These key technologies fall into the following categories:

- Tertiary treatment of final effluent for both non-potable and potable reuse.
- Alternative sludge treatment technologies with treated sludge fed to the EHTP process.
- Biomass processing technologies for further processing of hydrochar from the EHTP process into high-value products.
- CHP generation technology.

3.5.1.1 Water pathway

The contribution to the water pathway by wastewater treatment in the IWA framework is through wastewater reclamation and reuse (WRR). The reclaimed water can be reused in agriculture and industry and direct potable reuse. Tertiary treatment of final effluent is required to achieve the required quality for the intended reuse purpose. It involves suspended and colloidal solids removal, removal of dissolved solids and other micropollutants of concern and disinfection. Non-potable reuse often requires suspended and dissolved solids removal and disinfection, while potable reuse requires advanced treatment methods. Technologies are usually applied in series to achieve the desired reused water quality.

The EHTP process produces hydrochar that can be converted to activated carbon and used as adsorption media within the treatment processes. Figure 3.7 illustrates

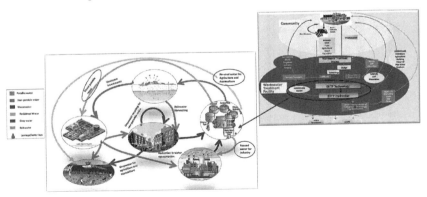

FIGURE 3.7 Illustration of potential coupling of EHTP process with other technologies within the water pathway (Adapted from IWA, 2016 and Musvoto & Mgwenya, 2022).

the potential coupling of the EHTP hydrochar activated carbon (HAC) with other tertiary treatment processes in WRR schemes within a CE.

A summary of the established and emerging technologies that can be coupled with the EHTP process for WRR as part of the water pathway is given in Table 3.6.

3.5.1.2 Energy pathway

The contribution to the energy pathway by WWTPs in the IWA framework is through the generation of energy from biosolids. The EHTP process produces hydrochar that can be used as a biofuel for CHP generation. In this regard, the EHTP technology can be coupled with the following technologies within the energy pathway, as illustrated in Figure 3.8, while associated CHP technologies are listed in Table 3.7.

- Anaerobic digesters to further treat DS combined with community biomass. CHP can be generated from the biogas produced during anaerobic digestion (AD) and EHTP hydrochar. The residual ash from hydrochar combustion can be beneficially used through the materials pathway in agriculture and the building industry for cement making. Metals can also be extracted from the ash.
- Gasification to process EHTP hydrochar to generate liquid and gaseous fuels, e.g., synoil and syngas. Similar to the above, the residual ash can be beneficially used through the materials pathway.
- EHTP hydrochar can be used to fabricate low-cost and high-performance air cathodes for MFCs. This can enable coupling with microbial fuel cell technology.

Depending on the available waste biomass from the community, other thermochemical conversion processes like hydrothermal liquefaction and pyrolysis can be incorporated at the WWTPs to process waste that cannot be processed in the EHTP reactor to increase the by-products for beneficial use.

3.5.1.3 Materials pathway

EHTP hydrochar and other by-products can be used within the materials pathway as follows:

- As a soil conditioner/fertiliser in agriculture.
- The building material for brick making. Ash from combusted biofuel can be used in cement making.
- Extraction of metals.
- Energy storage in hydrogen fuel cells.

Figure 3.9 shows the potential coupling of the EHTP process with other technologies within the material pathway. However, further research is required to develop some beneficial uses of EHTP hydrochar within the materials pathway.

TABLE 3.6

Water treatment technologies that can be coupled with the EHTP technology for wastewater reclamation and reuse as part of the water pathway

Technology	State of development	Brief description and application
Disinfection technologies		
Chlorination	Established	Chlorine-based disinfection for the removal of wastewater constituents and pathogenic microorganisms such as faecal coliforms, streptococci, Salmonella sp. and enteric viruses that are not removed by previous secondary treatments.
Ozonation	Established	Ozone applications involve oxidative reactions, where ozone can be used to disinfect or oxidise specific contaminants. Organic compounds that are difficult to oxidise include many solvents, pesticides, and compounds that cause tastes and odours.
Ultraviolet (UV) radiation	Established	UV disinfection transfers electromagnetic energy from a mercury arc lamp to an organism's genetic material (DNA and RNA) by penetrating through the cell wall and destroying the cell's ability to reproduce. UV disinfection destroys virtually all harmful pathogens, bacteria, viruses, spores and cysts. UV can also inactivate protozoa, notably Cryptosporidium and Giardia, that cannot be destroyed through chlorine-based disinfection.
Peracetic acid (PAA)	Emerging	PAA is an oxidising agent used as a routine wastewater disinfectant. It is a stronger oxidant than hypochlorite or chlorine dioxide but not as strong as ozone. PAA does not affect effluent toxicity, so it needs not to be removed as with chlorine. PAA does not explode. The solution is acidic (pH 2) and requires handling, transport, and storage care.
Adsorption technologies		
Granular- and powdered-activated carbon (GAC and PAC)	Established	Adsorption media for organic and inorganic pollutants removal. When a solution containing absorbable solute comes into contact with a solid with a highly porous surface structure, liquid–solid intermolecular forces of attraction cause some of the solute molecules from the solution to be concentrated or deposited at the solid surface. Removes heavy metals, colour, and some micropollutants of concern like EDCs.
Ion exchange (IX) resins	Emerging	IX resin technology has been used extensively as a practical and effective form of water treatment. The process removes soluble ionised contaminants such as hardness and alkalinity from the water via a reversible ionic interchange between a solid phase (resin beads) and a liquid phase (water). Selective resins have also been developed to remove heavy metals, nitrate, perchlorate, and other contaminants.

Novel green-activated carbons	Emerging	Green-activated carbons are made from renewable non-fossil fuel sources such as sawdust, waste tyres, prawn shells, mango seed kernels, wood chips, wheat straws, lemon peel, orange peel, tree barks, rice husks, maize cobs, and hazelnut husks. Activated carbon from EHTP hydrochar also falls into this category. Green-activated carbons remove heavy metals, colour, and micropollutants of concern, like EDCs. Some also remove other contaminants like nitrate, phosphorus, and ammonia.

Membrane liquid separation technologies

Ultrafiltration (UF)	Established	Pressure-driven ultrafine membrane media for solids removal. Removes particles 0.02–0.05 microns, including bacteria, viruses, and colloids. Usually applied in WRR to produce water for specific industrial reuse or as pre-treatment for reverse osmosis (RO) in direct and indirect potable reuse.
Nanofiltration (NF)	Established	Pressure-driven speciality membrane process that operates between UF and RO and rejects dissolved solutes in the range of 1 nanometre. These include organic molecules (molecular weight 200–400), metals and multivalent ions such as calcium chloride, sodium chloride, bacteria, and viruses. Application in WRR is usually for rejecting monovalent ions, such as chloride.
Reverse osmosis (RO)	Established	Pressure-driven separation process that employs a semipermeable membrane and the principles of crossflow filtration. Most effective separation process for all salts and inorganic molecules and organic molecules with a molecular weight greater than 100. Removes contaminants such as endotoxins/pyrogens, insecticides/pesticides, herbicides, antibiotics, nitrates, sugars, soluble salts, metal ions, bacteria, and viruses. Used as a polishing/further treatment stage in WRR for potable reuse or high-quality industrial reuse.

Advanced oxidation processes (AOP)

Hydroxyl radical and ozone-based AOPs	Established	Removal of recalcitrant organics that include non-biodegradable COD, TOC, VOC, dyes, surfactants, pesticides, herbicides, disinfection by-products, and EDCs.
Other novel AOPs (catalytic ozonation, photocatalysis)	Emerging	

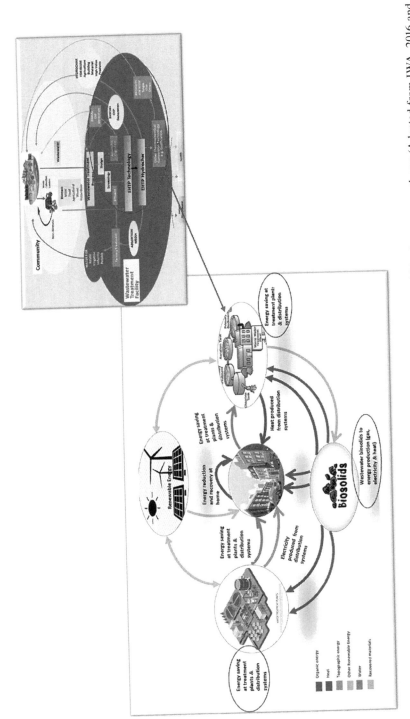

FIGURE 3.8 Illustration of potential coupling of EHTP process with other technologies within the energy pathway (Adapted from IWA, 2016 and Musvoto & Mgwenya, 2022).

TABLE 3.7

Technologies that can be coupled with the EHTP process as part of the energy pathway

Technology	State of development	Brief description and application
		Biochemical conversion processes
Anaerobic digestion (AD) and aerobic digestion	Established	The AD process consists of several sequential and parallel biochemical reactions that break down organic waste material to methane and carbon dioxide in the absence of oxygen to produce biogas containing mostly methane and carbon dioxide. Biogas can be burned directly for heat or steam or used in CHP generation. Aerobic digestion is the degradation of the organic sludge solids in the presence of oxygen. The microorganisms in the sludge convert the organic material to carbon dioxide and water, and the ammonia and amino species to nitrate. Sludge from both technologies can be processed in the EHTP technology with other waste biomass to generate hydrochar for CHP generation.
Microbial fuel cells (MFCs)	Emerging	MFC technology utilises microbes in the oxidation of organic substances to produce electricity. MFCs enable energy recovery from municipal wastewater while limiting both the energy input and excess sludge production. Good effluent quality and a low environmental footprint can be achieved from the process because of an effective combination of biological and electrochemical processes. The process is inherently amenable to real-time monitoring and control, which benefits good operating stability. EHTP hydrochar can be used to fabricate low-cost and high-performance air cathodes for MFCs.
		Thermochemical conversion processes
Gasification	Established	Thermochemical conversion process that converts biomass into gases, which are then synthesised into the desired chemicals or used directly. Thermal energy production is the main driver for this conversion route with five broad pathways: combustion, carbonisation, pyrolysis, gasification, and liquefaction. Hydrochar from the EHTP process can be further gasified with other waste biomass to produce synoil and syngas for CHP generation.
		Combined heat and power generation technologies
Various technologies	Established and Emerging	Include boilers, turbines and novel technologies like fuel microgrids that convert the hydrochar that is generated from the EHTP technology to heat and electric power.

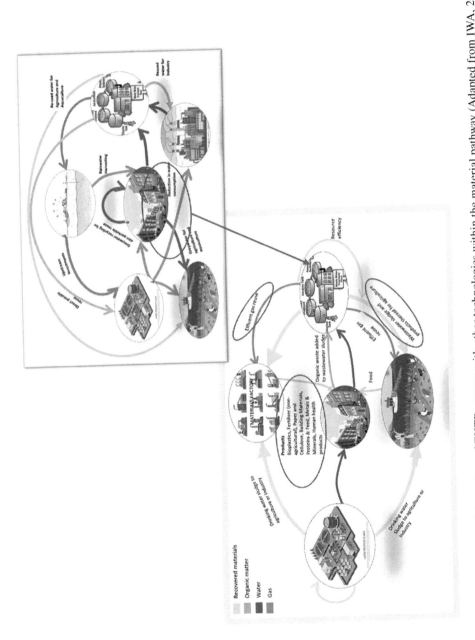

FIGURE 3.9 Illustration of potential coupling of EHTP process with other technologies within the material pathway (Adapted from IWA, 2016 and Musvoto & Mgwenya, 2022).

3.6 CONCLUSIONS

This chapter has reviewed the global status of CE adaptation globally and in LMICs like South Africa. A review of the IWA framework demonstrated that it covers all aspects of the water cycle and is considered the most appropriate framework and strategy for adoption by the water sector for transitioning to CE within LMICs. Junction opportunities presented by the three interrelated pathways of water, energy and materials are critical in achieving this transition in the water sector. Moreover, innovation has been noted to play a significant role as an accelerator in transitioning the water sector to a CE. In this regard, the emerging EHTP process has been demonstrated as a feasible technology for processing WWS in combination with FS and other waste biomass to a multiuse hydrochar useful as a biofuel, adsorption media, soil ameliorant and construction material. The possible coupling of the emerging technology with existing well-established technologies and other emerging technologies supporting the water, energy, and material pathways has also been outlined as feasible.

Furthermore, integrating waste management through the application of this multi-biomass processing emerging innovative technology as an accelerator to a CE within community settings is quite possible and achievable. Potential key social, economic, and environmental benefits of such an approach include creating new business models, jobs and new investment opportunities, developing new skills, and reducing the waste management carbon footprint. Overall, the chapter highlights the need for the wastewater services sector to rethink its wastewater management strategy, envisaging maximum benefits from resource recovery across the wastewater treatment cycle to support the CE implementation.

REFERENCES

Abu-Ghunmi, D., Abu-Ghunmi, L., Kayal, B. & Bino, A. 2016. Circular Economy and the Opportunity Cost of Not 'Closing the Loop' of Water Industry: The Case of Jordan. *Journal of Cleaner Production*, 131: 228–236.

Accenture. 2014. *Circular Advantage: Innovative Business Models and Technologies to Create Value in a World without Limits to Growth*. Available at: https://accntu.re/2kYSfPn.

Acsinte S and Verbeek A. 2015. Assessment of Access-to-Finance Conditions for Projects Supporting Circular Economy – Final report. Report prepared for DG Research and Innovation of the European Commission by InnovFin Advisory and European Investment Bank Advisory Services, Luxembourg.

Bačová, M., Böhme, K., Guitton, M., van Herwijnen, M., Kállay, T., Koutsomarkou, J., Magazzù, I., O'Loughlin, E. & Rok, A. 2016. *Pathways to a Circular Economy in Cities and Regions: A Policy Brief Addressed to Policy Makers from European Cities and Regions*. France: Interreg Europe Joint Secretariat.

Batstone, D.J., Hülsen, T., Mehta, C.M. & Keller, J. 2015. Platforms for Energy and Nutrient Recovery from Domestic Wastewater: A Review. *Chemosphere*, 140: 2–11.

Bowdish, L. 2016. *Trash to Treasure: Changing Waste Streams to Profit Streams*. Washington, DC.

de Jong, S., van der Gaast, M., Kraak, J., Bergema, R. & Usanov, A. 2016. *The Circular Economy and Developing Countries: A Data Analysis of The Impact of a Circular Economy on Resource Dependent Developing Nations*. Available at: https://hcss.nl/wp-content/uploads/2016/07/CEO_The-Circular-Economy.pdf

Ellen MacArthur Foundation. 2015. *Growth within: A Circular Economy Vision for a Competitive Europe*. Available at: https://www.ellenmacarthurfoundation.org/growth-within-a-circular-economy-vision-for-a-competitive-europe

European Commission. 2014. *Towards a Circular Economy: A Zero Waste Programme for Europe*. Brussels.

European Commission. 2015. *Closing the Loop - An EU Action Plan for the Circular Economy*. Brussels.

European Commission. 2017. *The Role of Waste-to-Energy in the Circular Economy. Communication from the Commission to the European Parliament, the Council, the European Economic and Social Committee and the Committee of the Regions*. Available at: https://eur-lex.europa.eu/legal-content/EN/TXT/HTML/?uri=CELEX%3A52017DC0034

European Commission. 2018. *Behavioural Study on Consumers' Engagement in the Circular Economy*. Luxembourg.

European Environment Agency. 2016. *Circular Economy in Europe - Developing the Knowledge Base*. Luxembourg.

Galvão, A.G.D., De Nadae, J., Clemente, D.H., Chinen, G. & De Carvalho, M.M. 2018. Circular Economy: Overview of Barriers. In *10th CIRP Conference on Industrial Product-Service Systems, IPS 2018*. Linköping: Elsevier B.V.: 79–85.

Geng, Y., Sarkis, J. & Bleischwitz, R. 2019. How to Globalize the Circular Economy. *Nature*, 565(7738): 5–7.

Geng, Y., Zhu, Q., Doberstein, B. & Fujita, T. 2009. Implementing China's Circular Economy Concept at the Regional Level: A Review of Progress in Dalian, China. *Waste Management*, 29(2): 996–1002.

Greyson, J. 2007. An Economic Instrument for Zero Waste, Economic Growth and Sustainability. *Journal of Cleaner Production*, 15(13–14): 1382–1390.

Herselman, J.E. & Moodley, P. 2009. *Guidelines for the Utilisation and Disposal of Wastewater Sludge: Volume 4 Requirements for the Beneficial Use of Sludge at High Loading Rates*. Pretoria.

Hislop, H. & Hill, J. 2011. *Reinventing the Wheel: A Circular Economy*. London: Green Alliance.

Humphris-Bach, A., Essig, C., Morton, G. & Harding, L. 2016. *EU Resource Efficiency Scoreboard 2015*. European Union. Available at: https://ec.europa.eu/environment/resource_efficiency/targets_indicators/scoreboard/pdf/EU Resource Efficiency Scoreboard 2015.pdf

International Water Association (IWA). 2016. *Water Utility Pathways in a Circular Economy*. Published by the International Water Association. Available at: https://iwa-network.org/water-utility-pathways-circular-economy-charting-course-sustainability/

Jazbec, M., Mukheibir, P. & Turner, A. 2020. *Transitioning the Water Industry with the Circular Economy, Prepared for the Water Services Association of Australia*, Institute for Sustainable Futures, University of Technology Sydney, September 2020. Available at: https://opus.lib.uts.edu.au/handle/10453/143397

Klimentov, M. 2018. *Coming Full Circle? The State of Circular Economies around the Globe*. Available at: https://www.greenbiz.com/article/coming-full-circle-state-circular-economies-around-globe

McDowall, W., Geng, Y., Huang, B., Barteková, E., Bleischwitz, R., Türkeli, S., Kemp, R. & Doménech, T. 2017. Circular Economy Policies in China and Europe. *Journal of Industrial Ecology*, 21(3): 651–661.

Musvoto, E. & Mgwenya, N. 2022. *The Role of Emerging Innovative Wastewater Sludge to Energy Technologies in Transitioning to a Circular Economy in the Water Sector: A South African Case Study*. Pretoria.

Musvoto, E., Mgwenya, N., Cingo, X. & Zvinowanda, C. 2019. *Application of Emerging Low Energy Technologies for the Removal of Endocrine Disrupting Compounds in Wastewater and Wastewater Sludge*. Pretoria.

Musvoto, E., Mgwenya, N., Mangashena, H. & Mackintosh, A. 2018. *Energy Recovery from Wastewater Sludge – A Review of Appropriate Emerging and Established Technologies for the South African Industry*. Pretoria.

Republic of South Africa. 2009. National Environmental Management Waste Act 59 (NEMA) of 2008. Government Gazette No. 32000, Vol. 525 (March 2009). Government Printer, Cape Town, Cape Town, p. 47.

Pearce, D.W. & Turner, R.K. 1991. Economics of Natural Resources and the Environment. *American Journal of Agricultural Economics*, 73(1): 227–228.

Peters, K.E., Xia, X., Pomerantz, A.E. & Mullins, O.C. 2016. Chapter 3- Geochemistry Applied to Evaluation of Unconventional Resources. In Y.Z. Ma & S.A.B.T.-U.O. and G.R.H. Holditch, eds. *Unconventional Oil and Gas Resources Handbook*. Boston: Gulf Professional Publishing: 71–126.

Puyol, D., Batstone, D.J., Hülsen, T., Astals, S., Peces, M. & Krömer, J.O. 2017. Resource Recovery from Wastewater by Biological Technologies: Opportunities, Challenges, and Prospects. *Frontiers in Microbiology*, 7(JAN): 1–23.

Rizos, V., Tuokko, K. & Behrens, A. 2017. *The Circular Economy a Review of Definitions, Processes and Impacts*. Report No 2017/08, CEPS, Brussels.

Snyman, H.G. & Herselman, J.E. 2006. *Guidelines for the Utilisation and Disposal of Wastewater Sludge Volume 1: Selection of Management options*. Pretoria.

Swartz, C.D., van der Merwe-Botha, M. & Freese, S.D. 2013. *Energy Efficiency in the South African Water Industry: A Compendium of Best Practices and Case Studies*. Pretoria.

Tesfamariam, E.H., Annandale, J.G., de Jager, P.C., Ogbazghi, Z., Malobane, M.E. & Mbetse, C.K.A. 2015. *Quantifying the Fertilizer Value of Wastewater Sludges for Agriculture*. Pretoria.

Vanner, R., Bicket, M., Withana, S., ten Brink, P., Razzini, P., van Dijl, E., Watkins, E., Hestin, M., Tan, A., Guilcher, S. & Hudson, C. 2014. *Scoping Study to Identify Potential Circular Economy Actions, Priority Sectors, Material Flows and Value Chains: Study Prepared for the European Commission, DG Environment's Framework*. Publications Office of the European Union, Luxembourg, 18 Sept 2014.

Van Vuuren, S.J., Dijk, M. Van & Loots, I. 2014. *Conduit Hydropower Pilot Plants*. Pretoria.

Zvimba, J.N. & Musvoto, E.V. 2020. Modelling Energy Efficiency and Generation Potential in the South African Wastewater Services sector. *Water Science and Technology*, 81(5): 876–890.

Zvimba, J.N., Musvoto, E. V., Nhamo, L., Mabhaudhi, T., Nyambiya, I., Chapungu, L. & Sawunyama, L. 2021. Energy Pathway for Transitioning to a Circular Economy within Wastewater Services. *Case Studies in Chemical and Environmental Engineering*, 4: 100144.

4 Transitional pathways towards sustainable food systems

Luxon Nhamo, Sylvester Mpandeli, Stanley Liphadzi, Samkelisiwe Hlophe-Ginindza, and Tafadzwanashe Mabhaudhi

4.1 INTRODUCTION

Food systems play an important role in sustainable development as they are at the centre of the nexus that links food and nutritional security, human health, provision of ecosystem services, climate change, and social justice (Caron et al., 2018; UNGA, 2015). However, the agriculture sector faces the challenge of meeting the food demands of a growing population without degrading the environment (Campbell et al., 2016; Misra, 2014). An increased world population of 2 billion people from the current 7 billion by 2050 will exert pressure on the agriculture sector to produce enough food to feed the increasing global population (Horton, 2017). As the population is projected to reach 9 billion people by 2050, agricultural production should increase by at least 70% during the same period to meet future food and nutritional requirements (Ehrlich and Harte, 2015; Krishna Bahadur et al., 2018). However, such changes will have to happen at a time when essential resources such as water, energy, and land are depleting and degrading, and at times compelling humankind to exceed planetary boundaries as demand and use exceed replenishment (Scoones et al., 2019; Whitmee et al., 2015). The challenges are compounded by climatic and environmental changes induced by unsustainable food systems (Misra, 2014). These adverse environmental changes result in the degradation of about 12 million hectares of fertile land globally per annum, sufficient to produce 20 tonnes of grain (Gibbs and Salmon, 2015; Higginbottom and Symeonakis, 2014). Besides, the intensity and frequency of droughts, cyclones, and floods have increased in recent years, further threatening food security (Nhamo et al., 2019a). The need to produce more food has witnessed an increase in the global cultivated area to more than a third (4.8 billion ha) of the total global surface area (13.5 billion ha) (FAO, 2020). As a result, the agriculture sector is now the second largest contributor of greenhouse gases after energy (IPCC, 2014) and the major contributor to land and water degradation (Borrelli et al., 2020).

In the case of southern Africa, agriculture contributes about 20.2% to the gross domestic product (GDP) and, thus, plays an important role in economic development (Nhamo et al., 2019b). However, the region has lost over 25% of its soil fertility over the years due to degradation and overexploitation, further exacerbating its vulnerability (FAO, 2020; Nkonya et al., 2016). This happens when the sector is expected

DOI: 10.1201/9781003327615-4

to produce more food to feed a population projected to reach 2 billion people by 2050 in southern Africa alone (Hall et al., 2017). There is, therefore, a need for transformational change in food systems through the adoption of smart and clean production systems that lead to a circular economy. Operationalising and implementing the circular economy model is anticipated to propel resource security and a cleaner environment (Hall et al., 2017). Adopting circular approaches in place of current linear models is the first transitional step towards sustainable food systems (Cosgrove and Loucks, 2015), as they provide pathways towards food and nutrition security for all and at all times without compromising the environment (Béné et al., 2019a). This is why food systems are at the heart of Sustainable Development Goals (SDGs) and are linked to at least 12 of the 17 goals (Chaudhary et al., 2018; UNGA, 2015).

A sustainable food system refers to an agricultural system that delivers healthy food to meet current food requirements while at the same time preserving healthy and sustainable ecosystems that are capable of providing food for generations to come with a controlled negative impact on the environment (Allen and Prosperi, 2016; UNGA, 2015). It is a system that encourages local production and knowledge, providing nutritious and healthy food which is available, accessible, and affordable to all at all times while protecting farmers, workers, consumers, and communities (Eakin et al., 2017).

A food system comprises sub-systems, including a farming system, waste management system, and input supply system. It is also intricately connected to other related systems such as energy, water, trade, and health systems (Figure 4.1) (Tomich et al., 2019). The interconnectedness of these systems indicates that any structural change in a food system might originate from a change in another system (Béné et al., 2019a). Thus, changes in a food system could be triggered by a policy that promotes more biofuel in the energy system, impacting the food system. Therefore, a food system is a complex system driven by intricately interlinked economic, social, cultural, and environmental factors, which require transformative thinking and integrated assessment tools to guide informed strategic policies that lead to sustainability in the whole agricultural value chain (Allen and Prosperi, 2016). Thus, food sustainability transitions include the transformation processes needed to drive changes in the food value chains towards sustainable food systems (El Bilali and Allahyari, 2018). Although it is complex, recent technological advances and digitalisation have enhanced ongoing transformation processes in global agriculture and food chains (El Bilali and Allahyari, 2018). Sustainability transitions refer to long-term, multi-dimensional, multi-sectoral, and structural transformational changes aimed at achieving shifts in socio-technical systems towards more sustainable modes of production and consumption (Klerkx and Rose, 2020). The term transition is associated with transitional pathways, a term referring to significant change processes in society (Geels et al., 2016). Sustainability transitions in the agriculture and food value chains facilitate changes towards novel production and consumption ways and practices that are more sustainable (El Bilali and Allahyari, 2018).

Therefore, transitioning towards sustainable food systems should be built around integrated strategic policies formulated around the intricately linked resources of water, land, environment and energy, nutrition, and health (Nhamo and Ndlela, 2021; Wittman et al., 2017). Transitional pathways concern a demarcated trajectory that

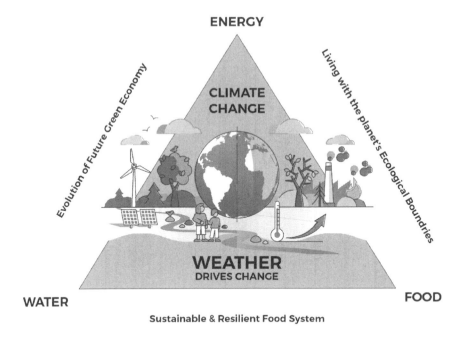

FIGURE 4.1 The impact of climate change on water, energy and food resources and how climate action drives the evolution towards the green economy and sustainable food systems and facilitates remaining within planetary boundaries.

leads from one situation to another through a particular territory. Transitions are evolutionary, open-ended, non-linear, and based on searching, learning, and experimentation (Geels et al., 2016). They are mainly supported by transformative and circular models, which are important in addressing today's challenges that cut across all sectors and require integrated, iterative and cross-sectoral interventions (Naidoo et al., 2021b). The pathways inform coherent, strategic policies that lead to sustainable adaptation and resilience.

Such informed policies provide transformative pathways towards national and regional targets like regional integration, employment creation, poverty alleviation, inclusive economic growth, climate action, and good health and well-being (Mabhaudhi et al., 2019; Nhamo et al., 2018). One such transformative approach is nexus planning, which is a catalyst for achieving the sustainability of food systems (Mabhaudhi et al., 2021; Nhamo and Ndlela, 2021). But nexus planning is also linked or informed by other transformative approaches, including scenario planning, just transitioning, circular economy, one health, strategic foresight, and horizon scanning (Nhamo et al., 2021). These circular models provide tools that inform investment decisions on agriculture infrastructure, climate-smart agriculture technologies, agriculture water management, and on-field decision-support tools to manage resource flow and implement and reduce losses (Adamides, 2020; Naidoo et al., 2021a). For example, smart systems and technologies that include product service systems and performance models are envisaged to guide the interlinkages between the circular economy and the Internet of Things (IoT) in food systems and accelerate the needed transformational change and achieve the green economy (Ingemarsdotter et al., 2019; Naidoo et al., 2021a). In this digital world of globalisation, the circular economy model is driven by digital technologies like the IoT, Big Data, and Data Analytics, which facilitate the smooth tracking and flow of products, components, and materials, allowing the derived data to be used to improve resource management and inform decision-making across various phases of the production cycle (Kristoffersen et al., 2020).

In particular, nexus planning and circular economy provide the decision-support pathways that lead to transformational change in the agricultural value chain and ensure socio-ecological sustainability (Rockström et al., 2017). Thus, this chapter aims to provide policy and decision-makers with tools that guide the transition towards sustainable food systems. Achieving sustainable food systems facilitates balancing social, economic, and ecological systems and sustainability (Lindgren et al., 2018). The rationale is to develop nexus planning and circular economy tools that guide the transitional pathways towards sustainable food systems, establishing the interlinkages between food system components, including producing, processing, packaging, distribution, retailing and consuming. This is essential for providing management solutions for both synergies and trade-offs and identifying priority areas for intervention.

4.2 THE CONCEPTUAL FRAMEWORK

As the concept of sustainable food systems is quite complex and cuts across many sectors and has various components, a conceptual framework was developed to guide the identification of pathways that drive towards sustainable food systems. The framework is based on the intricately interlinked but distinct components of a food system that include producing, processing, packaging, distribution, retailing, and consuming and how each connected system and component is impacted by climate change and other drivers of change. This is critical to understanding the socio-economic and environmental interactions and how they influence global environmental change. The derived knowledge facilitated the evaluation of societal outcomes such as food security, ecosystem services, and social welfare resulting from these interactions (Ericksen, 2008; Tendall et al., 2015). Figure 4.2 presents the developed framework, illustrating the interlinked processes of a food system and highlighting the role of nexus planning in transitioning towards sustainable food systems. Nexus modelling is preferred as it facilitates transformational change through its polycentric and circular modelling capabilities (Figure 4.2).

As food systems are complex social-ecological systems that include various interactions between humans (economic and political trends, food price volatility, population dynamics, changes in diets and nutrition, and advances in science and technology) and natural (landcover changes, land and soil degradation, climate change, biodiversity loss, sea-level rise, and air pollution) components (Béné et al., 2019b; Ericksen, 2008; Marshall, 2015), it is paramount to understand these relationships and assess them holistically. This is the initial phase in transitioning towards sustainable food systems. In between the social-ecological systems are external drivers (Figure 4.2), which include exposure and sensitivity, that also determine the

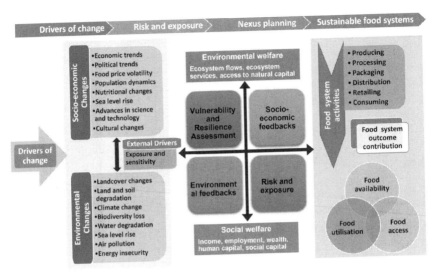

FIGURE 4.2 A nexus planning-based conceptual framework illustrating the connected processes and interactions needed to achieve a sustainable food system.

impact of change on human and environmental health. Knowledge of these drivers and how they influence activities and outcomes of food systems is important for informing policy decisions (Béné et al., 2019a). Food and nutritional security and sound human and environmental health are the main outcomes of any food system (Nemecek et al., 2016). Thus, a food system is considered vulnerable or resilient depending on its capability to deliver and ensure food security (Ericksen, 2008). According to Figure 4.2, nexus planning connects these interactions by defining, measuring, and modelling progress towards sustainability through indicators formulated around resource utilisation, accessibility, and availability (Nhamo et al., 2020). These developments facilitate modelling, monitoring, and simulating some aspects of sustainability.

The framework (Figure 4.2) emphasises the development of a food system that efficiently uses resources and reduces food waste at every stage, from primary production to transformation and consumption. An efficient food system is, therefore, built around circular models such as nexus planning, circular economy, one health, strategic foresight, horizon scanning and scenario planning (Jurgilevich et al., 2016), other than linear models that encourage the introduction of wastes into the environment, causing detrimental environmental and human health risks and climate change (Didenko et al., 2018). For example, nexus modelling develops knowledge-based tools that assess vulnerability and resilience, as well as recovery options and the potential of a food system (Nhamo et al., 2020). These tools facilitate the identification of pathways for simultaneous food security and resource conservation through an analysis of food system activities and outcomes, integrating environmental, social, political, and economic determinants summarised in socio-economic and global environmental change drivers (Figure 4.2). This is based on the understanding that food systems are socio-ecological systems comprising biophysical and social factors that are linked through feedback mechanisms (Binder et al., 2013; Ericksen, 2008; Marshall, 2015).

Identifying and modelling the intrinsic processes of a food system through nexus modelling ensures that food and nutritional outcomes are preserved or enhanced over time and across generations. This is achieved by promptly identifying priority areas for intervention, allowing decision-makers to trace progress towards sustainability and implement policies that foster positive transformations, and allowing humankind to remain within planetary boundaries in resource use. Thus, this chapter addresses the following identified thematic areas that drive towards sustainable food systems: (a) drivers of change, (b) risk and exposure, (c) nexus planning, and (d) pathways towards sustainable food systems.

4.2.1 GLOBAL DRIVERS OF CHANGE IMPACTING FOOD SYSTEMS

Achieving sustainability has become the guiding principle for transformational change and the main goal for human development (Mensah and Ricart Casadevall, 2019; UNGA, 2015). The current and closely interlinked grand challenges that transverse all socio-economic and ecological sectors (Figure 4.3) are prompting a shift from how humankind views the world from a linear view to a circular perspective (Geissdoerfer et al., 2017; Sariatli, 2017). A shock in one sector often triggers a host of interrelated but distinct challenges in the other sectors (Nhamo and Ndlela, 2021). For example, environmental degradation reduces the area under cultivation, causing low crop yields and triggering social distress, economic instability, food insecurity and price fluctuations (Gomiero, 2016).

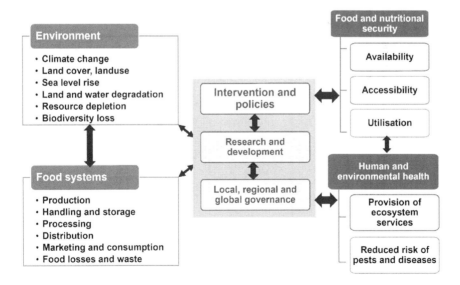

FIGURE 4.3 Interactions between environment and food systems as drivers of change and the pathways needed towards sustainable development, as well as human and environmental health.

Also compounding these existing socio-ecological challenges is the worsening climate change, which is causing biodiversity loss as new land is cleared for crop production, resulting in habitat loss for wildlife (Figure 4.3). Thus, the challenges are not only socio-ecological related but overlap with other sectors such as health and finance, resulting in huge job losses and economic decline (Béné et al., 2019b; Nhamo and Ndlela, 2021). Cross-cutting challenges must be addressed through circular models capable of informing coherent strategies that lead to resilience and adaptive management of resources (Nhamo and Ndlela, 2021; Velenturf and Purnell, 2021). Circular models acknowledge interlinked global environmental challenges' systemic and dynamic nature (Iacovidou et al., 2021).

Figure 4.3 presents the interactions between socio-ecological systems, food systems, and societal and economic factors. The interactions indicate that sustainability is only possible through cross-sectoral implementation of a range of changes within the agriculture value chain aimed at minimising environmental degradation, ensuring economic growth and equity, and improving human health and well-being (Lindgren et al., 2018). Failure to address these challenges holistically will only aggravate environmental degradation and increase exposure and sensitivity to the detriment of human health and resource security.

Food systems are, therefore, a predominant force behind environmental change, including climate change, environmental pollution, biodiversity loss, and degradation of land and freshwater resources (Malhi et al., 2020). Cross-sectoral interventions that include research and development and formulation of dedicated governance structures at different spatial scales will result in food and nutritional security and human and environmental health (Figure 4.3). An integrated assessment, as guided by a cross-sectoral governance framework, is critical to inform policy on pathways to reduce the environmental impact of food systems. The integrated assessment is possible through multi-criteria decision methods using sustainability indicators related to food and nutritional security and human and environmental outcomes (Nhamo et al., 2020). These outcomes include food availability, accessibility and utilisation, provision of ecosystem services, and reduced risk to human health from wildlife, respectively (Nhamo and Ndlela, 2021).

4.2.2 Increasing exposure and sensitivity of food systems

Agriculture faces the challenge of meeting the growing food demands of a growing population while at the same time reducing environmental degradation. The attainment of sustainability within a food system should be framed around developmental pathways that enhance agricultural intensification and increase crop production while reducing unsustainable use of water, nutrients, and chemicals that contaminate the environment. Nonetheless, significant trade-offs often accompany technological advances in the food supply. Processes in the agricultural value chain to food consumption often generate outputs other than consumable food that are returned to the natural environment, such as pollution and food waste (Allen and Prosperi, 2016). The most urgent need is to balance food systems with environmental health. However, this is a mammoth task due to the sensitivity and exposure of food systems (Béné et al., 2019a; Porter et al., 2014).

Despite the advances in increasing crop yields, chronic food insecurity persists in many world regions (Nhamo et al., 2019b). The situation could worsen without coherent and strategic policy responses aimed at transforming the agricultural processes, particularly with the world population projected to reach 9 billion people by 2050. The susceptibility of marginalised and impoverished communities to food insecurity is evident in many regions of the world (Misselhorn and Hendriks, 2017). The sensitivity and exposure of food systems are compounded by the increasing impact of extreme climate events, which also impact economies and the environment, requiring proactive global policies that enhance preparedness (Myers et al., 2017; Nhamo et al., 2019a). Furthermore, environmental trends that include changes in nutrient cycles, hydrological cycles, vegetation cover and composition, and pollution are eroding the capacity of ecosystems to continue providing sufficient services (Allen and Prosperi, 2016). These changes have seen significant spatial and temporal changes in the distribution of crop yields worldwide, further highlighting the sensitivity and exposure of food systems (Ray et al., 2015). Although efforts are being made to increase crop production, it has failed to keep pace with population growth, compounded by the depletion of natural fisheries, inefficient water management practices, actions threatening food systems, and food security and supply chains.

The evident multiple and often tight interlinkages between food system components, as shown in Figure 4.4, highlight the consequences of sector-based interventions to ease one type of vulnerability. Sector-based interventions often and unintentionally transfer those vulnerabilities to other sectors, hence the importance of nexus planning in cross-sectoral interventions. As vulnerability is a function of exposure and sensitivity, the adaptive capacity of food systems and food security outcomes depends on the responses to global environmental change (Porter et al., 2014).

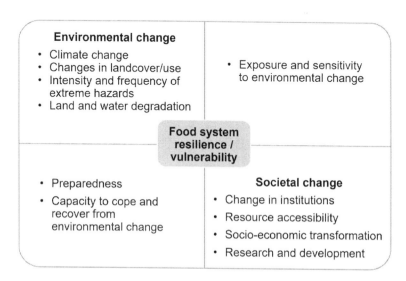

FIGURE 4.4 Food system vulnerability as a function of social and environmental change, exposure, preparedness, and adaptive capacity.

4.2.3   NEXUS PLANNING AS A PATHWAY TO ATTAIN SUSTAINABLE FOOD SYSTEMS

As nexus planning emphasises providing integrated solutions to distinct but inter-linked components, it is envisaged to provide integrated solutions to the intricately connected food system components (Freeman et al., 2015; Mercure et al., 2019; Nhamo et al., 2020; Nhamo and Ndlela, 2021). Its transformative and polycentric nature allows for an integrated assessment of food system components of produc-tion, processing, packaging, distribution, retailing, and consumption, allowing for integrated graphical visualisation of their relationships (Nhamo et al., 2020). This is facilitated by establishing numerical relationships of food system components through sustainability indicators. Sustainability indicators are essential for providing quantitative relationships between distinct components for informed resource man-agement and sustainable development. The sustainability indicators for food systems are related to food and nutritional security and human and environmental health (Nhamo and Ndlela, 2021). They include pillars such as availability, accessibility, and utilisation, as well as continued provision of ecosystem services and the reduc-tion of risk of pests and diseases (Figure 4.2) (Nhamo et al., 2020; Pérez-Escamilla and Segall-Corrêa, 2008).

Therefore, by considering the heterogeneity of the distinct components of food sys-tems over space and time and their repletion with non-linear societal and environmental feedback, nexus planning unpacks and addresses the complex and multi-causal chal-lenges within a food system (Bieber et al., 2018). A set of sustainability indicators related to food systems components are given in Table 4.1. The indicators are critical for provid-ing a form of measurement necessary to assess, monitor and evaluate performance, mea-sure achievement, and determine the system's accountability (Warhurst, 2002). These are the main elements that are critical in monitoring and evaluation. Therefore, sustainability indicators are basic decision-support tools for transforming complex relationships into simple formulations for easy interpretation, monitoring, and evaluation.

Regarding food systems, nexus planning balances competing needs against an awareness of humankind's environmental, social, and economic limitations (Nhamo and Ndlela, 2021). It is a transformative pathway that provides pathways that lead to resilience and adaptation while ensuring human and environmental health. Therefore, sustainability indicators are an integral part of nexus planning as they form the basic unit of measurement to understand complex interactions (Nhamo et al., 2020). The food systems nexus indicators (Table 4.1) are linked to related SDGs indicators, making nexus planning a relevant approach for assessing progress towards sustainable develop-ment over time. The indicators are used to develop indices that provide insights into the efficiency of processes within a food system (Chaudhary et al., 2018). The essence of establishing the numerical relationships of the components of a system is to indicate priority areas needing immediate intervention and to reduce risk and vulnerability.

4.3   PATHWAYS TOWARDS ACHIEVING SUSTAINABILITY OF FOOD SYSTEMS

As already alluded, food systems are complex, interlinked, and significantly con-tribute to the unsustainability of socio-ecological and economic processes. Thus,

TABLE 4.1

Sustainability indicators for assessing the sustainability of food systems

Food system component	Indicator	Units	SDG indicator
Producing	Proportion of agricultural area under productive and sustainable agriculture	%	2.4.1
	Proportion of land that is degraded over a total land area	%	15.3.1
Processing	CO_2 emission per unit of value added		9.4.1
	Manufacturing value added as a proportion of GDP and per capita	%	9.2.1
Packaging	Proportion of medium and high-tech industry value added in total value added	%	9.b.1
	Installed renewable energy-generating capacity	Watts/capita	12.a.1
Distribution	Proportion of the rural population who live within 2 km of an all-season road	%	9.1.1
	Passenger and freight volumes by mode of transport		9.1.2
Retailing	Number of companies publishing sustainability reports		12.6.1
	Material footprint, material footprint per capita, and material footprint per GDP	Tons/capita	12.2.1
	(a) Food loss index and (b) food waste index		12.3.1
Consuming	National recycling rate, tons of material recycled	Kg or %	12.5.1

building sustainable food systems has become a topical agenda at global conferences, particularly their role in achieving SDGs. The interconnectedness and the systemic nature of the interactions of food systems call for transformative and circular models that lead to integrated assessment to identify the intrinsic properties requiring timely interventions and guide progress towards sustainability. Therefore, providing a practical and action-based framework that guides policy and science towards food systems transformations is critical.

A comprehensive food systems framework that guides the transformational change should acknowledge that the sustainability of food systems entails long-term food and nutrition security in terms of availability, accessibility, utilisation, and stability dimensions (Figure 4.5) (Nhamo et al., 2020). The acknowledgement is based on meeting the food and nutrition security for the present and future generations; food systems components must be resilient, efficient, and sustainable (Béné et al., 2019a). The broad intricate interlinkages between food sustainability and food and nutrition security manifest at the global, national, local, and household levels (Mabhaudhi et al., 2016). Therefore, a multi-disciplinary approach that involves multi-stakeholders is needed to achieve sustainable food systems. It is never a one-way or linear approach but an iterative systemic and circular approach. The transitions, therefore,

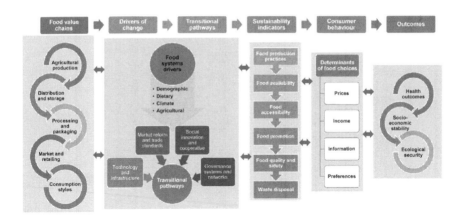

FIGURE 4.5 Transitional pathways towards sustainable food systems, representing an integrated and cross-sectoral intervention towards sustainability.

should include increasing efficiency (sustainable intensification), demand restraint (sustainable diets) and food systems transformations (alternative food systems). The transitional process towards sustainable food systems is a change from an agricultural-centred system to a food system policy and research framework. The stages are critical for integrating complex and holistic transformations in sustainable food systems, a precondition to achieving sustainable food and nutrition security.

The Food Systems Approach (Van Berkum et al., 2018) is widely used to guide strategic policy formulations towards sustainable food systems and support SDG 2 initiatives on achieving zero hunger. The approach attracts investment and supports innovations promoting healthier diets so that humankind does not exceed the planetary boundaries. Therefore, the food systems framework (Figure 4.5) outlines the pathways that provide insights into the structure, behaviour and performance of the interlinkages between food systems components. The framework differentiates the interrelationships, interlinkages and feedback between three fundamental components that include (Figure 4.5) (a) food system drivers (urbanisation, technology development, climate change, and economic growth), (b) food system components (production, distribution, packaging, retailing, and consumption), and (c) food system outcomes (health, sustainability, resilience and equity).

An important aspect of the framework (Figure 4.5) is that it identifies potential trade-offs between different dimensions of food systems (access, availability, safety, affordability, and resilience), how they can be addressed, and how synergies can be enhanced. The framework consists of three major phases that include (Béné, 2020; Brouwer et al., 2020; Fanzo et al., 2021; HLPE, 2017):

a. Societal demands emanate from diverse transitions in agriculture, demography, climate change, and changing diets.
b. Interventional strategies range from novel technologies and market transformations to social innovations and adaptive governance structures.
c. Interventions and leverage points with clear evidence showing the impact on key stakeholders.

Interactions between the three phases provide an overview of the local opportunities and constraints related to distinct interventions and the synergies and trade-offs between the food system components as informed by stakeholder engagement (Brouwer et al., 2020; HLPE, 2017).

4.3.1 UNDERSTANDING SOCIETAL TRANSITIONS

Achieving sustainable food systems requires an analysis of the different transitions in demography, diets, climate and agriculture that can either provide or undermine the healthy diets needed within planetary limits (Kimani-Murage et al., 2021; Lindgren et al., 2018). Previous studies provide valuable insights into these transitions, indicating an increasingly urban population, affluence and emerging middle class, and rising incomes that lead to rapidly changing diets towards more vegetables and animal-based proteins (IFPRI, 2017; Willett et al., 2019). Agricultural production is becoming more market-oriented, and agricultural intensification needs farm consolidation and employment shifts off-farm and non-farm activities (Giller et al., 2021). Challenges from the worsening climate change exacerbate vulnerability and risk and require additional investment for adaptation and mitigation (Eriksen et al., 2021).

4.3.2 IDENTIFYING INTERVENTION STRATEGIES

Coherent strategies are critical for guiding interventions that lead to improved food system performance. Intervention strategies and actions are extensive and varied, but the critical ones in achieving sustainable food systems include those that consistently influence agro-food system transitions (Kimani-Murage et al., 2021). This requires an understanding of the effectiveness of specific technological innovations, market-based incentives, or improved governance networks in accordance with the ongoing agri-food transitions (Giller et al., 2021). The interactions between these instruments are particularly relevant for concurrently achieving various key food systems outcomes (sustainable and healthier diets). An understanding of how sector or cross-sectoral interventions can result in feedback loops between distinct food system levels or produce dynamic spill-over impacts between public and private stakeholders that determine the overall performance of the food systems (Borman et al., 2022). The interventions, therefore, need to be guided by transformative systems modelling such as nexus planning, strategic foresight, scenario planning, horizon scanning, and one health that provide insight into opportunities for multi-stakeholder cooperation and public-private coordination (Naidoo et al., 2021b).

4.3.3 EVALUATION OF IMPACT AND LEVERAGE POINTS

An important phase in achieving sustainable food systems involves identifying the likely impacts of different interventions to attain stakeholders' desired food system outcomes. Rigorous and systematic data collection that provides real-time and reliable impact studies is critical. A systems approach to food provides valuable integrated solutions as it is both a practical and policy concept. The framework (Figure 4.5) provides the pathways to achieve the desired goals.

4.4 RECOMMENDATIONS

Transitioning towards a sustainable food system is a complicated process that requires improvements in land use and agricultural practices. Transformational and integrated approaches provide the pathways to sustainable food systems, but to achieve optimum results, we recommend the following guidelines:

a. Integrated pathways should emphasise critical biophysical and economic 'leverage points' in food systems, focusing on resource use efficiency and enhancing food production processes and the environment's performance with the least effort and cost. This calls for adopting modern technologies that enhance productivity in all domains.

b. Advances that are earmarked to improve agricultural productivity should also consider enhancing the food system's resilience. Although high-efficiency and mechanised agriculture have many benefits, it is also highly vulnerable to disasters that include extreme weather events, novel pests and diseases, and economic shocks (Calicioglu et al., 2019).

c. There is an urgent need to develop methods to evaluate trade-offs of agricultural practices and balance them with advances in technological developments. Research should develop decision-support tools to support management decisions, productivity, and environmental stewardship.

d. Sustainable development in the agriculture sector should be at par with technological development, as informed by circular and transformative modelling, which enhances transformational change, ensuring food security and environmental performance of food systems. Current linear models are generally sector-based and only exacerbate existing challenges by focusing on a single sector (Nhamo and Ndlela, 2021).

e. Transitioning towards sustainable food systems should be supported by coherent policies that create a strategic and enabling environment for agro-ecology. This is supported by a policy framework based on a holistic performance monitoring system that considers nutritional and environmental impacts and the system's long-term stability.

Agriculture is the key driver of environmental and climatic change. As a result, the sector requires a shift from the current linear approaches to circular modelling to enhance food production sustainably. The transformation should be accompanied by societal awareness to catalyse a change from current practices.

4.5 CONCLUSIONS

The systemic cross-sectoral nature and the intricate interdependencies and interactions of food systems require transformative approaches that address challenges in an integrated manner and simplify human understanding of complex socio-ecological connections. Nexus planning has been used to assess the food system's sustainability by identifying key properties that support life and healthy environments. The approach guides policy and supports decision-making to identify priority areas that need immediate intervention, a key step in ensuring food and nutrition security while promoting environmental sustainability. The essence of nexus planning is the capability to examine the multi-causality of dynamic processes within a complex system, including food systems. Indicating priority areas needing immediate intervention paves the way for using scenarios to evaluate various possibilities that lead to coherent strategies. This is critical to understanding and appreciating the change in food security and social and environmental outcomes. These outcomes depend on the decisions and actions taken in the activities practised during the food system but are also impacted by global socio-economic, political, and environmental drivers. Simplifying human understanding of the complex interactions among food system components provides pathways that reduce risk and exposure, an initial step towards sustainable development. This chapter has provided these pathways using nexus planning as today's challenges are complex, cut across sectors and interlinked. Sectoral interventions that do not consider the interlinkages and connectedness of sectors will only compound existing challenges.

REFERENCES

Adamides, G. (2020) A review of climate-smart agriculture applications in Cyprus. *Atmosphere* 11, 898.

Allen, T., Prosperi, P. (2016) Modeling sustainable food systems. *Environmental Management* 57, 956–975.

Béné, C. (2020) Resilience of local food systems and links to food security – A review of some important concepts in the context of COVID-19 and other shocks. *Food Security* 12, 805–822.

Béné, C., Oosterveer, P., Lamotte, L., Brouwer, I.D., de Haan, S., Prager, S.D., Talsma, E.F., Khoury, C.K. (2019a) When food systems meet sustainability–Current narratives and implications for actions. *World Development* 113, 116–130.

Béné, C., Prager, S.D., Achicanoy, H.A., Toro, P.A., Lamotte, L., Cedrez, C.B., Mapes, B.R. (2019b) Understanding food systems drivers: A critical review of the literature. *Global Food Security* 23, 149–159.

Bieber, N., Ker, J.H., Wang, X., Triantafyllidis, C., van Dam, K.H., Koppelaar, R.H., Shah, N. (2018) Sustainable planning of the energy-water-food nexus using decision making tools. *Energy Policy* 113, 584–607.

Binder, C.R., Hinkel, J., Bots, P.W., Pahl-Wostl, C. (2013) Comparison of frameworks for analyzing social-ecological systems. *Ecology and Society* 18(4), 26.

Borman, G.D., de Boef, W.S., Dirks, F., Gonzalez, Y.S., Subedi, A., Thijssen, M.H., Jacobs, J., Schrader, T., Boyd, S., Hermine, J. (2022) Putting food systems thinking into practice: Integrating agricultural sectors into a multi-level analytical framework. *Global Food Security* 32, 100591.

Borrelli, P., Robinson, D.A., Panagos, P., Lugato, E., Yang, J.E., Alewell, C., Wuepper, D., Montanarella, L., Ballabio, C. (2020) Land use and climate change impacts on global soil erosion by water (2015–2070). *Proceedings of the National Academy of Sciences* 117, 21994–22001.

Brouwer, I.D., McDermott, J., Ruben, R. (2020) Food systems everywhere: Improving relevance in practice. *Global Food Security* 26, 100398.

Calicioglu, O., Flammini, A., Bracco, S., Bellù, L., Sims, R. (2019) The future challenges of food and agriculture: An integrated analysis of trends and solutions. *Sustainability* 11, 222.

Campbell, B.M., Vermeulen, S.J., Aggarwal, P.K., Corner-Dolloff, C., Girvetz, E., Loboguerrero, A.M., Ramirez-Villegas, J., Rosenstock, T., Sebastian, L., Thornton, P.K. (2016) Reducing risks to food security from climate change. *Global Food Security* 11, 34–43.

Caron, P., y de Loma-Osorio, G.F., Nabarro, D., Hainzelin, E., Guillou, M., Andersen, I., Arnold, T., Astralaga, M., Beukeboom, M., Bickersteth, S. (2018) Food systems for sustainable development: Proposals for a profound four-part transformation. *Agronomy for Sustainable Development* 38, 1–12.

Chaudhary, A., Gustafson, D., Mathys, A. (2018) Multi-indicator sustainability assessment of global food systems. *Nature Communications* 9, 1–13.

Cosgrove, W.J., Loucks, D.P. (2015) Water management: Current and future challenges and research directions. *Water Resources Research* 51, 4823–4839.

Didenko, N.I., Klochkov, Y.S., Skripnuk, D.F. (2018) Ecological criteria for comparing linear and circular economies. *Resources* 7, 48.

Eakin, H., Connors, J.P., Wharton, C., Bertmann, F., Xiong, A., Stoltzfus, J. (2017) Identifying attributes of food system sustainability: Emerging themes and consensus. *Agriculture and Human Values* 34, 757–773.

Ehrlich, P.R., Harte, J. (2015) Opinion: To feed the world in 2050 will require a global revolution. *Proceedings of the National Academy of Sciences* 112, 14743–14744.

El Bilali, H., Allahyari, M.S. (2018) Transition towards sustainability in agriculture and food systems: Role of information and communication technologies. *Information Processing in Agriculture* 5, 456–464.

Ericksen, P.J. (2008) Conceptualizing food systems for global environmental change research. *Global Environmental Change* 18, 234–245.

Eriksen, S., Schipper, E.L.F., Scoville-Simonds, M., Vincent, K., Adam, H.N., Brooks, N., Harding, B., Lenaerts, L., Liverman, D., Mills-Novoa, M. (2021) Adaptation interventions and their effect on vulnerability in developing countries: Help, hindrance or irrelevance? *World Development* 141, 105383.

Fanzo, J., Haddad, L., Schneider, K.R., Béné, C., Covic, N.M., Guarin, A., Herforth, A.W., Herrero, M., Sumaila, U.R., Aburto, N.J. (2021) Rigorous monitoring is necessary to guide food system transformation in the countdown to the 2030 global goals. *Food Policy* 104, 102163.

FAO (2020) Land use and land cover statistics: Global, regional and country trends, 1990–2018. Food and Agriculture Organisation (FAO), Rome, Italy.

Freeman, O.E., Duguma, L.A., Minang, P.A. (2015) Operationalizing the integrated landscape approach in practice. *Ecology and Society* 20(1), 24.

Geels, F.W., Kern, F., Fuchs, G., Hinderer, N., Kungl, G., Mylan, J., Neukirch, M., Wassermann, S. (2016) The enactment of socio-technical transition pathways: A reformulated typology and a comparative multi-level analysis of the German and UK low-carbon electricity transitions (1990–2014). *Research Policy* 45, 896–913.

Geissdoerfer, M., Savaget, P., Bocken, N.M., Hultink, E.J. (2017) The circular economy–A new sustainability paradigm? *Journal of Cleaner Production* 143, 757–768.

Gibbs, H., Salmon, J.M. (2015) Mapping the world's degraded lands. *Applied Geography* 57, 12–21.

Giller, K.E., Delaune, T., Silva, J.V., Descheemaeker, K., van de Ven, G., Schut, A.G., van Wijk, M., Hammond, J., Hochman, Z., Taulya, G. (2021) The future of farming: Who will produce our food? *Food Security* 13, 1073–1099.

Gomiero, T. (2016) Soil degradation, land scarcity and food security: Reviewing a complex challenge. *Sustainability* 8, 281.

Hall, C., Dawson, T., Macdiarmid, J., Matthews, R., Smith, P. (2017) The impact of population growth and climate change on food security in Africa: Looking ahead to 2050. *International Journal of Agricultural Sustainability* 15, 124–135.

Higginbottom, T.P., Symeonakis, E. (2014) Assessing land degradation and desertification using vegetation index data: Current frameworks and future directions. *Remote Sensing* 6, 9552–9575.

HLPE (2017) Nutrition and food systems. A report by the High Level Panel of Experts on Food Security and Nutrition. HLPE, Rome, Italy, p. 152.

Horton, P. (2017) We need radical change in how we produce and consume food. *Food Security* 9, 1323–1327.

Iacovidou, E., Hahladakis, J.N., Purnell, P. (2021) A systems thinking approach to understanding the challenges of achieving the circular economy. *Environmental Science and Pollution Research* 28, 24785–24806.

IFPRI (2017) 2017 Global food policy report. International Food Policy Research Institute (IFPRI), Washington, DC, p. 148.

Ingemarsdotter, E., Jamsin, E., Kortuem, G., Balkenende, R. (2019) Circular strategies enabled by the internet of things—A framework and analysis of current practice. *Sustainability* 11, 5689.

IPCC (2014) Climate Change 2014: Mitigation of Climate Change. Contribution of Working Group iii to the 5th Assessment Report of the Intergovernmental Panel on Climate Change (IPCC), in: Edenhofer, O., Pichs-Madruga, R., Sokona, Y., Farahani, E., Kadner, S., Seyboth, K., Adler, A., Baum, I., Brunner, S., Eickemeier, P. (Eds.). Intergovernmental Panel on Climate Change (IPCC), Cambridge and New York, p. 1454.

Jurgilevich, A., Birge, T., Kentala-Lehtonen, J., Korhonen-Kurki, K., Pietikäinen, J., Saikku, L., Schösler, H. (2016) Transition towards circular economy in the food system. *Sustainability* 8, 69.

Kimani-Murage, E., Gaupp, F., Lal, R., Hansson, H., Tang, T., Chaudhary, A., Nhamo, L., Mpandeli, S., Mabhaudhi, T., Headey, D.D. (2021) An optimal diet for planet and people. *One Earth* 4, 1189–1192.

Klerkx, L., Rose, D. (2020) Dealing with the game-changing technologies of Agriculture 4.0: How do we manage diversity and responsibility in food system transition pathways? *Global Food Security* 24, 100347.

Krishna Bahadur, K., Dias, G.M., Veeramani, A., Swanton, C.J., Fraser, D., Steinke, D., Lee, E., Wittman, H., Farber, J.M., Dunfield, K. (2018) When too much isn't enough: Does current food production meet global nutritional needs? *PLoS One* 13 (10), e0205683.

Kristoffersen, E., Blomsma, F., Mikalef, P., Li, J. (2020) The smart circular economy: A digital-enabled circular strategies framework for manufacturing companies. *Journal of Business Research* 120, 241–261.

Lindgren, E., Harris, F., Dangour, A.D., Gasparatos, A., Hiramatsu, M., Javadi, F., Loken, B., Murakami, T., Scheelbeek, P., Haines, A. (2018) Sustainable food systems—A health perspective. *Sustainability Science* 13, 1505–1517.

Mabhaudhi, T., Chibarabada, T., Modi, A. (2016) Water-food-nutrition-health nexus: Linking water to improving food, nutrition and health in Sub-Saharan Africa. *International Journal of Environmental Research and Public Health* 13, 107.

Mabhaudhi, T., Nhamo, L., Chibarabada, T.P., Mabaya, G., Mpandeli, S., Liphadzi, S., Senzanje, A., Naidoo, D., Modi, A.T., Chivenge, P.P. (2021) Assessing progress towards sustainable development goals through nexus planning. *Water* 13, 1321.

Mabhaudhi, T., Nhamo, L., Mpandeli, S., Nhemachena, C., Senzanje, A., Sobratee, N., Chivenge, P.P., Slotow, R., Naidoo, D., Liphadzi, S. (2019) The water–energy–food nexus as a tool to transform rural livelihoods and well-being in Southern Africa. *International Journal of Environmental Research and Public Health* 16, 2970.

Malhi, Y., Franklin, J., Seddon, N., Solan, M., Turner, M.G., Field, C.B., Knowlton, N. (2020) Climate change and ecosystems: Threats, opportunities and solutions. *Philosophical Transactions of the Royal Society B*, 375(1794), 20190104.

Marshall, G. (2015) A social-ecological systems framework for food systems research: Accommodating transformation systems and their products. *International Journal of the Commons* 9(2), 881–908.

Mensah, J., Ricart Casadevall, S. (2019) Sustainable development: Meaning, history, principles, pillars, and implications for human action: Literature review. *Cogent Social Sciences* 5, 1653531.

Mercure, J.-F., Paim, M.-A., Bocquillon, P., Lindner, S., Salas, P., Martinelli, P., Berchin, I., de Andrade Guerra, J., Derani, C., de Albuquerque Junior, C. (2019) System complexity and policy integration challenges: The Brazilian Energy-Water-Food Nexus. *Renewable and Sustainable Energy Reviews* 105, 230–243.

Misra, A.K. (2014) Climate change and challenges of water and food security. *International Journal of Sustainable Built Environment* 3, 153–165.

Misselhorn, A., Hendriks, S.L. (2017) A systematic review of sub-national food insecurity research in South Africa: Missed opportunities for policy insights. *PLoS One* 12, e0182399.

Myers, S.S., Smith, M.R., Guth, S., Golden, C.D., Vaitla, B., Mueller, N.D., Dangour, A.D., Huybers, P. (2017) Climate change and global food systems: Potential impacts on food security and undernutrition. *Annual Review of Public Health* 38, 259–277.

Naidoo, D., Nhamo, L., Lottering, S., Mpandeli, S., Liphadzi, S., Modi, A.T., Trois, C., Mabhaudhi, T. (2021a) Transitional pathways towards achieving a circular economy in the water, energy, and food sectors. *Sustainability* 13, 9978.

Naidoo, D., Nhamo, L., Mpandeli, S., Sobratee, N., Senzanje, A., Liphadzi, S., Slotow, R., Jacobson, M., Modi, A., Mabhaudhi, T. (2021b) Operationalising the water-energy-food nexus through the theory of change. *Renewable and Sustainable Energy Reviews* 149, 10.

Nemecek, T., Jungbluth, N., i Canals, L.M., Schenck, R. (2016) Environmental impacts of food consumption and nutrition: Where are we and what is next? *The International Journal of Life Cycle Assessment* 21, 607–620.

Nhamo, L., Mabhaudhi, T., Modi, A. (2019a) Preparedness or repeated short-term relief aid? Building drought resilience through early warning in southern Africa. *Water SA* 45, 75–85.

Nhamo, L., Mabhaudhi, T., Mpandeli, S., Dickens, C., Nhemachena, C., Senzanje, A., Naidoo, D., Liphadzi, S., Modi, A.T. (2020) An integrative analytical model for the water-energy-food nexus: South Africa case study. *Environmental Science and Policy* 109, 15–24.

Nhamo, L., Matchaya, G., Mabhaudhi, T., Nhlengethwa, S., Nhemachena, C., Mpandeli, S. (2019b) Cereal production trends under climate change: Impacts and adaptation strategies in southern Africa. *Agriculture* 9, 30.

Nhamo, L., Ndlela, B. (2021) Nexus planning as a pathway towards sustainable environmental and human health post Covid-19. *Environment Research* 192, 110376.

Nhamo, L., Ndlela, B., Nhemachena, C., Mabhaudhi, T., Mpandeli, S., Matchaya, G. (2018) The water-energy-food nexus: Climate risks and opportunities in southern Africa. *Water* 10, 567.

Nhamo, L., Rwizi, L., Mpandeli, S., Botai, J., Magidi, J., Tazvinga, H., Sobratee, N., Liphadzi, S., Naidoo, D., Modi, A., Slotow, R., Mabhaudhi, T. (2021) Urban nexus and transformative pathways towards a resilient Gauteng City-Region, South Africa. *Cities* 116, 103266.

Nkonya, E., Johnson, T., Kwon, H.Y., Kato, E. (2016) *Economics of land degradation in sub-Saharan Africa, economics of land degradation and improvement–A global assessment for sustainable development.* Springer, Switzerland, pp. 215–259.

Pérez-Escamilla, R., Segall-Corrêa, A.M. (2008) Food insecurity measurement and indicators. *Revista de Nutrição* 21, 15s–26s.

Porter, J., Xie, L., Challinor, A., Cochrane, K., Howden, S., Iqbal, M., Lobell, D., Travasso, M. (2014) Food Security and Food Production Systems Climate Change 2014: Impacts, Adaptation and Vulnerability. Part A: Global and Sectoral Aspects. Contribution of Working Group II to the Fifth Assessment Report of the Intergovernmental Panel on Climate Change (IPCC). Cambridge University Press, Cambridge, pp. 485–533.

Ray, D.K., Gerber, J.S., MacDonald, G.K., West, P.C. (2015) Climate variation explains a third of global crop yield variability. *Nature Communications* 6, 1–9.

Rockström, J., Williams, J., Daily, G., Noble, A., Matthews, N., Gordon, L., Wetterstrand, H., DeClerck, F., Shah, M., Steduto, P. (2017) Sustainable intensification of agriculture for human prosperity and global sustainability. *Ambio* 46, 4–17.

Sariatli, F. (2017) Linear economy versus circular economy: A comparative and analyzer study for optimization of economy for sustainability. *Visegrad Journal on Bioeconomy and Sustainable Development* 6, 31–34.

Scoones, I., Smalley, R., Hall, R., Tsikata, D. (2019) Narratives of scarcity: Framing the global land rush. *Geoforum* 101, 231–241.

Tendall, D., Joerin, J., Kopainsky, B., Edwards, P., Shreck, A., Le, Q.B., Krütli, P., Grant, M., Six, J. (2015) Food system resilience: Defining the concept. *Global Food Security* 6, 17–23.

Tomich, T.P., Lidder, P., Coley, M., Gollin, D., Meinzen-Dick, R., Webb, P., Carberry, P. (2019) Food and agricultural innovation pathways for prosperity. *Agricultural Systems* 172, 1–15.

UNGA (2015) Transforming our world: The 2030 Agenda for Sustainable Development, Resolution adopted by the General Assembly (UNGA). United Nations General Assembly, New York, p. 35.

Van Berkum, S., Dengerink, J., Ruben, R. (2018) The food systems approach: Sustainable solutions for a sufficient supply of healthy food. Wageningen Economic Research, Wageningen, The Netherlands, p. 34.

Velenturf, A.P., Purnell, P. (2021) Principles for a sustainable circular economy. *Sustainable Production and Consumption* 27, 1437–1457.

Warhurst, A. (2002) Sustainability indicators and sustainability performance management. Mining, Minerals and Sustainable Development (MMSD), Warwick, p. 129.

Whitmee, S., Haines, A., Beyrer, C., Boltz, F., Capon, A.G., de Souza Dias, B.F., Ezeh, A., Frumkin, H., Gong, P., Head, P. (2015) Safeguarding human health in the Anthropocene epoch: Report of The Rockefeller Foundation–Lancet Commission on planetary health. *The Lancet* 386, 1973–2028.

Willett, W., Rockström, J., Loken, B., Springmann, M., Lang, T., Vermeulen, S., Garnett, T., Tilman, D., DeClerck, F., Wood, A. (2019) Food in the Anthropocene: The EAT–Lancet Commission on healthy diets from sustainable food systems. *The Lancet* 393, 447–492.

Wittman, H., Chappell, M.J., Abson, D.J., Kerr, R.B., Blesh, J., Hanspach, J., Perfecto, I., Fischer, J. (2017) A social–ecological perspective on harmonizing food security and biodiversity conservation. *Regional Environmental Change* 17, 1291–1301.

5 Strengthening the transformational implementation of national climate change adaptation plans to enhance agricultural resilience

Charles Nhemachena, Daniel Njiwa,
Mcloud Kayira Chirwa, Anabela Manhica,
Assan Ng'ombe and Protase Echessah

5.1 INTRODUCTION

Climate change, directly and indirectly, impacts food systems, food trade, food and nutrition security, and the attainment of the United Nations Sustainable Development Goals, such as achieving zero hunger, ending poverty, ensuring healthy lives, and promoting well-being. The Intergovernmental Panel on Climate Change (IPCC) sixth assessment report (AR6) states that evidence shows an increased intensity and occurrence of observed extreme climate changes such as heavy precipitation, agricultural and ecological droughts, heatwaves, and tropical cyclones since the AR5. The AR6 further highlights, with high confidence, that all regions are projected to experience further increases in hot climatic impact drivers (CIDs[1]) and decreases in cold CIDs. For example, extreme heat thresholds relevant to agriculture and health would be exceeded more frequently at higher global warming levels. Also, Africa is projected to experience increased frequency and or intensity of agricultural and ecological droughts with medium to high confidence (IPCC, 2021). These climate changes significantly impact food systems, ecosystem services, economic growth and development, disproportionately impacting vulnerable systems and communities.

Given the vulnerability of food systems to climate change variability and extremes, building resilience is crucial to help countries meet the growing demand for healthy and safe diets while achieving socio-economic and sustainability goals. The 2015

DOI: 10.1201/9781003327615-5

Paris Agreement, adopted by 196 Parties at the Conference of the Parties 21 on 12 December 2015 and entered into force on 4 November 2016, commits all Parties to engage in adaptation planning processes and implementing actions as well as developing or enhancing relevant plans, policies and/or contributions (Article 7.9) to contribute to the global goal to enhance adaptive capacity, strengthen resilience and reduce vulnerability (Article 7.1) (UNFCCC, 2015a). The United Nations Framework Convention on Climate Change (UNFCCC) secretariat's 2021 progress report on national adaptation plans (NAPs) indicates that as of November 2020, 125 of the 154 developing countries had undertaken activities to formulate and or implement NAPs (UNFCCC, 2021).

Article 7 of the 2015 Paris Agreement also calls for all Parties to implement, monitor, evaluate and learn from adaptation plans, policies, programmes and actions (UNFCCC, 2015a). Much of the focus on climate change adaptation progress has been on mainstreaming and effectiveness of planning of adaptation policies, plans and strategies with limited evidence on implementation and impacts of the adaptation plans (Bauer, Feichtinger, & Steurer, 2012; Leiter, 2021; Olazabal & De Gopegui, 2021; Runhaar, Wilk, Persson, Uittenbroek, & Wamsler, 2018; UNEP, 2021). The IPCC Fifth Assessment Report acknowledged that, at the global level, evidence of adaptation implementation remained limited and required overcoming resource, institutional and capacity barriers (Mimura et al., 2014). The extent of implementation, monitoring and evaluation remains limited across African countries, despite a series of technical and financial support by national and international partners assisting the countries in formulating the climate change adaptation plans/strategies/policies and implementing pilot projects. Often, the implementation ends at the pilot projects, and countries have not mainstreamed the allocation of resources in their national- and local-level planning and budgeting processes.

This gap makes indicators that assess whether a country has developed a national adaptation plan/strategy/policy, such as the SDG indicator "13.2.1 Number of countries with (…) national adaptation plans (…)" (UN, 2020) and the International Climate Fund (ICF) Key Performance Indicator (KPI) 13 scorecard on mainstreaming climate change in national agriculture plans/strategies/policies misleading to policymakers and the public as they assume climate change adaptation and resilience are being addressed. The lack of evidence on climate change adaptation implementation affects the ability to understand whether countries are effectively preparing their populations and economic sectors to better cope with climate change shocks (Binet et al., 2021). The planning of climate change adaptation and resilience is important; however, translating the plans into implementation is critical to building the adaptative capacity to respond to the increasing number and intensity of climate change shocks.

The governments of Malawi and Mozambique recognise the critical role of climate change adaptation in their medium- and long-term development plans and in a range of other strategies and policies, to the extent that they have developed and are mainstreaming programmes addressing climate change through, for example, the Malawi National Climate Change Management Policy (Government of Malawi, 2016, 2017) and the Mozambique National Climate Change Adaptation and Mitigation Strategy (NCCAMS) 2013–2025 (Government of Mozambique, 2012). With technical and sometimes financial support from development partners, the governments of Malawi and Mozambique have extensively invested in national adaptation plans/strategies/policies. The national plans, strategies and policies highlight the importance of responding to the impacts of climate change and building adaptive capacity to better prepare for future risks and shocks. However, despite the favourable national framework and adaptation plans/strategies/policies developed to guide adaptation investments in these countries, implementing agricultural sector adaptation priorities remains a challenge. Furthermore, despite several projects implemented across the countries with support from national and international partners, there is scant empirical evidence of the implementation outcomes; and at a higher level, there is limited evidence of monitoring and learning of the national climate change adaptation plans/strategies and policies.

The chapter's main objective was to assess the extent and challenges of implementing, monitoring, and evaluating climate change adaptation plans/strategies/policies to enhance agricultural resilience at national and sub-national levels in Malawi and Mozambique. This chapter contributes to the need for more empirical evidence on implementing, monitoring, and evaluating climate change adaptation and resilience policies and plans beyond stated intentions in national planning documents and country submissions to the UNFCCC. Relying on stated intentions in NAPs leads to over-estimation of countries implementing, monitoring and evaluating the progress of their plans (Leiter, 2021). The need for more empirical research on climate change policies/plans implementation (Rykkja, Neby, & Hope, 2014) is confirmed by the AGRA ICF KPI 13 scorecard results (AGRA, 2019, 2020, 2021) and the 2019 CAAP Biennial Review. The empirical findings of this chapter contribute to climate change advisory reports to engage stakeholders in the respective countries to strengthen the implementation of national climate change plans/strategies/policies.

5.2 LITERATURE REVIEW

5.2.1 OVERVIEW OF THE STATUS OF CLIMATE CHANGE ADAPTATION PLANNING AND IMPLEMENTATION

The UNFCCC was established in 1992 to help countries formulate and implement national adaptation strategies. Least developed countries developed National Adaptation Programmes of Action (NAPA) that documented the country's perceived urgent and immediate needs to adapt to climate change (UNFCCC, 2011). In addition to the NAPs addressing medium- to long-term impacts of climate change, after the 2015 Paris Declaration, countries have developed Intended Nationally Determined Contributions (INDCs). The IPCC Special Report on the impacts of global warming of 1.5°C above the pre-industrial levels shows the need for the urgency of greater ambition in NDCs if the global mean temperature is to be limited to 1.5°C (IPCC, 2018). Pauw and Klein (2020) argue that the ambition of the intended NDCs before or shortly after the 2015 Paris Climate Conference is not enough, and there is a need for countries to improve the effectiveness of the plans and policies underpinning their NDCs. This can be achieved through improved transparency, coherence and implementability of the NDCs (Pauw, Castro, Pickering, & Bhasin, 2020; Pauw & Klein, 2020; Pauw et al., 2018).

Mainstreaming climate change in policy development across sectors is important (England et al., 2018) to ensure countries reduce the adverse impacts of climate change risks and better prepare to respond to projected future changes. Climate change adaptation is increasingly integrated into national planning and policy processes. Röser, Widerberg, Höhne, and Day (2020) argue that the process of preparing the NDCs contributes positively to national climate policy processes by raising awareness, catalysing institutional change, and improving political buy-in across government and non-government stakeholders. However, the process of preparing and implementing NDCs in developing and emerging countries faces challenges such as political support, financial, human and technical resources, and analytical capabilities (Röser, Widerberg, Höhne, & Day, 2020). Despite evidence of delays due to the COVID-19 pandemic on NAP development processes in some countries, especially least developed countries, there is considerable progress on NAP agendas. As of August 2021, more than 75% of African countries had adopted at least one national-level adaptation planning instrument (such as a plan, strategy, policy, or law) (see Figure 5.1) (UNEP, 2021).

Leiter (2021) conducted an evidence-based global stocktake of monitoring and evaluation systems of national climate change adaptation plans to determine whether

National plan, strategy, law or policy in place

FIGURE 5.1   Status of adaptation planning in Africa, as of 5 August 2021.

Note: Territories marked as N/A are those which are recognised as disputed by the United Nations or whose status has not yet been agreed upon.

Source: UNEP (2021).

governments track their implementation. The study highlighted that despite more than 70 countries adopting national climate change adaptation plans, there was little evidence of the extent of implementation of these plans. Leiter (2021) found that less than 40% of the 70 countries that adopted a NAP reported on implementation progress or evaluated them. Similarly, the Adaptation Gap Report (AGR) 2021 indicated that on monitoring and evaluating NAPs, only 8% of the countries had evaluated their adaptation plans, 26% have M&E systems in place, and 36% are still developing M&E systems (UNEP, 2021). The main constraints reported are technical, human and financial resources.

Climate change and the COVID-19 pandemic compounded the risks and vulnerability, adversely impacting the adaptive capacity of governments, communities and societies, especially in developing countries (UNEP, 2021). The COVID-19 pandemic triggered economic slowdowns and loss of income and livelihood sources that disproportionately affected vulnerable populations, further reducing their capacity to adapt to extreme climate change events (UNEP, 2021). There is a need to enhance the implementation of adaptation actions and effective mainstreaming of climate change risks in decision-making processes and the COVID-19 recovery investments. Although COVID-19 stimulus recovery investments present opportunities for mainstreaming resilient and green recoveries, the opportunities have not been seized (UNEP, 2021). Learning from the experiences of the COVID-19 pandemic, the 2021 AGR highlights the need for governments to address compound risks through integrated risk management approaches, such as prioritising green recovery investments that achieve economic growth and climate change resilience. Also, developing countries can increase the resilience of fiscal frameworks to address compound risks through flexible disaster finance frameworks to ensure predictable, timely and cost-effective finance availability for immediate responses to emergencies such as the COVID-19 pandemic and extreme climate events (UNEP, 2021).

Figures 5.2 and 5.3 show the outlook on the African continent's adaptation progress and countries, including selected adaptation interventions in stimulus packages, as of 31 January 2021. Development partners and international and multilateral organisations have provided technical, development and financial support to countries across the continent to formulate national climate change adaptation and mitigation policies and plans, including mainstreaming climate change into sector strategies and plans such as in agriculture. In many African countries, the implementation of climate change adaptation plans/strategies/policies has been supported by national and international partners working with government line ministries and agencies to pilot climate change adaptation projects and programmes. The pilot implementation periods range from 1 to ±5 years. In many cases, these have not been sufficient to build long-term sustainable adaptive capacity and adaptation activities for the target communities.

5.2.2 CHALLENGES IN THE IMPLEMENTATION OF CLIMATE CHANGE ADAPTATION POLICIES

Climate change adaptation is considered an integral part of national policy planning. In the past decade, countries (such as in Africa) have made significant progress in developing NAPs/NASs and policies that outline the strategic goal of addressing the

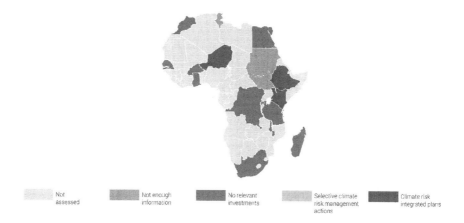

FIGURE 5.2   Countries including selected adaptation interventions in stimulus packages, as of 31 January 2021.

Source: UNEP (2021).

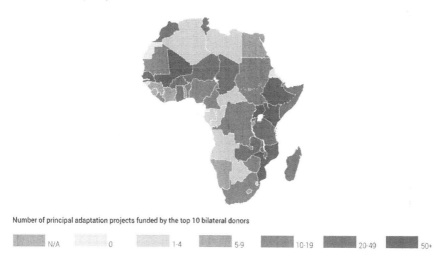

FIGURE 5.3   Geographic distribution of principal adaptation projects funded by the top ten bilateral donors.

Source: UNEP (2021).

impacts of climate change and building adaptive and resilience capacity. Despite adaptation being set on the political agendas of countries, several studies (Alves et al., 2020; Ampaire et al., 2016; Dupuis & Knoepfel, 2013; Leiter, 2021; Pauw & Klein, 2020; Totin et al., 2015) identify that across countries, the implementation of climate change adaptation policies are lacking. For example, the climate change adaptation policy implementation deficit remains a challenge, especially across many African countries, adversely impacting efforts to respond to the impacts of climate change and better prepare for future shocks.

The implementation of adaptation actions involves a complex interaction between the framing of the problem in specific places (place-based framing) and the utilisation of the strengths in places (key determinants for action) to promote ownership and progressive action (Barnett et al., 2015; Dewulf, 2013; Eriksen, Nightingale, & Eakin, 2015; Mackay, Hennessey, & Mackey, 2019). The challenges that hinder the implementation of climate change adaptation actions emerge over time and are often a result of poor framing of the problem and planning (Barnett et al., 2015; Dupuis & Knoepfel, 2013). Furthermore, formulating national climate change adaptation plans is often based on approaches prescribed at the international level, such as from the UNFCCC, which might not be representative of local contexts at national and sub-national levels.

The main climate change adaptation implementation challenges include a lack of resources (technology, finance, and knowledge) and institutional characteristics (Alves et al., 2020; Biesbroek, Klostermann, Termeer, & Kabat, 2013; Klein et al., 2014; Mackay, Hennessey, & Mackey, 2019; Mataya, Vincent, & Dougill, 2020; Totin et al., 2015). Some of these challenges are discussed in detail below. These are collaborated by discussion from empirical studies that have analysed climate change implementation barriers, particularly in Africa and other developing countries.

5.2.2.1 Lack of knowledge, information and policy awareness

Knowledge of climate change adaptation implementation remains low globally despite extensive knowledge on framing adaptation problems and planning actions to address them, indicating the extent of the implementation deficit (Fünfgeld, Lonsdale, & Bosomworth, 2019; Mackay, Hennessey, & Mackey, 2019). The limited awareness and knowledge by decision-makers and implementers of the policy processes, the contents of the policies and how to translate them into activities to achieve the planned outputs and outcomes hamper effective policy implementation. For example, in Ghana, Mali and Senegal, the implementation of climate change policies was hampered by the disconnect between the national climate policy development processes and information at the sub-national (district) level. The staff expected to implement the policies and stakeholders who required the information in climate development work, despite being aware of the developed climate change policies, were either not involved in the development process or had not seen the policy documents (Totin et al., 2015). Strengthening the implementation of climate change policies requires ensuring the flow of information and sufficient knowledge on climate change policies to sub-national-level staff members. This includes awareness and policy literacy across all governance levels through effective dissemination and communication (including translating into local languages) on climate change policies. Creating a better understanding of the climate change policies is critical for sub-national-level staff to translate the policies into context-specific problem questions and actions to achieve expected results and outcomes.

5.2.2.2 Financial resources

Despite formulating and adopting climate change adaptation plans/strategies/policies, many African governments still fail to allocate resources within their planning and budgeting processes. For example, the analysis of implementation challenges and emerging lessons on Uganda's NAPA in 2015 found that the government was not allocating funding for local-level adaptation (Nyasimi, Radeny, Mungai, & Kamini, 2016). The implementability of NDCs heavily relies on whether they are conditional upon receiving support and whether the requested support for implementation is available (Pauw & Klein, 2020). The NDCs of most developing countries are partly conditional upon receiving financial support; however, the financial support requested to implement the conditional NDCs far exceeds the available funding pledges (Pauw, Castro, Pickering, & Bhasin, 2020).

The NDCs risk not being implemented when the expected financial support from developed countries is unavailable. The adaptation finance gap grew in 2021 compared to 2019 and 2020 despite a gradual increase in international public adaptation finance for developing countries. However, the current financing allocated to adaptation is expected to decline due to the COVID-19 pandemic as both developed and developing countries prioritise limited financial resources to address pressing health and financial needs for their economies (UNEP, 2021). The COVID-19 pandemic adversely impacted adaptation planning and further constrained available financial resources. Constrained financial resources further hinder the implementation of national climate change adaptation plans. However, there are opportunities for mainstreaming adaptation, green growth and climate resilience in the COVID-19 public rescue and recovery financing streams (UNEP, 2021). The pandemic also disrupted and, in some cases, eroded livelihood and income sources in many developing countries, increasing their vulnerability to other shocks, such as climate change variability and extreme events thereof. Scaling up and increasing public adaptation finance for direct investment and overcoming barriers to private sector adaptation is an urgent priority in implementing adaptation policies (UNEP, 2021).

5.2.2.3 Lack of coordination and coherence in planning and implementation

Some key constraints include the lack of harmonised sectoral planning and inconsistencies between national and sub-national adaptation policies and strategies (Ampaire et al., 2017; Hisali, Birungi, & Buyinza, 2011). Often coordinated planning and implementation of national policies/strategies is challenging in many countries. Traditionally, the primary mandate of climate action in many countries is the Ministry of the Environment. However, implementing NAPs often includes other institutional actors such as sectoral line ministries, government agencies, nongovernmental organisations and the private sector. The extent to which the desired policy, such as the NDCs, can be successfully implemented (implementability) requires agreed and well-defined roles and responsibilities for implementation (by whom, how, what scope, stakeholder involvement, etc.), as well as public and political acceptability of implementation needs and consequences (costs involved, need for support, equality of process and outcomes) (Pauw & Klein, 2020; Alves et al., 2020; Mackay, Hennessey, & Mackey, 2019).

The lack of clear roles and responsibilities between the national government ministries and the sub-national implementation authorities also constrains the implementation of national climate change adaptation plans/strategies/policies. Other barriers include the absence of statutory obligations and the interactions with other policies and development plans. Also, implementing national climate change adaptation plans/

strategies/ policies suffers from the lack of resources transferred from the national treasury to sub-national and local levels. These findings call for strengthened coordination and coherence in implementing climate change adaptation plans/strategies/ policies. Another challenge for implementing national adaptation strategies is that the impacts of climate change occur at multiple levels (from national, sub-national, and local). Adaptation planning and implementation need to consider local dynamics and complexities across all levels of governance that determine the failure or success of implementation (Mackay, Hennessey, & Mackey, 2019; Hupe & Hill, 2016; Dupuis & Knoepfel, 2013).

Further, as most climate impacts and responses happen at the local level, the decentralisation and devolution policies contribute positively towards the development and implementation of adaptation programmes and investments. In Kenya, where the country is pursuing a devolved governance and development framework, implementing climate adaptation programmes is beginning to see increased locally developed adaptation projects. For example, county governments are beginning to operationalise the County Climate Change Funds, developing county-integrated development plans that include climate change adaptation initiatives.

5.2.2.4 Institutional capacity

Institutional constraints, including failure to elevate adaptation as a political priority; considering adaptation as an isolated task of a sector/governance unit; and lack of horizontal and vertical coordination between different administrative levels and between formal agencies and private stakeholders, are the main hindering factors to climate change adaptation implementation (Biesbroek, Klostermann, Termeer, & Kabat, 2013; Mimura et al., 2014; Calliari, Michetti, Farnia, & Ramieri, 2019). The poor cross-level coordination often contributes to the lack of implementation of national climate change adaptation plans/strategies/policies. The multi-level layers of implementation require engagement with stakeholders at the different levels to design context-specific and responsive action plans to translate the national adaptation plans/strategies/policies into relevant and implementable projects. Despite national and international investments supporting developing countries to develop institutional capacities and improve cross-sectoral collaboration and coordination mechanisms between ministries and relevant stakeholders (FAO & UNDP, 2018), empirical evidence still indicates that implementation of national climate change adaptation plans/strategies/policies suffers from lack of the above factors.

Successful formulation and implementation of national climate change adaptation plans/strategies/policies require effective institutional capacity across all levels to identify and integrate climate change adaptation actions in sectoral planning and budgeting processes (FAO & UNDP, 2018). The UNFCCC NAP Technical Guidelines (Steps 2 and 3) and the NAP-Ag Guidelines (FAO, 2017) focus on assessing gaps and weaknesses in undertaking the NAP process and "enhancing capacity for planning and implementation of adaptation in the agriculture sector" (FAO & UNDP, 2018). Figure 5.4 presents the main elements of the NAP institutional capacity assessment. The elements show all the critical institutional capacity needs that are addressed in preparing countries for NAP formulation, implementation and monitoring. However, one of the main factors that hamper the implementation of

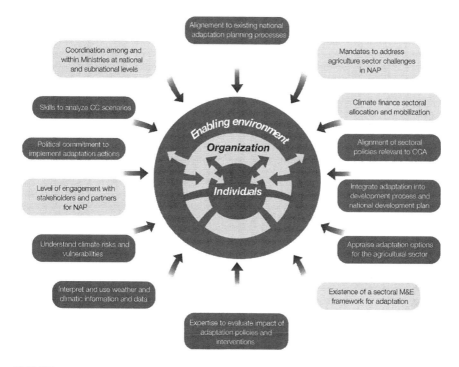

FIGURE 5.4 Main elements of the NAP institutional capacity assessment.

Source: FAO and UNDP (2018).

climate change adaptation plans/strategies/policies is the lack of institutional capacity across different levels. This calls for exploring the challenges beyond the formulation of the NAP to strengthen the translation of the climate change adaptation plans/strategies/policies to outputs and outcomes. Institutional capacity strengthening and harmonised cross-sectoral collaboration and coordination mechanisms between line ministries and relevant stakeholders across all governance levels are crucial for the sustainable implementation of climate change adaptation plans/strategies/policies (FAO & UNDP, 2018).

5.2.3 EMPIRICAL EVIDENCE ON CHALLENGES IN THE IMPLEMENTATION OF CLIMATE CHANGE ADAPTATION POLICIES

Leiter (2021) conducted a systematic review and outreach to country representatives and international organisations and developed an inventory of NAP monitoring and evaluation systems that documented practices from over 60 countries. The stocktake relied on the evidence from the monitoring and evaluation reports instead of stated intentions to conduct monitoring and evaluation. The results showed that compared to the baseline of the 2017 Adaptation Gap Report of the United Nations Environment Programme, the number of countries developing or using NAP monitoring and evaluation systems and with published NAP evaluations increased by 40%. Leiter's (2021) stocktakes also found that systematic assessment of NAP implementation was lacking in more than 60 percent of the countries that adopted NAPs, making it difficult to understand the impacts of the NAPs. The findings above call for increased efforts in ensuring countries implement the commitments in their NAPs and conduct systematic monitoring and evaluation of the implementation and impacts in the respective economies.

A study on ways to effectively build capacity to adapt to climate change in Malawi found that capacity building from long-term and short-term training complementarily influences the design and implementation of successful adaptation practices (Mataya, Vincent, & Dougill, 2020). This includes designing and implementing short-term training participatory workshops customised to the needs of the trainees and using context-specific examples as well as on-the-job training, action planning and mentoring after the training (Mataya, Vincent, & Dougill, 2020). The study also reiterated the importance of coordinated design, implementation and monitoring of adaptation capacity-building activities and ensuring appropriate institutional support after the training sessions to improve adaptation planning across the continent.

The Mozambique Government, in collaboration with national and international partners, has implemented several climate change-related projects and programmes. The gaps and barriers identified in the implementation of climate change adaptation actions in Mozambique include insufficient coordination and governance mechanisms leading to policy coherence at the national, provincial and district levels; lack of technical capacity to mainstream climate change at national, provincial and district planning and budgeting systems; and poor climate change and gender-sensitive data and information (UNDP, UNEP, & GEF, 2020). The National Adaptation Plan Global Support Programme (NAP-GSP) identified the following opportunities to strengthen the NAP formulation and implementation processes in Mozambique: define precise coordination mechanisms; operationalise the implementation mechanisms of the NCCAMS; elaborate and implement the capacity plan to conduct research in relevant areas; increase the capacity to lead the climate change adaptation cycle; strengthen relevant institutions to collect and manage data and information, run climate models and elaborate scenarios at provincial levels; develop and implement strategies for climate change education, awareness-raising, communication and public participation; assess adaptation technology needs; update sectoral policies; develop or improve monitoring and evaluation tools; strengthen capacities to mainstream other cross-cutting issues such as gender or biodiversity; build national technical and institutional capacities to design and manage projects to access climate financing; and establish climate insurances (UNDP, UNEP, & GEF, 2020).

Alves et al. (2020) analysed implementation challenges of climate change policies and agendas in 13 countries. They found that despite accounting for different non-governmental stakeholders, the NAPs/NASs remain largely state-centred, with the steering and implementation responsibilities assigned to each country's Ministry of Environment. The other finding from the same study was that the objectives of the NAPs reflected a more global agenda with less focus on national/regional contexts and vulnerabilities.

The experiences from Niger indicate that most of the NAPA priorities were addressed through pilot projects supported by bilateral or multilateral cooperation arrangements. However, the challenge that remains in most developing countries is to scale these into the medium and long term (UNDP, UNEP, & GEF, 2018). The experiences of the NAPA process in Niger also highlight similar constraints observed in other countries, such as coordination; institutional and technical capacity; data availability, reliability and management; integrating climate change adaptation into planning and budget processes. Also, mobilising financial resources remains critical to scaling up and sustaining the pilot projects co-implemented with bilateral and multilateral partners (UNDP, UNEP, & GEF, 2018).

Ampaire et al. (2017) and Ampaire, Happy, Van Asten, and Radeny (2015) analysed policy development and implementation gaps in Rakai District, Uganda, focusing on institutional challenges to climate change adaptation. The studies were based on literature reviews across multiple governance levels, spatial scales, and field assessments. The policy development processes were centralised at the national levels, and led by central government agencies with the insufficient engagement of other actors, and local stakeholders (communities) were excluded. In addition, the study found the main constraints to climate policy implementation included a disconnect in communication across all governance levels (national, district and community), limited technical capacity and finances, political interference and absence of functional implementation structures across all levels (Ampaire, Happy, Van Asten, & Radeny, 2015; Ampaire et al., 2017). The study recommended measures to enhance linkages across all governance levels and among actors to improve policy formulation, implementation and adaptation by smallholder farmers.

Ampaire et al. (2016) analysed barriers to the successful implementation of climate change policy in Tanzania. They found that there have been considerable efforts to support resilience-building actions in the agriculture sector. The main barriers to implementing climate change actions included limited climate change knowledge across levels, lack of effective national finance mechanism to direct climate funds and poor coordination of climate change actions from national to local levels (disconnect between national and local governments).

Uittenbroek (2016) analysed the role of organisational routines in constraining the mainstreaming of climate change adaptation at the implementation stage. The study found that despite the relative ease of mainstreaming climate change adaptation in national policies, the problem is with implementation. Often, policies are implemented by actors other than the policymakers, whose actions are guided by organisational routines, which, if not adjusted, hamper the implementation of new policy goals such as climate change adaptation (Uittenbroek, 2016). Ensuring appropriate changes in organisational routines across all levels (national and sub-national) is important to strengthen the implementation of climate change adaptation plans/strategies/policies. Some required changes include reallocating resources and adapting existing practices to implement the priorities and actions in climate change adaptation plans/strategies/policies.

Totin et al. (2015) found that despite progress in formulating national climate change policies and action plans in Ghana, Mali and Senegal, district-level staff and the general public at the regional and local levels lacked awareness and understanding of the climate policy implementation processes. The common barriers to policy development and effective implementation in the three countries included: a lack of awareness and funding, a lack of operational capacity at lower administrative levels and little involvement from stakeholders. Furthermore, the effective implementation of climate policy was hampered by a lack of information flows on existing climate policy processes between national and local levels. The study recommended supervised knowledge-sharing platforms for national, regional and local policymakers and other stakeholders to strengthen information flows and support policy development and implementation. Other constraints that hamper the translation of climate change policies and plans into concrete actions and implementations in Ghana, Mali, and Senegal include lags in the policy planning, development and approval processes. Furthermore, the development of climate change policies in Ghana, Mali and Senegal was not comprehensive in the participation of all relevant stakeholders, especially at the sub-national levels. Effective participation is important in ensuring the policy development process integrates context-specific inputs to create awareness and understanding of the priorities to be mainstreamed in development activities, especially at the sub-national levels.

5.3 METHODS OF THE STUDY

Building on other empirical studies on the implementation of climate change adaptation policies, such as those presented by several researchers (Ampaire et al., 2016, 2017; Alves et al., 2020; Leiter, 2021), the chapter is based on a review of the literature and qualitative data collected from key informant interviews with identified key national and sub-national stakeholders. The systemic desktop review focused on climate change adaptation planning and policy documents from national government ministries/departments (such as the Ministries of Environment, Agriculture, and Trade) responsible for designing and implementing climate change adaptation plans/strategies/policies. The review assessed the availability of monitoring and evaluation systems and reports for the country's national adaptation plans/policies as indicated in the NAP technical guidelines (UNFCCC, 2012) and Article 7 of the 2015 Paris Agreement (UNFCCC, 2015a). We also explored agriculture sector monitoring and evaluation systems and published literature like the global stocktake of NAP monitoring and evaluation systems such as Leiter (2021), UNFCCC NAP progress reports, and UNEP adaptation gap reports (UNEP, 2021), reports from academics, national and international organisations. In addition, the review identified gaps and challenges in climate change policy implementation processes, such as technical capacity and budget provisions to translate the policy actions into outputs and outcomes.

For key informant interviews, the identified stakeholders included policymakers, farmers, scientists, and non-state actors such as development partners and the private sector working on climate change adaptation and resilience in the respective focus countries. The analysis triangulated the findings from the systemic literature review through outreach to key stakeholders from government and national and international partner organisations working on climate change adaptation in the respective countries. The findings from the review and outreach to key informant stakeholders from the respective countries helped develop climate change adaptation advisory reports to inform stakeholder engagement.

5.4 RESULTS AND DISCUSSION

This section discusses findings from stakeholder engagements in Malawi and Mozambique on the extent of implementation of climate change policies and priorities. as highlighted above. The two countries have comprehensive sets of policy frameworks developed to address climate change issues. The stakeholders from both countries reported that they have adequate climate change policy frameworks that, if implemented, would significantly contribute to building climate change adaptative and resilience capacity across all levels and sectors. The respective Ministries of Environment coordinate the national policy frameworks. The stakeholder engagements showed that the challenges affecting the implementation of climate change adaptation priorities/actions in national policies included the following: lack of financial and technical resources, implementation coordination challenges, and lack of awareness of the policy frameworks, especially at sub-national levels. The stakeholder engagements showed that much of the efforts have been on mainstreaming climate change adaptation in national planning documents and policies. There is limited evidence of significant traction on the implementation of these policies. Malawi and Mozambique remain vulnerable to climate change variability and extremes. In early 2022, both countries experienced tropical storms, Anna and Gombe, respectively, which significantly destroyed livelihood sources and infrastructure in the affected communities.

5.4.1 LIMITED FINANCIAL RESOURCES

The stakeholder engagements in both Malawi and Mozambique showed that the low and often limited allocation of financial resources remains a significant constraint to implementing climate change adaptation priorities across sectors such as agriculture. Despite the comprehensive national climate change frameworks in both countries, without adequate financial resources, many policy documents get to their end dates without considerable implementation. The respective climate change departments reported limited budget allocations to operationalise their annual plans. Malawi and Mozambique have budget challenges and significantly depend on donor support; even if they have the political will, limited financial resources hamper their ability to operationalise their climate change adaptation policies/plans. Some government stakeholders highlighted that the limitations in financial resources leave the countries largely dependent on development partners who often drive their agenda, which sometimes does not align with government priorities.

Another challenge is the priorities regarding budgeting allocations; for example, in Malawi, close to 50% of the agriculture budget is allocated to the input subsidy programme, leaving minimal resources for other activities, including implementing climate change adaptation policies/plans.

The other challenge regarding financial resources stakeholders highlighted in Malawi is balancing public good programmes (such as food and nutrition security) and commercial programmes in allocating public resources. Because governments are constrained in resources, development partners and NGOs drive their own agenda. This affects the sustainability of such programmes beyond the funding programmes if the government considers them primarily donor/NGO driven without effective partnership in designing, planning and implementation.

Innovative financial approaches are critical to driving the implementation of climate change adaptation from domestic and international sources. The capacity of government and other domestic institutions should be improved to help them access international climate change adaptation finances to implement their adaptation plans and policies. Advocacy for increased investments in climate change adaptation is critical to ensure the strengthened implementation of adaptation policies and plans. Climate change adaptation should not be taken as an extra in the planning of the Ministry's annual plans. Still, it should be embedded in the ongoing activities to bring transformative adaptation outcomes that help the countries develop and build resilience to climate-related risks.

5.4.2 Limited Awareness of the Policy Frameworks, Especially at Sub-National Levels

The stakeholder engagements in both countries also showed limited awareness of climate change policy frameworks at the sub-national levels where implementation occurs. Despite some of the officials at sub-national levels being consulted in developing these documents, when completed, copies are often not shared with them. In some cases, the policy documents were reported to be too long and difficult to read and understand easily. The stakeholders highlighted that these need to be simplified into easy-to-read and useable versions to facilitate easy reading, understanding and use in the planning and implementation of sub-national development plans. The stakeholders involved in implementing climate change adaptation at sub-national levels highlighted that in some cases, the officials have either not seen the national climate change policies/plans or they have not read them. Stakeholders in both Malawi and Mozambique reported that, in some cases, the climate change adaptation documents remain in national offices and are never seen at the sub-national level.

The above findings contribute to weak mainstreaming and implementing climate change adaptation at the sub-national level as the officials mandated to develop and oversee sub-national development activities have either been limited or are unaware of climate change policies/plans. Some of the stakeholders in Malawi argued that in some cases, the climate change adaptation policies/plans are known to the officials actively involved in their development. The limited awareness in other line ministries beyond the staff engaged in consultations during the development of the climate change adaptation policies/plans affects the integration of adaptation and resilience in broader national programmes and activities. Furthermore, some non-state stakeholders argued that although there are efforts to engage various actors in developing climate change adaptation policies/plans, more needs to be done to ensure effective and inclusive participation. The stakeholders highlighted that engagement should not only validate already developed policies/plans but facilitate participation in actively shaping their development.

5.4.3 Coordination Challenges in Planning and Implementation of Adaptation Priorities

Engagements with government, development partners, NGOs, farmer organisations, etc., reiterated the lack of coordinated planning and implementation of climate change adaptation and resilience activities in both countries. Government departments (including in the same ministry) still work in silos on climate change adaptation issues despite efforts and structures to coordinate efforts. Similarly, NGOs and Development Partners (DPs) were reported to implement their own programmes/projects sometimes without the effective involvement of the government. The government stakeholders argued that NGOs sometimes get money in the name of helping the government implement adaptation priorities; however, there is no accountability to the government and sometimes reported outputs and impacts are not what is on the ground. Some stakeholders highlighted a disconnect between the results and impacts in institutional reports and what can be verified in the target communities.

Furthermore, coordination of DPs (among themselves and with or by the government) on implementing climate change adaptation and resilience activities was limited. Stakeholders highlighted the urgent need to improve coordination among DPs and also with the government in planning and implementing climate change adaptation priorities for the respective countries. For example, in Malawi, some stakeholders reported that whoever funds the Technical Committee on Climate Change called the shots, and the committee's focus ended up with the focus of the funding agency. Some of the stakeholders highlighted that the coordination of the Technical Committee on Climate Change should be strengthened to mirror the effectiveness of the DCAFS in coordinating and driving the implementation of climate change adaptation and resilience priorities in the country. Furthermore, these technical committees should move beyond discussing projects to focus on the country's thematic climate change adaptation/resilience priorities.

The lack of effective and inclusive coordination in the planning and implementation of climate change adaptation and resilience at sub-national levels results in staff at these levels receiving multiple and different uncoordinated climate change adaptation and resilience information and projects. The climate change messaging and programming can be overwhelming to sub-national staff and end-users and fail to achieve the desired outputs and impact. Coordination can help streamline climate change adaptation, resilience messaging, and implementation across all levels. The coordination of the committee needs to remain broad to cover national priorities and drive their implementation across all sectors. Some stakeholders highlighted that due to the lack of national coordination, different institutions focus on getting as many resources as possible in the name of climate change adaptation and resilience; however, there is no evidence to demonstrate the impact. Despite the mandate of respective Ministries of Environment to coordinate climate change issues in each country, current efforts are inadequate as different institutions continue to plan and implement their own activities within the climate change space.

5.4.4 Limited transparency in the implementation of national adaptation policies/plans

While some stakeholders engaged in Malawi reported that the National Climate Change Resilience Strategy developed by the government was not being implemented, engagements with other government departments showed that implementation started with a pilot in six districts, and plans are to scale to other districts and the rest of the country. One of the challenges highlighted during the stakeholder engagements is that the National Resilience Strategy is now housed at the Department of Disaster Management (DoDMA) without much awareness and reach to stakeholders across the country beyond disasters. Without awareness and visibility of the strategy to other sectors and stakeholders who are expected to implement some of the priorities,it is challenging to improve implementation, monitoring and evaluation significantly.

Stakeholders in Malawi highlighted that multi-sectoral climate change adaptation/resilience policies/plans should not be housed in a department or line ministry. The experience in the country has been most of this ends up being plans for that respective department or line ministry. The engagements highlighted that departments or line ministries sometimes act as rivals because each needs access to climate change adaptation/resilience resources. This results in adaptation activities being implemented piecemeal without coordinated planning to scale up the implementation. Structures such as the Office of the President are ideal for driving multi-sectoral efforts such as climate change adaptation. However, the limitation is that anything under the statehouse will live as long as the President is in power. There is a need for structures with convening power to bring different ministries, DPs, and NGOs together to strengthen coordination and alignment.

5.4.5 LACK OF NATIONAL MONITORING AND EVALUATION OF ADAPTATION PROGRESS

This is linked to the limitations in coordinated planning and implementation despite ongoing efforts in Malawi, such as developing a management information system (MIS) to monitor and track all climate change investments, outputs and impacts in the country. When fully operational, the information systems being developed by the Department of Environment will help the country monitor and track all climate change-related investments and progress. However, neither country could provide documented evidence of monitoring and evaluation reports on national climate change adaptation priorities during the stakeholder engagements. This is despite climate change adaptation policies and frameworks being developed with monitoring and evaluation plans. The finding also highlights the limited capacity to monitor and evaluate the implementation of climate change policies and plans in relevant institutions. The individual investments by different actors usually have monitoring and evaluation of results for specific projects, and there is no readily available data at the national level on project performance. Monitoring and evaluating climate change adaptation activities at the national level is important to ensure that countries identify success stories to scale to other parts of the countries and learn from the implementation to improve future adaptation programmes.

5.4.6 RECURRENT CLIMATE CHANGE SHOCKS AND RESPONSES TO EMERGENCIES

Malawi and Mozambique have been hit by several tropical cyclones, storms and droughts that have increased in frequency and intensity in recent years. The latest IPCC report shows that this trend will continue in the future due to climate change and variability. The recurrent climatic extreme events that often hit both countries reduce the capabilities of the respective governments to always respond to emergencies that significantly impact medium- and long-term planning and implementation of programmes. The severe impacts of the shocks mean that government budgets are always inadequate as available resources are channelled to respond to emergencies.

5.4.7   LIMITED INSTITUTIONAL CAPACITY

The chapter also undertook an institutional capacity assessment focusing on implementing climate change policies in the agriculture sector. Stakeholder engagements highlighted that several national and sub-national government departments mandated to implement climate change adaptation have inadequate institutional capacity to deliver on their goals. The institutional capacity challenges reported include a lack of laws, regulations and frameworks to ensure the department gets a budget from the national treasury, limited human and technical capacities (such as the number of skilled officials and representation at sub-national levels) and competing institutional mandates. For example, in Malawi, the Department of Climate Change and Meteorological Services and the DoDMA highlighted that no legal frameworks exist to guide their operations. As such, they have no budget votes.

Furthermore, expertise is needed to translate the scientific information in national climate change policies and climate forecasts into easy-to-use forms for end-users at different levels. Climate change adaptation and resilience should be mainstreamed in sub-national-level extension services to strengthen access to climate information for improved decision-making that builds adaptive capacity and resilience to future shocks. The decentralisation of government in Malawi and ongoing efforts in Mozambique require the institutional capacity to mainstream, implement and monitor climate change adaptation and resilience policies and plans at sub-national levels, which often is not there.

5.5   CONCLUSIONS AND RECOMMENDATIONS

Planning climate change adaptation and resilience is important; however, translating the plans into implementation is critical to building the adaptative capacity to respond to climate change shocks. The assessment showed that despite progress in mainstreaming climate change considerations in national policies and strategies, the extent of implementation, monitoring and evaluation remains limited. Often, the implementation ends at the pilot projects, and countries have not mainstreamed the allocation of resources in their national- and local-level planning and budgeting processes. The lack of evidence on climate change adaptation implementation affects the ability to understand better whether countries are effectively preparing their populations and economic sectors to better prepare for climate change shocks. The stakeholder engagements showed that the challenges affecting the implementation of climate change adaptation priorities/actions in national policies included the following: lack of financial and technical resources, implementation coordination challenges, and lack of awareness of the policy frameworks, especially at sub-national levels.

The recommendations to address some of these challenges include the following:

Design and implement innovative financing mechanisms and strengthen technical capacity and resources: The results showed that Malawi and Mozambique, like many developing countries, lack viable financing mechanisms to operationalise adaptation policies and plans. There is a need to create innovative financing options leveraging public (especially financial support from public funds in the national budget) and private sector sources (domestic and international). This includes integrating adaptation financing in budgeted development interventions to ensure transformative adaptation outcomes that help the countries develop and build resilience to climate-related risks. Other measures include expanding and strengthening the capacity of government and other domestic institutions to help them access international climate change adaptation finances to implement their adaptation plans and policies. Countries should also continuously develop the technical capacity of their staff in translating climate change adaptation policies into action and innovative financing options to ensure the policies and plans are operationalised.

Strengthen advocacy and awareness of climate change adaptation policy frameworks, especially at sub-national levels: The climate change adaptation policies and plans must be packaged in user-friendly formats for dissemination to diverse stakeholders across the countries. Deliberate efforts must ensure climate change adaptation policies are widely disseminated beyond the national offices coordinating their development. Increased climate change adaptation policy advocacy should be strengthened, including inclusive development, planning and implementation of these policies and plans, especially at the sub-national levels, other ministries and departments and sector-wide stakeholders. Inclusive climate change adaptation stakeholder participation should be beyond validating policy/planning documents to active engagement in their development, implementation, monitoring and evaluation.

Strengthen coordination in planning and implementation of adaptation priorities in national and sub-national development programmes: The evidence from the review and stakeholder engagements calls for an urgent need to strengthen coordination in planning and implementing climate change adaptation activities at national and sub-national levels. This includes coordination within government ministries and departments and with sector stakeholders (development partners, private sector, farmer organisations, NGOs, etc.). There is also a need to strengthen coordination among other stakeholders themselves, such as within the development partners and NGOs, to better plan and coordinate climate change adaptation interventions. This would help to coordinate climate change adaptation and leverage resources to scale the implementation of national priorities and bring transformational change. Also, strengthening coordination would help streamline climate change adaptation and resilience messaging and implementation across all levels.

Improve transparency in implementing national adaptation policies/plans: Deliberate efforts are required to ensure the visibility of progress with climate change policies to sector-wide stakeholders. Implementing national climate change policies and plans should not be closed within some departments but visible to other departments, ministries and stakeholders. This is also important to ensure accountability for action and results in climate change adaptation interventions. There is a need for national structures with convening power to bring different ministries, DPs, and NGOs together to strengthen coordination and alignment.

Strengthen national monitoring and evaluation of adaptation progress: There is an urgent need to develop and/or strengthen monitoring and evaluation systems (and management information systems) of climate change adaptation activities at the national level. This would help to ensure that countries document their investments in climate change adaptation and track the impacts, identify success stories to scale to other parts of the countries and learn from the implementation to improve future adaptation programmes. The institutional and individual capacity to monitor and evaluate the implementation of climate change adaptation policies and plans should be developed and strengthened at national and sub-national levels.

Build climate change forecasting capacity to improve planning and decision-making in responding to recurrent climate change shocks and emergencies: There is an urgent need to strengthen the capacity of the respective departments and ministries working with partners to produce close to real-time, medium- and long-term forecasting of climate changes, including shocks such as the recurrent tropical storms and droughts. The information should be readily accessible to sector-wide stakeholders to inform appropriate planning beyond reactive responses to emergencies when there is a shock. This would also help avoid diverting significant budget allocation to other development programmes to attend to climate change emergencies.

NOTE

1 "Climatic impact-drivers (CIDs) are physical climate system conditions (e.g., means, events, extremes) that affect an element of society or ecosystems. Depending on system tolerance, CIDs and their changes can be detrimental, beneficial, neutral, or a mixture of each across interacting system elements and regions" (IPCC, 2021).

REFERENCES

AGRA. (2019). AGRA International Climate Finance (ICF) KPI 13 scorecard. Regional Food Trade and Resilience Programme. Alliance for a Green Revolution in Africa, Nairobi, Kenya.

AGRA. (2020). AGRA International Climate Finance (ICF) KPI 13 scorecard. Regional Food Trade and Resilience Programme. Alliance for a Green Revolution in Africa, Nairobi, Kenya.

AGRA. (2021). AGRA International Climate Finance (ICF) KPI 13 scorecard. Regional Food Trade and Resilience Programme. Alliance for a Green Revolution in Africa, Nairobi, Kenya.

Alves, F., Leal Filho, W., Casaleiro, P., Nagy, G., Diaz, H., Al-Amin, A., & Saroar, M. (2020). Climate change policies and agendas: Facing implementation challenges and guiding responses. *Environmental Science & Policy*, 104, 190–198.

Ampaire, E., Happy, P., Van Asten, P., & Radeny, M. (2015). The role of policy in facilitating adoption of climate-smart agriculture in Uganda. Copenhagen: CGIAR Research Program on Climate Change, Agriculture and Food Security (CCAFS).

Ampaire, E., Jassogne, L., Providence, H., Acosta, M., Twyman, J., Winowiecki, L., & Van Asten, P. (2017). Institutional challenges to climate change adaptation: A case study on policy action gaps in Uganda. *Environmental Science & Policy*, 75, 81–90. doi: 10.1016/j.envsci.2017.05.013.

Ampaire, E., Okolo, W., Acosta, M., Jassogne, L., Twyman, J., Muindi, P., & Mwongera, C. (2016). Barriers to successful climate change policy implementation in Tanzania. CCAFS Info Note. Copenhagen: CGIAR Research Program on Climate Change, Agriculture and Food Security (CCAFS).

Barnett, J., Evans, L., Gross, C., Kiem, A., Kingsford, R., Palutikof, C., ... Smithers, S. (2015). From barriers to limits to climate change adaptation: Path dependency and the speed of change. *Ecology and Society*, 20(3), 5. doi: 10.5751/ES-07698-200305.

Bauer, A., Feichtinger, J., & Steurer, R. (2012). The governance of climate change adaptation in 10 OECD countries: Challenges and approaches. *Journal of Environmental Policy & Planning*, 14(3), 279–304.

Biesbroek, G., Klostermann, J., Termeer, C., & Kabat, P. (2013). On the nature of barriers to climate change adaptation. *Regional Environmental Change*, 13(5), 1119–1129.

Binet, S., De Bruijn, M., Horikoshi, D., Kim, R., Lee, B., Markrich, M., ... Uvarova, G. (2021). Independent evaluation of the adaptation portfolio and approach of the green climate fund. Evaluation Report No. 9, February 2021. Songdo: Independent Evaluation Unit, Green Climate Fund.

Calliari, E., Michetti, M., Farnia, L., & Ramieri, E. (2019). A network approach for moving from planning to implementation in climate change adaptation: Evidence from southern Mexico. *Environmental Science & Policy*, 93, 146–157.

Dewulf, A. (2013). Contrasting frames in policy debates on climate change adaptation. *Wiley Interdisciplinary Reviews: Climate Change*, 4(4), 321–330.

Dupuis, J., & Knoepfel, P. (2013). The adaptation policy paradox: The implementation deficit of policies framed as climate change adaptation. *Ecology and Society*, 18(4), 31. doi: 10.5751/Es-05965-180431.

England, M., Dougill, A., Stringer, L., Vincent, K., Pardoe, J., Kalaba, F., ... Afionis, S. (2018). Climate change adaptation and cross-sectoral policy coherence in Southern Africa. *Regional Environmental Change*, 18(7), 2059–2071.

Eriksen, S., Nightingale, A., & Eakin, H. (2015). Reframing adaptation: The political nature of climate change adaptation. *Global Environmental Change*, 35, 523–533.

FAO. (2017). Addressing agriculture, forestry and fisheries in National Adaptation Plans (NAPs)-Supplementary guidelines. Rome: Food and Agriculture Organisation of the United Nations.

FAO & UNDP. (2018). Institutional capacity assessment approach for national adaptation planning in the agriculture sectors. Briefing Note. Rome and New York: Food and Agriculture Organization of the United Nations and United Nations Development Programme.

Fünfgeld, H., Lonsdale, K., & Bosomworth, K. (2019). Beyond the tools: Supporting adaptation when organisational resources and capacities are in short supply. *Climatic Change*, 153(4), 625–641.

Government of Malawi. (2016). National climate change policy. Ministry of Environment and Climate Change Management. Lilongwe: Government of Malawi.

Government of Malawi. (2017). Malawi growth and development strategy III. Lilongwe: Government of Malawi.

Government of Mozambique. (2012). National Climate Change Adaptation and Mitigation Strategy 2013–2025. Maputo: Government of Mozambique.

Hisali, E., Birungi, P., & Buyinza, F. (2011). Adaptation to climate change in Uganda: Evidence from micro level data. *Global Environmental Change*, 21(4), 1245–1261.

Hupe, P., & Hill, M. (2016). 'And the rest is implementation.' Comparing approaches to what happens in policy processes beyond Great Expectations. *Public Policy and Administration*, 31(2), 103–121.

The references should be as follows:

IPCC, 2018: Global Warming of 1.5°C. An IPCC Special Report on the impacts of global warming of 1.5°C above pre-industrial levels and related global greenhouse gas emission pathways, in the context of strengthening the global response to the threat of climate change, sustainable development, and efforts to eradicate poverty [Masson-Delmotte, V., P. Zhai, H.-O. Pörtner, D. Roberts, J. Skea, P.R. Shukla, A. Pirani, W. Moufouma-Okia, C. Péan, R. Pidcock, S. Connors, J.B.R. Matthews, Y. Chen, X. Zhou, M.I. Gomis, E. Lonnoy, T. Maycock, M. Tignor, and T. Waterfield (eds.)]. Cambridge University Press, Cambridge, UK and New York, NY, USA, 616 pp. https://doi.org/ 10.1017/9781009157940.

IPCC. (2021). Summary for policymakers. In V. Masson-Delmotte, P. Zhai, A. Pirani, S. Connors, C. Péan, S. Berger, B. Zhou, Climate Change 2021: The Physical Science Basis. Contribution of Working Group I to the Sixth Assessment Report of the Intergovernmental Panel on Climate Change (pp. 3–32). Geneva: Intergovernmental Panel on Climate Change.

Klein, R., Midgley, G., Preston, B., Alam, M., Berkhout, F., Dow, K., & Shaw, M. (2014). Adaptation opportunities, constraints, and limits. In C. Field, V. Barros, D. Dokken, K. Mach, M. Mastrandrea, T. Bilir, … L. White, Climate Change 2014: Impacts, Adaptation, and Vulnerability. Part A: Global and Sectoral Aspects. Contribution of Working Group II to the Fifth Assessment Report of the Intergovernmental Panel on Climate Change (pp. 899–943). Cambridge and New York: Cambridge University Press.

Leiter, T. (2021). Do governments track the implementation of national climate change adaptation plans? An evidence-based global stocktake of monitoring and evaluation systems. *Environmental Science & Policy*, 125, 179–188.

Mackay, S., Hennessey, N., & Mackey, B. (2019). *Barriers to the implementation of climate change adaptation plans and action: Considerations for regional Victoria*. Brisbane: Griffith University.

Mataya, D., Vincent, K., & Dougill, A. (2020). How can we effectively build capacity to adapt to climate change? Insights from Malawi. *Climate and Development*, 12(9), 781–790.

Mimura, N., Pulwarty, R., Duc, D., Elshinnawy, I., Redsteer, M., Huang, H., … Sanchez Rodriguez, R. (2014). Adaptation planning and implementation. In C. Field, V. Barros, D. Dokken, K. Mach, M. Mastrandrea, T. Bilir, … L. White, Climate Change 2014: Impacts, Adaptation, and Vulnerability. Part A: Global and Sectoral Aspects. Contribution of Working Group II to the Fifth Assessment Report of the Intergovernmental Panel on Climate Change (pp. 869–898). Cambridge and New York: Cambridge University Press.

Nyasimi, M., Radeny, M., Mungai, C., & Kamini, C. (2016). Uganda's National Adaptation Programme of Action: Implementation, challenges and emerging lessons. Copenhagen: CGIAR Research Program on Climate Change, Agriculture and Food Security (CCAFS).

Olazabal, M., & De Gopegui, M. (2021). Adaptation planning in large cities is unlikely to be effective. *Landscape and Urban Planning*, 206, 103974.

Pauw, W., Castro, P., Pickering, J., & Bhasin, S. (2020). Conditional nationally determined contributions in the Paris Agreement: Foothold for equity or Achilles heel? *Climate Policy*, 20(4), 468–484.

Pauw, W., & Klein, R. (2020). Beyond ambition: Increasing the transparency, coherence and implementability of Nationally Determined Contributions. *Climate Policy*, 20(4), 405–414.

Pauw, W., Klein, R., Mbeva, K., Dzeb, A., Cassanmagnago, D., & Rudloff, A. (2018). Beyond headline mitigation numbers: We need more transparent and comparable NDCs to achieve the Paris Agreement on climate change. *Climatic Change*, 147(1), 23–29.

Röser, F., Widerberg, O., Höhne, N., & Day, T. (2020). Ambition in the making: Analysing the preparation and implementation process of the Nationally Determined Contributions under the Paris Agreement. *Climate Policy*, 20(4), 415–429.

Runhaar, H., Wilk, B., Persson, A., Uittenbroek, C., & Wamsler, C. (2018). Mainstreaming climate adaptation: Taking stock about "what works" from empirical research worldwide. *Regional Environmental Change*, 18(4), 1201–1210.

Rykkja, L., Neby, S., & Hope, K. (2014). Implementation and governance: Current and future research on climate change policies. *Public Policy and Administration*, 29(2), 106–130.

Totin, E., Traoré, S., Zougmoré, R., Homann-Kee, S., Tabo, R., & Schubert, C. (2015). Barriers to effective climate change policy development and implementation in West Africa. CCAFS Info Note. Copenhagen: CGIAR Research Program on Climate Change, Agriculture and Food Security (CCAFS).

Uittenbroek, C. (2016). From policy document to implementation: Organizational routines as possible barriers to mainstreaming climate adaptation. *Journal of Environmental Policy & Planning*, 18(2), 161–176. doi: 10.1080/1523908X.2015.1065717.

UN. (2020). Global indicator framework for the Sustainable Development Goals and targets of the 2030 Agenda for Sustainable Development. New York: United Nations. Retrieved from https://unstats.un.org/sdgs/indicators/indicators-list/.

UNDP, UNEP, & GEF. (2018). National Adaptation Plan process in focus: Lessons from Niger. New York: United Nations Development Programme.

UNDP, UNEP, & GEF. (2020). National Adaptation Plans in focus: Lessons from Mozambique. New York: United Nations Development Programme.

UNEP. (2021). The Gathering Storm Adapting to climate change in a post-pandemic world. Adaptation Gap Report 2021. Nairobi: United Nations Environment Programme.

UNFCCC. (2011). National Adaptation Programmes of Action. Rio de Janeiro and New York: United Nations Framework Convention on Climate Change.

UNFCCC. (2012). National Adaptation Plans, Technical guidelines for the national adaptation plan process. Rio de Janeiro and New York: United Nations Framework Convention on Climate Change.

UNFCCC. (2015a). Paris Agreement. Rio de Janeiro and New York: United Nations Framework Convention on Climate Change.

UNFCCC. (2015b). Report of the Conference of the Parties on its twenty-first session, held in Paris from 30 November to 13 December 2015. Addendum, Part two: Action taken by the Conference of the Parties at its twenty-first session. Rio de Janeiro and New York: United Nations Framework Convention on Climate Change. Retrieved from https://unfccc.int/documents/9097.

UNFCCC. (2021). Progress in the process to formulate and implement national adaptation plans. Revised note by the secretariat. Rio de Janeiro and New York: United Nations Framework Convention on Climate Change.

6 Progress towards the circular economy
Case studies of sanitation and organic waste–derived resource recovery technologies in South Africa

Taruvinga Badza, William Musazura, Mendy Zibuyile Shozi, Alfred Oduor Odindo, and Tafadzwanashe Mabhaudhi

6.1 INTRODUCTION

6.1.1 DEFINING THE CIRCULAR ECONOMY

Globally, the circular economy (CE) concept has been receiving increasing attention for the past decade. Although this is not a new term, as it was first introduced into the literature by Pearce et al. (1990), its practical implementation in businesses and industries has not been quantified and reported until recently. It is still ambiguous what precisely CE is, and this confusion is evident through the diverse ways in which this economic model is being approached by different practitioners, as well as by the existence of so many definitions given to this concept by various schools of thought (Nikolaou and Tsagarakis, 2021). Nevertheless, Merli et al. (2018) noted this as evidence of a strong and quickly evolving field, signifying that this is a concept of undefined boundaries with ever-changing actors of various perceptions.

The CE is defined in various ways depending on the field one looks at (Korhonen et al., 2018). Kirchherr et al. (2017) identified approximately 114 definitions of the CE concept, and Merli et al. (2018) described it as a non-static concept definition, meaning that CE is still undergoing an evolutionary path (Velenturf and Purnell, 2021). This pluralistic definition has left CE wide open to multiple interpretations associated with various principles. Some schools of thought say this is an approach of principles for saving the environment and driving sustainable development (Mathews and Tan, 2011; Naustdalslid, 2014), while others observed this as a concept for industrial or economic growth through industrial symbiosis (Ellen Macarthur Foundation, 2013; Kirchherr et al., 2017; Velenturf and Purnell, 2021). Thus, good waste management

DOI: 10.1201/9781003327615-6

101

and recycling mitigate environmental degradation while presenting some economic opportunities. For example, about 0.75% of the European Union's GDP was ascribed to waste management and recycling in Europe. More so, the recycling sector reportedly had a turnover of 24 billion Euros and created employment for about half a million persons (Mathews and Tan, 2011). This industrial symbiosis is achieved by finding synergistic interactions between or amongst industries where waste products from one industry are turned into inputs for another industry's production processes (Mathews and Tan, 2011). Others think CE is defined as industrial ecology, a concept combining engineering and natural sciences (Murray et al., 2017). However, the consensus derived from these multiple definitions is that CE is an approach entirely focusing on increasing resource use efficiency, minimising resource exploitation and maximising waste reduction (Figure 6.1) (Velenturf and Purnell, 2021). In other words, it is fair to claim that CE is the foundation of the world's current and future economic development framework. It must be embraced, as it is our best option to curb irreversible damages likely to be brought about by the overexploitation of world resources and limit the environmental consequences of waste accumulation (Mhatre et al., 2021).

6.1.2 TRANSITIONING TO A CIRCULAR ECONOMY: WHAT IS THE ORIGIN OF THE CURRENT STATUS, AND WHY?

Globally, the growth in natural resource use tripled between 1970 and 2010, from 23.7 to 70.1 billion tons (Velenturf et al., 2019), and later to 89 billion tons by 2017 (OECD, 2019). This far exceeds the 82 billion tons previously projected by 2020 (Figure 6.2) (Ellen Macarthur Foundation, 2013). This clearly indicates that, in recent years, global natural resource extraction has been happening faster than previously forecasted. Further expansion in resource extraction is expected to continue, mainly driven by population growth, economic development, and changes in consumption patterns (UNEP, 2016). According to Lim et al. (2022), primary resource consumption is expected to double the consumption rate of 2017 and reach about 167 billion tons by 2060. The continuous rise in natural resource consumption has been driven by the current linear economic model, which is predominant internationally. Take-make-use-dispose (Figure 6.3) is a linear economy (LE) model (Ellen Macarthur Foundation, 2013) and has been the preferred economic development model for over 150 years, particularly since industrialisation (Principato et al., 2019).

Overall, the LE model is characterised by mass production and consumption patterns associated with large waste generation volumes in the production cycles. Recent statistics indicate that with the current production and resource consumption magnitude, the world would need about 1.7 times the Earth's size to replenish exploited natural resources, maintain current economic rates, and absorb the pollution generated by the LE model (World Economic Forum, 2019). Observing the situation from a distance, as we are currently propelling ourselves, and calling it 'business as usual' would deplete many resources, resulting in vulnerable economies (World Economic Forum, 2019). For instance, the current projections forecast a rise in the global resource use, demand and extraction of energy (Fossil fuels: oil, gas and coal), non-metallic minerals, metals and water. However, non-metallic materials

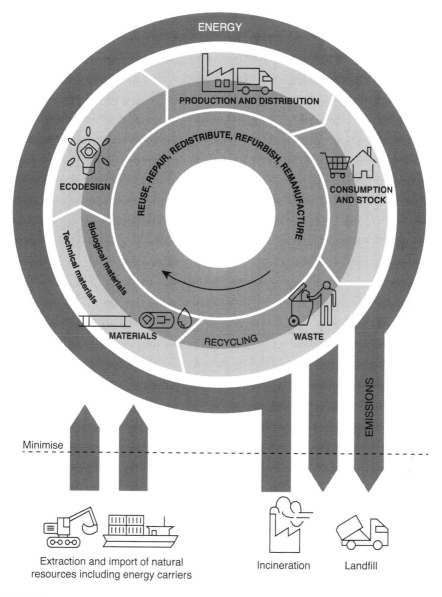

FIGURE 6.1 Natural resource flows in a CE for enhanced resource use efficiencies.

Source: European Environment Agency (2019).

demand is likely to remain above all other classes of resources, and their extraction will significantly increase with time from 2030 (Hatfield-Dodds et al., 2017; Lopez et al., 2020).

The world's current state of economic development originated and developed through the current LE framework; however, this has neglected sustainability principles and is being increasingly challenged by practitioners globally (Ellen Macarthur

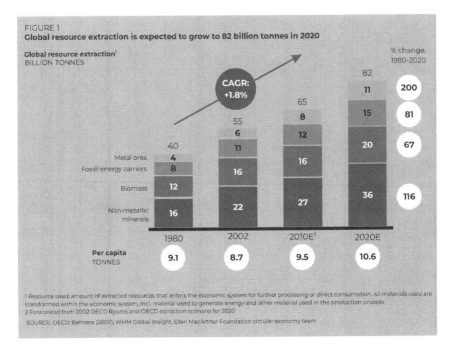

FIGURE 6.2   Forecasted growth and increase in global resource extraction from the year 1980–2020.

Source: Ellen Macarthur Foundation (2013).

Linear Economy

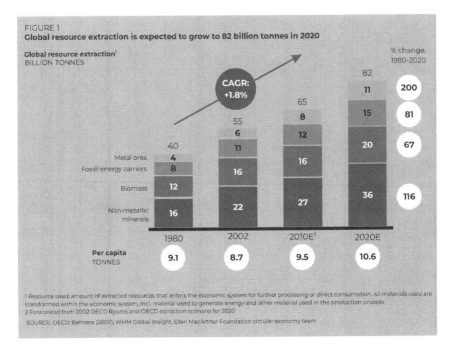

FIGURE 6.3   Linear economy resource flow.

Foundation, 2013). This linear framework has placed many countries at risk of resource overexploitation and depletion, potentially disrupting the continuous international economic growth. These are signs indicating an impending plateau in terms of economic development.

Considering the wide range of challenges and limitations associated with LE (Box 6.1), there is no doubt that migrating from LE to CE is critical. This migration would be a positive transition from the current Cradle-to-Grave economic system towards an alternative Cradle-to-Cradle system, returning as many resources to their original state as possible or replenishing resources where possible, thus supporting sustainable development (Burchard-Dziubinska, 2017; Drabe and Herstatt, 2016; Özkan and Yücel, 2020). The challenges in balancing industrial development, economic growth, and environmental and human health have strengthened the need to support the CE concept. For instance, sanitation and organic waste products management focus on environmental and human health services provision and drives CE; product management promotes reduced pollution and a cleaner environment. Clean environments are critical for human health. In addition, CE needs to move from a theoretical development phase to successful implementation across multiple sectors and all countries. This concept can be partially implemented sector by sector depending on a sector's potential to implement CE, or what is also called a "closed-loop" economy. However, for a wholistic realisation of CE, regions or countries must have sectoral engagement so that none of the sectors is left behind in playing its part in this noble economic model to drive sustainable development (Owojori and Okoro, 2022; Sharma et al., 2021).

Literature and emerging research show CE as the best economic model to replace the LE (Table 6.1); however, it is clear that CE is still in an infant stage in terms of implementation. In the European Union, programmes, regulations, and directives were implemented in support of the CE model (Camilleri, 2020; European Commission, 2017, 2018; European Union, 2020; Oakdene, 2018); however, little is observed in other regions, especially in the developing world (Desmond and Asamba, 2019). There is little or no clear action to support transformative efforts and

TABLE 6.1
Differences between linear and circular economies

Category	Linear economy	Circular economy
Attitude towards nature	Extensive resource extraction from the Earth	Decoupling economic activities from the consumption of scarce resources, keeping products and materials in use
Attitude towards production	Take-make-use-waste	6Rs – Reduce, Reuse, Recycle, Rethink, Repair, Recover
Closing loops and flows	One lifetime use of products, materials and energy	Materials and renewable energy flow infinitely in circles through the economy
Product attributes	Products become obsolete while still usable	Product life extension and used materials serve as valuable inputs for other products
The ecosystem services	Best simply on efficiency and one-size-fits-all approach	Many connecting nodes and scales show a greater resistance in the face of external stresses and shocks
Economic key values	Money and efficiency are the dominant values in linear businesses	Highly focused on economic, ecological, and social aspects

Source: Adapted from Shevelov (2020).

policies facilitating the migration from LE to CE in developing countries, especially in Africa (Desmond and Asamba, 2019; Negi et al., 2021).

This chapter traces previous and current CE efforts in South Africa. The focus is on the historical development of resource recovery from sanitation, particularly organic waste materials. As highlighted before, the economic and developmental sector or industry drives the perception of a CE, which is different in mining, agriculture, energy, water, human settlement, and waste management, to name a few. This chapter evaluates CE progress in South Africa from the perspective of resource reclamation from organic waste. Several technologies and projects in this field that have been or are running are described. Here, these projects are identified, and their strategies elucidated, including opportunities for scaling. Challenges, as well as projects that were halted, are also discussed. This work also describes the potential sustainable integration of these approaches into the wider South African economy and remedial pathways for such projects and technologies to succeed. In short, we assess the progress made so far in organic waste management and its contribution to accelerating the CE model in South Africa.

6.1.3 An overview of the South African circular economy
FRAMEWORK: POLICIES AND INVOLVEMENT

South African government and practitioners have put extensive resources and effort into developing rigorous guidelines for waste reuse. Risks become a critical factor whenever the CE is considered and wastes are re-used. Any waste destined for reuse must be classified as safe before being slotted back into the CE. Local guidelines support this effort, making it simple for practitioners to test and classify waste as safe for reuse. For instance, in organic waste management and reuse, elegant and clear regulations and guidelines have been developed in South Africa, mainly focusing on municipal sewage and water treatment residual sludge reuse in agriculture (Herselman, 2013; Snyman and Herselman, 2006).

Similarly, the testing and application of general organic waste are well-described under the National Environment Management: Water Act of 2009 (Godfrey et al., 2021). As much as this is a noble and useful attempt to regulate the use of organic wastes, the current guidelines focus on quality standards for sludges targeted for reuse in the agriculture sector, protecting the receiving environment from contaminants. This risk-focused approach can potentially limit rather than support or facilitate the transition to a CE. Ideally, considering this problem's magnitude, policy should describe risk mitigation and create directives for this transition. The specific policies necessary to support and mandate the transition to CE do not currently exist in South Africa. Although proposals have been tabled for consideration, they are yet to be promulgated into government policies and legislation (Desmond and Asamba, 2019). Current and historical efforts in strengthening the move to CE, particularly regarding organic waste and resource recovery, are driven mainly through individual sectorial efforts, either by researchers or private companies.

Due to these disintegrated efforts, the model has been marred with several challenges hindering a smooth take-off into a viable and sustainable transformative effort from LE to CE. In recent years, the South African Department of Science and

Innovation (DSI) launched the 'Science, Technology, and Innovation for Circular Economy' initiative. The department pledged to support the country's transition to CE through these avenues. This initiative has made significant strides in CE policy formulations, paving the way for waste research and developing an innovation roadmap towards a functional transition. To align its efforts with the national and international priorities for CE, the Waste Research, Development and Innovation plan was initiated, a ten-year waste management plan aimed at growing and transforming South Africa's waste sector (Department of Science and Technology, 2014). The government also included CE as a paramount option for sustainable growth. In general, policy documents like the White Paper on Science, Technology and Innovation (Department of Science and Technology, 2019), the Decadal Plan (Department of Science and Innovation, 2021), and the National Waste Management Strategy (Department of Environment Forestry and Fisheries, 2021). Although these are noble initiatives in support of the drive towards CE, the country's policy framework on CE remains fragmented across various government institutions and departments, like the Department of Water and Sanitation (DWS), the DSI, the Department of Science and Technology (DST), and others (Nahman et al., 2021). This lack of a consolidated approach does not acknowledge the urgency of the situation and is one of the primary factors inhibiting the progress of the transition.

Despite limited national strategic planning towards a CE, South Africa is working towards a transformative migration. In the regional and international space, South Africa has assumed a leading role in pushing the CE forward through its current role as co-chair of the African Circular Economy Alliance, which aims to redress these consolidation challenges, linking up continental projects and programmes, and facilitating collaboration to drive the transformation to a CE. This alliance was jointly launched by South Africa, Rwanda and Nigeria at COP23, the annual United Nations Climate Change conference held in 2017 in Bonn, Germany. South Africa is also a co-founder of the African Circular Economy Network (ACEN), which was formed in June 2016 by a group of CE professionals in Cape Town. This CE network envisions strategies for a restorative continental economy that generates social cohesion and community success through economic production and consumption that supports the regeneration of environmental resources (GRID-Arendal, 2021). It doubles as an active participant in the World Circular Economy Forum and a member of the Global Alliance on Circular Economy and Resource Efficiency (Nahman et al., 2021). Its involvement in these regional and global CE organisations allows the country to share challenges and learn from the world's transformative ideas that will help shape and align the national strategic policy framework with the global sustainable development goals.

6.1.4 ORGANIC WASTE MANAGEMENT AS A CIRCULAR ECONOMY PLATFORM IN SOUTH AFRICA

Like in any other developing country, waste management in South Africa is still a challenge. The published statistics show that approximately 80% of the 108 Mt of 2017 waste generated in South Africa has been landfilled (Department of Environmental Affairs, 2018). In 2020, a notable proportion of waste was reported as mismanaged,

with almost 37.4% of households having no access to refuse removal, necessitating illegal dumping (Chitaka and Schenck, 2022). This has been compounded by
the rise in population, rural urban migration, and industrialisation (Department of
Environmental Affairs, 2017). These demographic and social challenges give rise to
increasing waste generation, which is challenging to manage. This increases pressure
on already-limited water supply infrastructure, environmental health and sanitation,
and many other services, especially in cities and peri-urban areas. In South Africa,
a notorious management challenge is associated with water and waste infrastructure in mushrooming unplanned community settlements, especially in peri-urban
areas. This was evidenced by the cholera outbreaks reported in South Africa from
1980 to 1986 (Sidley, 2001), in 2003 and from 2008 to 2009 (National Institute for
Communicable Disease, 2009). Cholera outbreaks are a sign of public health system failure, associated with a lack of access to running water and proper functional
sanitation services (Ali et al., 2011; Ismail et al., 2013). This was the case with South
Africa during those outbreaks, when about 80% of informal settlement residents had
no regular access to clean water, and close to 18 million rural South African citizens
had no access to municipal sanitation services (Sidley, 2001).

To address these challenges, soon after attaining independence in 1994, the government of South Africa initiated large-scale sanitation infrastructure programmes.
A national sanitation programme called the National Sanitation Policy White Paper,
was developed and launched in 1996, which defined the basic sanitation technologies
fit for households (Bhagwan et al., 2019). Many communities had limited or no access
to dignified sanitation services during the apartheid era. However, since water access
and availability have always been a major challenge, especially in rural and informal
settlement communities, and considering the scale of addressing this inequality that
needed redressing, these communities received primarily on-site sanitation systems,
which are more cost-effective than flush toilets. Although on-site technologies come
in different forms, most households were installed with ventilated improved pit (VIP)
latrines. Since the inception of this sanitation programme and after 1994, over two
million VIPs and other on-site toilets were installed (Bhagwan et al., 2019).

As much as these technologies were deemed adequate, ideal in line with United
Nations (UN) standards, and cost-effective for rural communities, they also come
with challenges. The associated limitation observed with the VIPs is the high rate of
saturation, which then demands either decommissioning of the full toilets or intermit-
tent emptying of the faecal matter if the same toilet is to be used continuously (Mjoli,
2010; Still et al., 2012). Either of these choices is associated with expenses. When
the former is chosen, the household would need to rebuild another toilet in a different location, which demands land and space. Emptying and reusing latrines involves
handling the faecal matter and disposal costs of the human excreta. Generally, sludge
disposal in South Africa from wastewater treatment plants (WWTPs) and on-site
systems is a challenge to municipalities (Pillay and Bhagwan, 2021). Proper sludge
handling is fundamental to reducing illegal sludge dumping and consequent environmental contamination.

Large amounts of faecal matter are generated from these on-site sanitation systems in communities. Therefore, several options for waste handling were proposed
as waste management strategies. Initially, the focus was managing these organic

materials as waste. However, with the help of emerging academics and industrial research, the idea of resource recovery from human excreta material was developed. The CE concept was born, gaining prominent attention in the waste industry (Department of Environmental Affairs, 2017; Still et al., 2012). With the need to implement successful strategies for resource recovery from human excreta materials, different sanitation technology prototypes were designed and piloted in South Africa by stakeholders like researchers, municipalities and public-private companies. Ideally, these are advanced sanitation technologies, as they should have the capacity to enhance the value chain of resource recovery more effectively than VIP latrines. These advanced technologies include urine diversion dry toilets (UDDTs), decentralised wastewater treatment systems (DEWATS), organic waste compositing and biochar material production. The details of how these technologies work and the products produced for reuse are well documented in the literature (Gutterer et al., 2009; Kvarnström et al., 2006; Mkhize et al., 2017; Mnkeni and Austin, 2009; Musazura et al., 2018; Vinnerås, 2001; Vinnerås and Jönsson, 2002).

There is adequate evidence that recycling and reusing human excreta-derived material benefits agriculture. This knowledge led to pilot projects targeting generating resources out of organic wastes (wastewater, sludge, human faecal matter and urine, food, or green waste). Although many strategies have focused on waste disposal, the new paradigm focuses more on resource recovery, deriving fertiliser materials, and harnessing municipal effluent for agricultural use. This doubles as a positive strategy for sustainable environmental protection (Sharma et al., 2022). It is a sustainable intervention because it reduces the amount of waste to be channelled into the environment by diverting wastes from landfills into reusable materials like organic fertilisers and irrigation water.

6.1.5 ORGANIC WASTE–DERIVED RESOURCES RECOVERY TECHNOLOGIES AND INNOVATIONS SUPPORTING THE CIRCULAR ECONOMY IN SOUTH AFRICA

Like any other country, South Africa faces environmental degradation and pollution challenges from municipal sludges, landfills and dumping sites for food and other organic waste. In addition, rapid resource extraction and depletion are prevalent across the country. A wealth of literature has shown that CE has the potential to address these challenges (Sehnem et al., 2019; Tahulela and Ballard, 2020; Wijkman and Skånberg, 2015). Various technologies are available to facilitate resource recovery and reuse of organic waste–derived materials from waste streams. These waste products are as broad as food waste (food market dumping sites and household waste), wastewater, and municipal sludge (from centralised or on-site sanitation facilities) like faecal sludge or human excreta and urine. Urine can be harvested from on-site sanitation facilities like UDDTs, improved urine diversion toilets and VIP latrines. Treatment is always aimed at improving the materials' quality regarding physical, chemical and biological properties while maintaining their beneficial value (e.g., nutrient content). Risk is an important consideration, and processing must render it safe and pleasant for handling, agricultural use and consumption of the associated products (i.e., crops grown in waste-fertilised soils). This section details the strategies and technologies currently used for resource recovery from organic wastes in

South Africa. We also highlight how these strategies support the CE concept and are linked to sustainable waste management through organic waste treatment.

6.1.6 FAECAL SLUDGE TREATMENT TECHNIQUES FOR RESOURCE RECOVERY AND REUSE

In the context of promoting access to sanitation services for all in South Africa, coupled with resource recovery, technologies have been developed and installed around the country. For example, over 80,000 UDDT were installed in eThekwini municipality, Durban (Bhagwan et al., 2019). This type of sanitation technology was an upgrade from the general VIP, as it separates faecal matter and urine, facilitating easy drying and making emptying, collection and transport manageable. However, there was some resistance to adopting the technology as people were familiar with their traditional VIPs (Roma et al., 2013). The study by Roma et al. (2013) showed that the rejection was attributed to reasons such as smell and malfunctioning of pedestals. As a result, the authors recommended that adopting such technologies can be improved by educating the users on the potential benefits of using such technologies, especially regarding the nutrient recovery aspect. Etter et al. (2015a) reported that user acceptance was increased with adequate education, contributing to increased urine collection from households for valorisation. Faecal sludge from existing sanitation technologies like pit latrines in most rural communities of South Africa is considered unsafely managed. This poses both environmental and health risks (Bishoge, 2021; Kalulu et al., 2020; Mamera et al., 2020), necessitating improved treatment and collection strategies and awareness campaigns, hence making the UDDT an important technology to consider during the transition towards CE. To reduce faecal sludge's pathogenicity, toxicity and odour for use as soil fertilising material, sludge must be treated and stabilised before use. Several techniques are applied for faecal sludge treatment in South Africa and globally to improve the sludge quality for agricultural reuse. The existing treatment techniques in South Africa include the use of composting and co-composting, wastewater treatment using a DEWATS, black soldier fly larvae (BSFL), latrine dehydration and palletisation (LaDePa), and pyrolysis. The next section details these faecal sludge treatment techniques.

6.1.6.1 Composting, co-composting, and vermicomposting

In faecal sludge management (FSM), composting and co-composting are heat and microbial stabilisation processes used to sanitise organic waste materials, making them fit for handling, use in agriculture, and lowering human health risks. Sánchez et al. (2017) define composting as an aerobic, thermophilic microorganism-mediated solid-state fermentation process, transforming organic waste materials into more stable organic compounds. Co-composting is a simultaneous composting process of two or more types of organic waste materials, which are sources of N or C to enhance microbial activity (Das et al., 2011; Petric et al., 2012). The processes increase the potential for improved and enriched compost quality for agricultural use (Paredes et al., 1996). With the global increase in the generation of organic waste, for example, garden and food wastes, which are being disposed into landfills, leading to environmental pollution through greenhouse gas emissions, it is important to consider innovative and ecologically sustainable waste management strategies. Technologies such as co-composting and subsequent agricultural use of the compost materials minimise volumes of organic wastes entering landfills and environmental pollution. This ensures environmental sustainability, as physical waste volumes or released organic compounds from such organic wastes are reduced, limiting their transfer into groundwater and surrounding trophic food chains. Composting processes have proven to be useful (Körner et al., 2003), as they transform organic wastes into nutrient-rich fertilising materials (Scheutz et al., 2011) and are used as soil conditioners (Iqbal et al., 2010). Such strategies for waste resource recovery and reuse in agriculture close the loop of an originally linear system and create a CE in the sanitation or organic waste management system.

Faecal sludge and municipal sludge can be composted too or co-composted with other agro- or green waste or animal manure to produce a compost suitable for use as a soil conditioner or fertilising material (Iqbal et al., 2010; Petric et al., 2012). Generally, compost materials are typically low-value products regarding plant nutrient content and can be primarily used as soil conditioners. However, co-composting or fortification with municipal sludge or human urine can enrich compost materials and increase their fertiliser value (Cofie et al., 2016). Fortification can also be done by adding a fraction of chemical fertilisers, usually using nitrogen or phosphorus fertilisers. In addition, most composting techniques reduce pathogen loads in faecal matter (Dumontet et al., 1999; Grantina-Ievina and Rodze, 2020). Common composting or co-composting is done in windrows or piles. However, sometimes composting can be done at the household level, especially when communities are provided with designed composting toilets as on-site sanitation services.

Vermicomposting is one of the alternative methods used for degrading organic matter. According to Singh et al. (2011), vermicomposting is the decomposition of solid organic waste facilitated synergistically by microbes and earthworms. The authors state that, even though microbes primarily facilitate waste degradation, earthworms are the true foundational drivers of the decomposition process. They fragment and condition the substrate and enhance microbial degradation. Vermicomposting produces a more nutrient-rich compost material than the traditional composting process (Suthar, 2009). There is some mixed information on the ability of vermicompost to deactivate pathogens. Ndegwa and Thompson (2001) reported that the vermicomposting process cannot deactivate pathogens from organic wastes such as faecal sludge. However, this contrasts with earlier studies by Eastman (1999), which indicate that vermicomposting can deactivate pathogens more than general composting. According to Samal et al. (2022), vermicomposting can deactivate pathogens if the process is done properly, and from their review, it was stated that pathogen deactivation takes 60 days under optimal conditions.

South Africa is one of the Sub-Saharan African countries facing food insecurity due to degraded soils resulting from minimal use of organic fertilisers (ten Berge et al., 2019). The recovery of nutrients from organic wastes or co-compost and reuse in agricultural fields as a soil conditioner helps improve soil properties by enhancing microbial activity for nutrient recycling, increasing soil moisture and nutrient retention capacity and increasing soil aggregate stability (Cofie et al., 2016; Fuhrmann et al., 2022; Iqbal et al., 2010). This recovery and reuse create a closed-loop circular system in a way that intertwin food security while addressing sustainable waste management in line with responsible consumption and production (SDG 12) (Drangert et al., 2018; Harder et al., 2020).

Although these processes have gained momentum as resource recovery technologies from solid organic waste such as faecal or municipal sludges, the development of business models for implementation in South Africa is still lacking. However, sewage co-composting is currently underway as a pilot project through a multidisciplinary project called **R**ural **U**rban **N**exus: Establishing a Circular Economy for Resilient city-region food systems (RUNRES) implemented by the University of KwaZulu-Natal's (UKZN) Crop Sciences team. It is executed in collaboration with some private companies and the uMngeni WWTP in the Msunduzi Municipality, KwaZulu-Natal, South Africa. In this case, shredded green waste (garden waste) is mixed with municipal sludge and co-composted on windrows over time. Periodic sampling and analyses are employed to continuously monitor the composting process, focusing on the changes in microbial pathogens, chemical composition, and the final compost quality. The compost is not yet sold on the formal market due to the unavailability of certification of faecal sludge-derived compost products. As a result, compost production is still under research and not yet produced at a large, economically viable scale. Until then, operated as a business entity, this is still limited in supporting the CE approach. However, with relevant support and policies in place, the demonstration of such composting case studies could support this technology to drive CE in sanitation and organic waste materials.

6.1.6.2   Decentralised wastewater treatment system (DEWATS)

The DEWATS is a robust waterborne package that treats various types of wastewater close to the generation source. The DEWATS is a low-cost, decentralised, community-based wastewater treatment technology that can be made from low-cost, locally available materials and operates on a low energy demand (Gutterer et al., 2009). By design, DEWATS works the same way as the conventional system. The treatment follows the common four processes, i.e., primary, secondary, tertiary (advanced secondary treatment) and post-treatment phases (Gutterer et al., 2009; Singh et al., 2019). The DEWATS anaerobically degrade organic compounds from various wastewater types into inorganic compounds, producing effluent that contains mineral nutrients and some pathogens. In hybridised DEWATS, the planted gravel filters (PGFs) (horizontal flow constructed wetlands; HFCW and vertical flow constructed wetlands; VFCW) have sand filters to further polish the effluent to remove pathogens and other nutrients (Singh et al., 2019). The effluent enters the VFCW and vertically flows down the sand filters, and during seepage, the oxygen promotes nitrification processes. The difference between HFCW and VFCW is that in the former, effluent moves horizontally, but in both systems, pathogens are captured onto gravel particles, where they are eventually deactivated (Gutterer et al., 2009). The deactivation of pathogens is hastened by several processes, including predation by protozoa and other bacteria such as *Bdellovibrio bacteriovorus* (Wand et al., 2007). The advantages over conventional treatment include easy operation in small communities (decentralised), simplicity (minimum operational skills, no to low energy requirements) and reusable product generation (resource recovery from treated wastewater) (Singh et al., 2019; Varma et al., 2022). DEWATS have been used widely in developing countries like India

(Singh et al., 2019), Nepal (Bright-Davies et al., 2015), Brazil (Dariva and Araujo, 2021), Indonesia (Kerstens et al., 2012) and South Africa (Reynaud and Buckley, 2015), among others. However, in South Africa, this technology is still at the pilot scale at Newlands Mashu Ecological Centre. In 2018, the eThekwini municipality planned to scale the DEWATS to other areas such as Banana City and kwaDabeka (Tuyens et al., 2018). However, the same idea has been adopted for rural schools, whereby the suitability of DEWATS technology is being piloted at iNtapuka primary school (H_2O Sanitation Services, 2022). The RUNRES project operating in Msunduzi has selected various innovation platforms. One is the DEWATS plant connected to urine diversion toilets and will be piloted for wastewater treatment at a rural school in Howick. The innovation will include the recovery of wastewater and urine for agricultural use; if successful, the project plans to scale out the innovation. This implies that South Africa is still in the transitional phase when it comes to the implementation of DEWATS technologies in Ces.

Generally, conventional centralised municipal sewage systems are the most common treatment technologies installed in towns and cities. Although their sizes may vary in design and preferences, they are considered most suitable due to the large volumes of wastewater they can handle and treat at a time and their efficiency in organic compound removal. However, they are associated with several challenges as they are considered resource intensive, for example, (i) high initial capital investment requirement, (ii) high operations & maintenance (O&M) costs, (iii) high technical capacities requirement, and (iv) high energy requirements, among others (Bhagwan et al., 2019; Gutterer et al., 2009). Thus, despite the quality of effluent, the mentioned set of challenges of these systems often make them not feasible for rural communities and unplanned peri-urban and informal settlements. Designing such plants for undulating and mountainous locations also involves extensive additional costs.

Despite these challenges, every community deserves to have a dignified sanitation system at its service. To overcome the hurdles of large-scale conventional treatment systems, DEWATS can be used instead. DEWATs have been identified as an alternative wastewater treatment approach to conventional systems. They can be used for wastewater treatment with the on-site sanitation technologies commonly employed in small townships and rural communities of South Africa. This system is considered economically feasible as a Community-Based Sanitation framework for small, densely populated communities in rural and peri-urban settlements (Water and Sanitation Program, 2013). Although this technology is decentralised, best suited for small communities, and capable of treating low wastewater flows ranging from 1 m^3–1,000 m^3 per unit per day (Gutterer et al., 2009), it can potentially be used to complement the conventional treatment system when needed. Evidence-based information shows that the DEWATS technology supports CE by allowing water and nutrient recovery for agricultural use (Bame, 2012; Busari et al., 2020; Magwaza et al., 2020).

Furthermore, studies by Musazura and Odindo (2021) showed that the use of DEWATS effluent from both the anaerobic filter (AF) section and after the PGFs have no negative effects on soils, crops, environment and irrigation equipment. However, in South Africa, the effluent originating from the DEWATS is not being used for agriculture despite evidence that it is suitable for the purpose and following the existing World Health Organization guidelines (Reynaud and Buckley, 2015). The pilot scale

DEWATS package in Newlands, South Africa, is solely used for research purposes. However, the eThekwini municipality plans to scale out the technology to other areas around its jurisdiction. The municipality also envisions a reuse component, but the existing South African policies are unclear on producing food crops using effluent. Recent efforts by the Water Research Commission (WRC) were to establish a South African specific practical guideline for using DEWATS effluent for agricultural use (Odindo et al., 2022). The current project, WRC C2021/2022–00603, investigates traditional and advanced technologies for eliminating pathogens for unrestricted agricultural use. Implying that, at the research level, the future of DEWATS technologies in CE is bright in South Africa. DEWATS construction and use as a sanitation technology has been tested in South Africa and has proven valuable in alleviating pressure on sanitation services. However, there is low general uptake as a resource recovery technology, likely due to the lack of information on technology availability and safety and pairing waste with crops and soil types. Lack of education, conscientisation and limited stakeholders' involvement at the planning level of an innovation or technology can stagnate implementation of these technologies to support CE. Many relevant stakeholders left out at the initial stages were not aware of this technology. This makes adoption and implementation challenging, requiring a transformational approach to ensure success in such projects. As explained earlier, there is a need for certification of these products and strategies to endorse them as safe for handling and use for agriculture and create consumer confidence. The end users need assurance from the relevant authorities that the product is approved for such purposes to facilitate technology transfer from piloting to public implementation.

6.1.6.3   Black soldier flies larvae (*Hermetia illucens* L.)

BSFL is another technology used for faecal sludge treatment for resource recovery in organic waste management, especially in low-income countries (Diener et al., 2011). According to Otoo et al. (2015), BSFL technology is a composting process that allows BSFL (maggots) to feed on solid organic materials like faecal sludge or food waste. After the larvae are grown, especially at a prepupa stage, the larvae would be harvested and fed to animals like chickens as a protein source. In such instances, where sludge residues or frass and larvae themselves are both the input waste and products from the technology, the use of BSFL would be of a double benefit: firstly, as a waste management technique for resource recovery, and secondly, as animal feed for protein provision. Through this process, a CE approach can be achieved as the larvae effectively contribute to closing some nutrient cycles (Fuhrmann et al., 2022). After degradation by maggots, the organic materials' residues can then be used as is for soil conditioning, can be further composted before use, or may be used as feedstock for biochar production (Nkomo et al., 2021). This biochar can be applied back into the VIP to further treat the faecal sludge material and to control or limit the leaching of pollutants from the faecal sludge (Mamera et al., 2021). Since faecal sludge is potentially harmful due to microbial pathogens and organic pollutants, after being stored in a UDDT or VIP latrines (Austin, 2001), it is critical to ensure that the communities using these on-site sanitation technologies have a faecal sludge treatment plan in place. The use of BSFL is one of the relevant and tested technologies that can be employed, and its residues could produce biochar that, in turn, could be applied back to the VIP toilets. If the BSFL residues are tested and meet national land application standards, they could be used as soil conditioners on agricultural lands. However, if the residual sludge is deemed unfit for use, it is recommended to be incinerated and used as biochar, a potential soil conditioner. Biochar is not as nutrient-rich but has many soil benefits (Mutsakatira et al., 2018).

The growth cycle of the BSFL is briefly detailed here. The BSFL is an insect that consumes biodegradable organic waste and consists of four life cycles (Figure 6.4). The larvae feed voraciously on a range of organic substrates but stop feeding as they reach the prepupal stage. Once they stop feeding, they rely on the stored fat from the larval stage and migrate out of the food source in search of a dryer pupation site. The larval form takes two to three weeks until it reaches the final larval stage, the prepupae (Maleba et al., 2016; Peguero et al., 2021). The larvae grow at different rates and sizes depending on the diet and environmental conditions. The BSFL needs a warm environment to grow, which becomes a problem during winter. However, this is far less limiting in most African countries than in northern hemisphere countries.

Based on the target business model one would choose, the BSFL can be used to feed on the faecal waste/sludge. However, it has been found that faecal sludge alone is not a good diet for producing bigger larvae if one targets larvae for animal feed. It would then need to be mixed with food market wastes to produce bigger BSFL in a shorter period. The BSFL can convert organic biomass from waste into protein (approximately 40%–44% crude protein) and fat (dependent on a diet) (Shumo et al., 2019). This makes the larvae, once dried, suitable to be used as animal feed. The extracted oil, during drying, can be used to make other products, such as cosmetics (Maleba et al., 2016; Mutsakatira et al., 2018). This wide range of products that could be potentially produced from these larvae makes this a promising technology, in addition to its quick turnover time (two to three weeks larval stage) relative to other composting technologies.

The current status of this technology in South Africa is promising, as there is already a commercial endeavour. AgriProtein is a Cape Town-based company that has conducted treatment of biodegradable waste using BSFL for several years, which

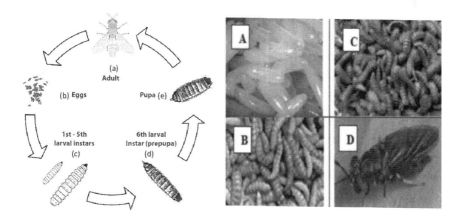

FIGURE 6.4 Developmental stages (life cycle) of BSFL (*Hermatia illucens*); (A) eggs of the BSFL; (B) larvae stage; (C) pupa stage; and (D) adult fly (Oliveira et al., 2015).

has been done in collaboration with Stellenbosch University and BioCycle. In the eThekwini Municipality, the feasibility of this method was assessed by Khanyisa Projects, with funding from the Bill & Melinda Gates Foundation. However, these are still run as research and development innovations. There are no business models to transition to large-scale production relative to other composting technologies. According to Cofie et al. (2016), this takes up to six months, depending on the feedstock used. However, Sustainability through Entomology is a private company based in Cape Town which is producing insect protein from BSFL reared using agricultural by-products (Innovus, 2022). The company produces a wide range of products from BSFL, including compost from the BSFL residues, animal feed, BSF breeder's feed and BSF eggs. However, they are not using faecal sludge, meaning that BSFL CE business models currently taking place in South Africa are impeding the adoption of human excreta as feedstock.

6.1.6.4 Latrine Dehydration and Pasteurisation (LaDePa) process

In South Africa, the LaDePa machine and process were developed in response to the need for an environmentally friendly FSM strategy. The VIP latrines installed in the eThekwini municipality communities were filling up without a clear sludge disposal plan (Gounden et al., 2006; Septien et al., 2018; Zuma et al., 2015). The LaDePa process is a faecal sludge treatment option developed by the eThekwini Water and Sanitation (EWS) in conjunction with Particle System Separation (Harrison and Wilson, 2012). Latrine dehydration and pasteurisation technology is an innovative process of sludge drying and pasteurisation using a LaDePa machine (Mirara et al., 2020). The faecal sludge is a valuable resource in the LaDePa process (Figure 6.5A). During processing, the LaDePa machine extrudes sludge as pellets that are dried and then exposed to infrared radiation for pasteurisation (Septien et al., 2018). As they leave the LaDePa machine (Figure 6.5B), the final pellets are dry and hygienically safe to handle.

After sludge treatment in the LaDePa process, the sludge material can be used as a soil conditioner or fertiliser for agricultural production in the form of dried pellets. This closes the loop of nutrient cycling. Generally, pellets are a product characterised by a relatively high phosphorous and carbon content in addition to substantial amounts of other trace elements needed for crop growth. Alternatively, these final products can be used for biofuel energy production (Diener et al., 2014). In this process, critical sludge treatment occurs when the pellets are exposed to infrared radiation that reduces the microbial pathogens found in the sludge, like faecal coliforms, Ascaris, and helminth ova (Septien et al., 2018). This method has been piloted in eThekwini Municipality in Durban, South Africa, and has the potential for expansion. The machine has a relatively low capital cost and employs basic mechanical and electrical technology. It is robust, simple to operate, and can be placed in containers for mobility (Harrison and Wilson, 2012; Mirara et al., 2020).

Considering that most rural communities and informal settlements in and around peri-urban areas of South Africa use VIP latrines, the LaDePa machine and process presents an opportunity to facilitate an environmentally friendly FSM strategy that can bring significant national impact. Currently, most communities in South

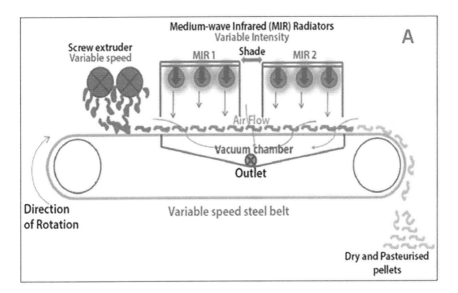

FIGURE 6.5 Schematic diagram of the LaDePa process (A) and the pellets that are produced (B) (Mirara et al., 2020).

Africa have no sludge emptying services when their VIP latrines fill up. Instead, they abandoned the full pit and installed a new latrine in another location. This rotational system not only presents the challenge of land shortage but also exposes the groundwater systems to pollution risk due to leachate from the untreated faecal sludge, which is increasingly concentrated as latrines displace land mass. In addition to being a feasible sludge disposal option, fertiliser pellets are a promising income stream with well-designed business models. Since the LaDePa machine is shown to be an inexpensive and movable product (Harrison and Wilson, 2012), there is a possibility that it can be operated at the local level and the produced pellets can be used for agricultural production locally, limiting transport and logistical costs and encouraging local nutrient circularity.

6.1.6.5   Urine and urine-derived products

Earlier, we highlighted BSFL and LaDePa processes dealing with mixed sewage waste, including faecal sludge and sewage from urine-diverting toilets. However, urine is another valuable sanitation product, considered a "golden liquid" if efforts are made towards beneficiation and nutrient recovery for agricultural use. Urine can be a source of several products applicable to agricultural productivity. The inception of UDDT in South Africa brought in urine as another valuable resource besides faecal sludge. Installation of UDDT opened an opportunity for source separation of human excreta, which would allow urine valorisation and nutrient recovery and reuse in agriculture.

As the Durban metropolitan city expanded in 2002, the municipality was mandated to provide safe and dignified sanitation services to everyone, including the indigent group. One of the ideas was to introduce ecological sanitation solutions; however, expanding the waterborne sewer network system was beyond their scope since it is expensive. As a result, urine diversion toilets were deemed the minimum sanitation technology to put in place. However, there were no strategies to deal with human excreta. The initial plan was to bury faecal sludge on-site, while the urine was drained in the soak pits, although this was not environmentally sustainable.

In response to this complex sanitation-environment challenge, in 2010, the Valorisation of Urine Nutrients in Africa (VUNA) project took a multidisciplinary approach to investigate methods for managing urine from the UDDT. As eThekwini is one of the cities in the Sub-Saharan region facing massive food insecurity, the EWS and Swiss Federal Institute of Aquatic Science and Technology (EAWAG) collaborated on this pilot project to (i) promote the use of urine diversion toilets, (ii) produce agricultural fertilisers from urine, and (iii) reduce environmental pollution by recovering and valorising urine nutrients. This multidisciplinary pilot project was done by a team of social scientists, engineers and crop scientists from the Water, Sanitation & Hygiene Research and Development Centre (WASH R&D) (Pollution Research Group), the Centre for Development and Cooperation at EAWAG, Environmental Chemistry Lab, Plant Nutrition Group, and the School of Agriculture, Earth and Environmental Sciences Crop Science team from the UKZN.

Since the use of urine in agriculture has some limitations, including pharmaceuticals, pathogens and the loss of nitrogen through volatilisation, treatment technologies must focus on minimising these risks while maximising potential benefits. As a result, various urine valorisation technologies were considered, including struvite, nitrified urine concentrate (NUC) and urine storage as recovery methods to ensure that the urine and its products are free from pharmaceuticals and pathogens and maintain high concentrations of nutrients (Udert et al., 2016). Urine storage is recommended to reduce pathogens, but the process is associated with excessive nitrogen losses (Ouma et al., 2016). The nitrification and distillation process also effectively reduces pathogens and pharmaceuticals while stabilising the urine into a concentrated product rich in nutrients (Etter et al., 2015a). The VUNA project investigates a complete recovery process that involves removing the water to minimise nutrient loss. The urine is first stabilised to prevent ammonia from volatilising through nitrification. Bacteria oxidise half the ammonia into non-volatile nitrate, and as the pH decreases, the other half is stabilised as ammonium. Once the urine is nitrified, water is removed through distillation. Distillation concentrates urine nutrients in a liquid fertiliser (a form of NUC). Nitrogen loss during this step is minimal (Etter et al., 2015b).

The struvite production process might not completely remove pathogens, but Bischel et al. (2016) recommended that initial heating under moist conditions followed by desiccation can be done. Precipitation of struvite is a known process for recovering phosphorous from urine. The precipitation process produces solid struvite from the urine solution during a chemical reaction. Struvite is subsequently dried at ambient conditions. Struvite precipitation allows ammonium recovery, which is an environmental advantage and an economic benefit, as it can be sold as a fertiliser (Etter et al., 2015b; Udert et al., 2015). During the VUNA pilot project, the above-mentioned products (NUC, struvite, struvite effluent and stored urine) were produced. Their potential agricultural use as fertilisers was investigated, and substantial evidence shows that struvite is good as common as a P fertiliser source, while NUC provides the required crop N (Bonvin et al., 2015).

Recovery of nutrients from urine as a business model canvas for the urine value chain from collection to the final product showed that large-scale urine production is not expensive, but the transport costs are very high. It was further reported that concentrated urine production provides more revenue than solid bulk fertiliser (Etter et al., 2015a). The system produces distilled water that can be marketed profitably. To see this being implemented country-wide, scaling up was suggested, and the EAWAG acquired an operating licence for urine fertiliser production from the Swiss Federal Office for Agriculture. To date, Aurin, a concentrated fertiliser, has been endorsed by the Swiss government for use on flowers and vegetables (Halbert-Howard et al., 2021). In South Africa, its use is still limited to research only, and it is not scaled to commercial endeavours yet.

6.1.6.6 Biochar

Biochar is a product of pyrolysis, biomass (wood, manure, leaves, sludge) degradation process at high temperatures ranging between 350°C and 800°C (Krueger et al., 2020) in the absence of oxygen, resulting in the production of charcoal (solid), bio-oil (liquid) and gas (Demirbas and Arin, 2002; Lai et al., 2018; Lehmann, 2009; Wang et al., 2017). It is a carbon-rich material that has been widely used to improve soil properties, reduce soil nutrient losses and promote agricultural production (Cui et al.,

2020a). In addition, studies have highlighted that biochar has the potential to improve soil carbon capture, climate change mitigation, soil pollution remediation, wastewater treatment and energy storage (Ahmad et al., 2014; Chen et al., 2016; Inyang and Dickenson, 2015; Leng et al., 2015; Yan et al., 2022).

Technical details like feedstock types and the associated pyrolysis temperatures have been highlighted and summarised in a review by Ahmad et al. (2014). The feedstocks range from woody material like tree plantation residues, crop residues, and animal waste/manure to sludge waste, grass and saw-dust. Biochar quality and stability depend on the biochemical composition of the feedstock, temperature level and time taken to heat the feedstock. Although biochar use has gained global attention, its production from faecal sludge feedstock has not been adopted at a large scale; rather, it is still limited to laboratory-scale research (Krueger et al., 2020). However, there is evidence of potential benefits from faecal sludge biochar. Besides being used as a soil conditioner, faecal sludge can be made into biochar and be returned to on-site sanitation technologies like VIP toilets to help reduce microbial pathogens and leaching of pollutants into underground waters (Ahmad et al., 2014; Mamera et al., 2021, 2022) and other soil and sludge conditioning (Bai et al., 2018; Deng et al., 2022).

Biochar has the potential to address sanitation and soil fertility challenges. However, there is a need for a transdisciplinary approach to transition from experimental-based scales to viable commercial scales. Although this technology is promising to address the limitations, its implementation, especially regarding faecal sludge biochar production, is still confined to research and has not been expanded to commercial-scaled endeavours.

6.1.7 CHALLENGES LIMITING THE SUCCESS OF CIRCULAR ECONOMY AND PROGRESS IN RESOURCE RECOVERY TECHNOLOGIES

Although the CE approach is beneficial relative to the current LE, it is still derailed at the interface between research, design and implementation. Several hindrances impede the implementation of CE approaches, either at the sectoral or national level. Meanwhile, 193 countries from the United Nations globally attended and proposed a set of actions, including strategies to sustainably boost their economies by 2030 (UNGA, 2015). However, with eight years to go to achieve these goals, there is limited evidence of achieving 100% progress on several set goals. Understanding the hurdles impeding this transition is crucial (Jensen, 2022). The authors have identified several challenges in implementing the CE approach in the sanitation and organic waste management sector in South Africa.

6.1.7.1 Technological and financial challenges

Successful implementation of the CE approach in the organic waste management sector requires a well-planned investment in technology. Developed technological infrastructure is critical at various waste management value chain stages to realise substantial benefits of waste-derived materials and recycling. For off-site centralised wastewater treatment, lack of financial support for O&M, upgrading and maintaining ageing infrastructure are some of the challenges highlighted by the DWS's Green Drop National Report, a regular national report investigating the reuse of wastewater in agriculture (Department of Water and Sanitation, 2022). Several WWTPs scored below 31%, based on the assessment criteria put in place, revealing the dismal state of wastewater management in the country in meeting the Green Drop Status (Department of Water and Sanitation, 2022). Green Drop Status essentially investigates the nationwide attempts to transform waste into a product that can be utilised in agriculture, much like many of the technologies discussed in this report. While the upgrade and expansion of these plants are imperative, the difficulties WWTPs face include challenges in sourcing the required funds, skills to access such funds, and the time it takes to develop new financing mechanisms.

Many technologies are available to transform waste into useful and safe outputs, but particularly in developing countries, even maintaining the current infrastructure is financially limiting. Funding is a primary problem in implementing these novel ideas that are gaining international support. Additionally, centralised wastewater treatment technologies typically treat household influents from flush toilets. This approach is not sustainable in terms of both water and sanitation security. Mixing the faecal material with water in flush toilets requires additional and expensive treatment during resource recovery. On-site sanitation treatment technologies, with waste isolation and separation capabilities, are a viable alternative, as described above. Such options allow no or minimal use of water resources (no-flush toilets) relative to conventional flush toilets, which is particularly critical in a drought-ridden country like South Africa. Operations and maintenance remain one of the key challenges for on-site sanitation. The WRC of South Africa has driven some work investigating the reasons for these challenges, including a lack of O&M budgets, poor revenue collection and limited capacity to manage toilet facilities. Competing needs from other sectors, such as tertiary education, the social grant systems, and the country's low economic growth, have placed tremendous strain on budgets allocated for these on-site sanitation technologies (Akinsete et al., 2019).

6.1.7.2 Business-as-usual mindset for waste treatment

Conventional wastewater treatment techniques for removing organics, nutrients, heavy metals and pathogens have a long local legacy, and the effluent is typically discharged into the receiving environment. This is an important step for water recovery, to either be returned into the water cycle or later treated to potable standards at water treatment plants. However, this is more common in neighbouring Namibian cities than in South Africa. Although this method seeks to reduce health risks and protect the environment, treating wastewater to appropriate discharge standards is an energy-intensive process. Natural fertiliser resources, such as nitrogen and phosphorous, quickly decline but can be recovered from wastewater or general human waste. Additionally, energy can be simultaneously generated from these wastes through the co-digestion of sludge (Junior et al., 2021). This whole process requires a systemic shift from *wastewater treatment plants* to *waste resource recovery facilities*.

However, a major stumbling block is the current linear model of treating wastewater instead of exploring closed-loop or circular modes of treatment. Additionally, perceptions are critical, and a public mindset shift is necessary, encouraging a popular view of wastewater as a resource that requires specific economic resource recovery management strategies. This includes carefully assessing the fate of waste-derived products that can be generated from wastewater treatment systems as part of the overall decision-making processes for designing and operating relevant technologies. It also includes commercial stakeholders, bridging municipal role players (waste managers), researchers and companies, and evaluating economic models for transitioning WWTPs to waste recovery facilities.

6.1.7.3 Social hesitancy in accepting the product

There has been great headway in conducting research that evaluates user acceptance of sanitation technologies from VIP latrines (Gounden et al., 2006; Mkhize et al., 2017) to low-flush toilets (Akinsete et al., 2019) and various urine-diverting dry and flush toilets (Devkota et al., 2020). According to Akinsete et al. (2019), waterless dry toilet systems have not been accepted. The chance of a new or innovative sanitation technology being deemed acceptable is higher when there have been in-field experiments and testing technologies within the communities that would benefit from implementation. Action research is a promising strategy, bridging the interface between science and social dimensions. This process involves simultaneously acting (implementing a strategy) and doing research, linked by critical reflection steps, which have been shown to influence the uptake of these technologies, as described by Owojori et al. (2022).

This facilitates feedback to the technical team about the technology's user design and experience, allowing for changes to be made that would suit the receiving community (Kabundu et al., 2022). However, due to a lack of funding, some technology developers cannot always financially support the lengthy redesign process and its reiterations. In terms of perception, in previous studies, although hesitancy was reported on acceptance of the sanitation technologies, there was positive attitude and perceptions on the use of human waste–derived fertilising products (Gwara et al., 2021) due to perceived economic benefit from these fertilising materials relative to chemical fertilisers. The 'yuck' factor (for example, utilising a fertiliser derived from human waste for food crops) is one aspect that would need to be overcome through social behaviour change, education and awareness.

6.1.7.4 Stakeholder involvement hesitancy

Successful transitioning from an LE to a CE model in organic waste management requires participatory water and waste governance as an enabling mechanism. While there is legislation and strong support in theory, the implementation of policies and engagement with principles for community and stakeholder participation in waste and water resource management has not been effective. The involvement of stakeholders in these CE initiatives remains a challenge, particularly at the level of priority-setting, planning, decision-making and implementation (Hove et al., 2021).

6.1.7.5 Lack of supportive institutional regulatory and policy frameworks

As stated by Desmond and Asamba (2019), specific policies supporting the CE do not currently exist in South Africa, although they have been tabled for consideration. This is a challenge when attempting to valorise organic waste or waste-derived products. If there are no policies, regulations, or standards that determine the acceptable quality of waste-derived products (closely related to those made for fertilisers, soil conditioner materials, animal feed products, and others), then it becomes more difficult to develop a market for the CE in the country, particularly since perception is such a difficult hurdle. Strong quality evaluation systems and concepts similar to "Organic" labelling could tap into an environmentally aware market. According to Montwedi et al. (2021), the policies that could support this initiative include:

i. Participation of all stakeholders to provide clear mandates and roles.
ii. Supporting the provision of free basic services for all citizens, particularly the provision of water and sanitation services by local or district municipalities.
iii. Reintroducing recovered waste products as safe end-use products through implementing standards to provide legal certainty for market uptake.
iv. Limiting the discharge of waste sludge to landfills.
v. Finding effective links between water, sanitation and resource recovery through policies and national frameworks that touch on environmental protection and health.

Resource recovery should not be at the expense of human or environmental health.

6.1.8 FUTURE DIRECTION FOR THE SUCCESS AND GROWTH OF CE IN SANITATION AND ORGANIC WASTE MANAGEMENT

Although researchers have demonstrated the potential of technologies supporting CE in organic waste management, there is still a very limited implementation of these waste materials feeding into the CE, especially in developing countries (Negi et al., 2021). Several "wicked" challenges have been highlighted to hinder the expected progress and implementation of CE in organic waste materials. This section, in contrast, details the potential enabling avenues which could foster the success of a CE in sanitation and organic wastes.

6.1.8.1 Creation of public-private partnerships (PPPs)

Collaboration is critical to the successful commercial implementation of such endeavours. Although independently run or funded business entities exist, the CE approach in organic waste management through resource recovery and reuse requires a collaborative business model due to the nature of the system, including a municipally managed waste stream and numerous decentralised waste treatment options proposed here. This is a unique business model that largely depends on the primary service type that an entity will offer. In the case of organic waste–derived materials, services may target resource recovery and reuse (fertilising material, water, or energy source), as well as sanitation or environmental protection services. Planning and implementation require a broad knowledge and skill set unavailable to a single individual or sector in the organic waste management sector and sanitation service chain. For example, accessibility and availability of quality raw materials and transportation, especially when dealing with faecal sludge from on-site sanitation systems, could be associated with high costs that might not meet initial investments, especially for young sanitation waste management businesses (Mallory et al., 2020a).

Generally, it requires a large initial capital investment that could be difficult to finance if one is a sole proprietor, in addition to the challengingly diverse socio-economic dynamic around the actors involved along the organic waste value chain. Other than the non-monetary benefits of organic waste management, like sanitation services and mitigating environmental pollution, the return on investment (ROI) may not be positive if one is to consider the fertiliser value of organic waste–derived materials, which is relatively lower than synthetic fertilisers. Therefore, these challenges could be addressed by forming public-private partnerships (PPPs) and building collaboration at various levels along the value chain. PPPs are utilised to generate synergistic collaborations between public and private sectors, during which the private sector's somewhat public operations are efficiently carried out, allowing co-financing and capitalising of its innovations (Yescombe, 2011).

These PPPs are critical in the waste management services of local authorities or any entity hoping to venture into waste management as part of their business model. Private companies are believed to have the capacity to implement waste management businesses successfully, while public institutions like the local authorities primarily lack financial and institutional capacities and the required technological skills (Khajuria and Rudra, 2016). Developing PPPs as a pathway to enhance the success of CE in waste management business entities could benefit both the local authorities and private companies in facilitating better, new and improved technologies. Such would put in place relevant infrastructural development, increased financial support, job and product market creation, and increased cost efficiencies (Khajuria and Rudra, 2016). Therefore, implementing some long-term PPP contracts could be highly beneficial in driving forward the CE in sanitation and organic waste in South Africa. The public sector would use the private sector's financial strength, technological know-how, flexibility and innovation to improve service delivery on sanitation services provision, waste management and other shared initiatives (Cui et al., 2020b).

6.1.8.2   Collaborations, stakeholder engagement, and co-designing of circular economy projects

Transitioning towards a CE business model in sanitation and waste-derived materials and reuse is now central; after years of research and development, we are ripe for implementation. However, although this business approach and concept dominates the current discussions within corridors of various sectors and actors, most of these conversations are still being conducted in "silos". The individualism in these non-collaborative discussions fragments the confidence needed to encourage the implementation of CE projects amongst stakeholders. The "silos" prevent holistic planning and cause duplication of activities between institutions. Multi-stakeholder collaborations and engagements are still not functioning well in South Africa, as conversations in the sanitation and organic waste management value chain are individually held or converge at best at a local level. For example, the strongest footprint is at research institutions and universities, disseminated as research papers or at national conferences.

In South Africa, collaborations have been limited to universities and the WRC providing funding for developing and testing prototype technologies initiated by various universities. A few industrial actors like WWTPs and some local municipalities are encouragingly engaging with the research, especially successful collaborations between the UKZN and the eThekwini municipality. Although there seem to be some strong collaborations, these initiatives and engagements are research-focused, developing prototypes and piloting technologies. Noble technologies promising to drive the CE in sanitation organic waste management have been developed; however, no single technology has been scaled up and implemented as a functional and revenue-generating commercial entity. There is a need to strengthen the existing initiatives and facilitate productive collaborations and engagements among relevant stakeholders within and outside the sanitation and organic waste management value chain to ensure successful dialogues regarding transitioning towards a CE in South Africa. Engagement of various stakeholders of different backgrounds for a common cause could be between research institutions, institutes of higher learning (universities and colleges), private sectors, government departments and local authorities (public sector), including representatives from the local communities.

Collaborations among relevant stakeholders facilitate progressive engagement for strategic project design, sharing ideas regarding how project implementation could be initiated (Mishra et al., 2021). Such collaborative engagements would allow for designing, redesigning and co-designing project strategies (Nakakawa et al., 2010). Collaborations are integral to any business entity's success, and role players must be carefully selected for successful engagement. Particularly, astute and experienced commercial role players are necessary to rigorously consider the financial feasibility of these technologies.

To ensure a smooth flow of organisational arrangement, planning and coordination among the stakeholders, to convene the necessary meetings and to facilitate rational dialogues, there is a need for a **lead organisation** that chairs the whole platform and keeps the itinerary and inventory of relevant stakeholders. Although there are already ongoing initiatives, the discussions are still under the corridors of a few institutions, like research institutes, academic institutions, or municipalities, without centralising national activities and collaborations among stakeholders. There is a need for multi-stakeholder platform development from now onwards, upon which stakeholder engagement strategies can be designed to move towards the co-implementation of CE's sustainable and integrated development scenarios in the sanitation and organic waste management sector. This multi-stakeholder platform can be the vehicle to enhance stakeholder engagement, involvement, and participation in the creation of sustainable CE in sanitation and organic waste–derived materials and reuse. Recently, after becoming aware of the work the UKZN (Department of Crop Sciences) is doing on innovations around resource recovery and reuse derived from various organic waste streams, the DWS directorate came on board to gain an understanding of the work being done by UKZN Crop Science team.

The team also initiated the dialogue on formulating FSM guidelines. It is also developing potential business models on CE in sanitation products that will be used as a national guide on how communities could realise some revenue flows (learning new financial techniques and improved local economic benefits) from sanitation and organic waste resource recovery programmes. The involvement of government departments is an encouraging stance and the beginning of a long-term planning trajectory in fostering CE in the sanitation and organic waste–derived materials and reuse sector. These government stakeholders are the first representatives in the relevant forums for policy formulations needed to support the success of these CE projects. We, therefore, propose that as the deliberations and dialogues on driving CE in sanitation and organic waste management procedures, the DWS should be at the centre as the chair of these dialogues and facilitate stakeholder engagement for this cause.

The department is a sector leader and shoulders a responsibility to guide the sector to take steps towards transitioning into the CE, unlocking potential sanitation and resource recovery and reuse economic opportunities. This must be done in conjunction with other sister departments, which include the Department of Agriculture, Land Reform and Rural Development, the Department of Environment, Forestry and Fisheries (Department of Environmental Affairs), among others. Extensive research institutions and scientists are nationally involved in this endeavour and willing to support such initiatives with quality testing strategies, technology design and action research bridging science and perceptions. The dialogues initiated in KwaZulu-Natal are critical towards achieving a positive drive towards CE in sanitation and organic waste resource recovery at the national level. The challenges of resource shortages and organic waste generation are a nationwide concern. It will be important if such platforms and deliberations are replicated across all provinces of the Republic of South Africa.

6.1.8.3   Designing policy, standards, and regulation framework promoting circular economy

South Africa is a member of the ACEN, a network of 14 African countries. Like in any other developing country, South Africa still has no clear CE policy framework to regulate and guide companies or individuals to venture into resource recovery and reuse of organic waste–derived materials. However, there are some existing supporting policies. Although not directly focused on the CE, they can be used as a springboard to create policy frameworks specifically targeting CE in sanitation and organic waste resource recovery.

The National Sanitation Policy of 2016 (Department of Water and Sanitation, 2016) acknowledges that sanitation is economically valuable. It also recognises that the demand for resources derived from human excreta, such as plant nutrients, can create self-sustaining sanitation businesses and encourage investment in sanitation, thereby reducing dependence on public and donor funding. Although no policy currently speaks directly to the CE per se, as highlighted above, the DWS recently initiated consultations. A process is underway with the UKZN Crop Science team to produce a project document that will be a baseline for drafting the FSM guidelines towards CE in sanitation and organic waste resource recovery. The Department shares the sanitation sector research vision that a CE replaces the current linear food consumption model, followed by human excreta secretion into sanitation facilities. The sludge would be funnelled into septic tanks or pits for disposal.

Therefore, designing the FSM guide to regulate the CE in the sanitation value chain would facilitate transforming sanitation products, components, and materials, such as faecal sludge, into products of the highest utility and value possible. At this juncture, faecal sludge can now be treated for beneficial use or resource recovery, which can be utilised in various sectors. As such, it is an urgent need and DWS' responsibility to facilitate the dialogue and drafting of guidelines that support the development of financial mechanisms and business models to support the CE in the sanitation value chain to ensure economically and financially sustainable sanitation services following the National Sanitation Policy (2016).

6.1.8.4 Technical and financial support

Currently, venturing into organic waste resource recovery does not seem to be a viable business model due to the low fertiliser value of most of the products, and they are often costly activities along the value chain. For example, a study by Mallory et al. (2020b) revealed that very few case studies exhibited a value above $5/person/year from municipal sludge reuse, indicating very low returns on investment. Therefore, to successfully promote implementing the CE approach in organic waste resource recovery as a viable business model, other drivers, such as socio-economic benefits, should be highlighted besides financial gains. Although the benefits do not directly translate to monetary gains, their effects are valuable. For example, reusing organic waste contributes to environmental protection and sanitation services within communities, which brings human dignity and potentially motivates investment.

Additionally, fertilisers will soon become a limited resource, driving the value of soil conditioners up as demand for fertilisers continues to rise (Bumb and Baanante, 1996; Heffer and Prud'homme, 2016; Mogollón et al., 2018). As such, the waste management and sanitation sectors need some funding mechanisms to assist interested parties intending to venture into resource recovery from organic waste. As of now, several technological innovations have been successfully tested to generate resources from waste; however, they have not been able to be scaled into communities because they are technically complex or require hefty monetary investments for their production and operations at a large scale. Examples include the LaDePa process that pelletises treated municipal sludge and the urine and urine products processing technologies tested in eThekwini municipality, KwaZulu-Natal, South Africa. However, these are still yet to be developed for large-scale production, for example, the VUNA project. Investment and financial plans, subsidies and incentives are required nationally through government and private sector collaborations and interventions.

6.1.8.5 Education, conscientisation and awareness

Attitude and perceptions studies demonstrate hesitancy in accepting human waste–derived fertilisers for reuse in agriculture, which has been reported by several scientists globally (Chen et al., 2015; Guo et al., 2021; Gwara et al., 2021; Msaki et al., 2022; Nancarrow et al., 2008; Simha et al., 2017, 2020). As much as the participants in the studied locations ascribed hesitancy to social, cultural, or religious beliefs, generally, lack of information, awareness, and knowledge has been central across all studies (Guo et al., 2021). In their study, Simha et al. (2017) showed that the farmers' position in society influenced hesitancy, with those in high positions in society fearing ridicule for using human waste–derived fertilising materials. A review by

Gwara et al. (2021) highlighted that the level of education and lack of awareness of the benefits of these materials negatively influenced the decision to accept the recycling and reuse of human waste–derived fertilising materials.

There is a need for evidence-based information and action research to inform relevant stakeholders about the importance, benefits and limitations associated with recycling and reuse of sanitation and organic waste materials in agriculture. Not only the consumers of the final product seem hesitant due to a lack of knowledge about these materials. Some high-profile stakeholders relevant within the sanitation value chain, especially those who might not have been involved in the initial phases of planning and inception of these innovations, are also often unaware of the benefits. This, therefore, shows the importance of cross-sectoral planning and co-designing of development projects so that relevant to the project is involved from initiation, educated, well-informed and conscientious. This will help in imparting knowledge and reduce misinformation, particularly for the more vulnerable who might associate waste with low status. Equipping the stakeholders with relevant knowledge and awareness will help them understand that recycling and reuse of human waste–derived materials are not about status, but about the choices, values and benefits attached to these materials, including potential economic and definite environmental benefits.

6.2 CONCLUSION

This work explored the current progress towards achieving a CE in the sanitation and organic waste management sector in South Africa. The focus was on the need for transitioning towards a CE and gave a detailed inventory of the current resource recovery technologies and associated opportunities for transitioning to a CE in sanitation products and organic waste. The work highlighted that the country faces myriad challenges with its current linear economic model, which has strained several resources nationally. However, we identified different technologies currently at various development and implementation phases and detailed how these potentially contribute to achieving CE through resource recovery. The identified technologies include composting and co-composting, using DEWATS, the LaDePa process, BSFL, biochar and urine valorisation products. These were seen as noble FSM strategies, with the potential to recover resources for agricultural use and potentially improve livelihoods within rural and peri-urban informal communities. However, it was seen that although the existing technologies have been scientifically evaluated and proven effective in treating organic wastes and turning them into safe-to-use fertilisers, the implementation of these technologies as a feasible business model is still not evident. These technologies' general uptake and scaling are limited, limiting their use to local research institutes and a few pilot projects in collaboration with the government. It was also further revealed that several barriers impede achieving a CE in the sanitation and organic waste value chain. The authors propose strategic pathways to navigate the identified challenges limiting the progress, particularly consolidating collaboration and strategic national management. If these steps are put in place, and the environment to do business around these technologies and their products is improved, particularly in terms of communication between parties, this would enhance the transformative approach and may be the next step in the transition to a CE in South Africa.

REFERENCES

Ahmad, M., Rajapaksha, A.U., Lim, J.E., Zhang, M., Bolan, N., Mohan, D., Vithanage, M., Lee, S.S., Ok, Y.S. (2014) Biochar as a sorbent for contaminant management in soil and water: A review. *Chemosphere* 99, 19–33.

Akinsete, A., Bhagwan, J., Hicks, C., Knezovich, A., Naidoo, D., Naidoo, V., Zvimba, J.N., Pillay, S. (2019) The sanitation economic opportunity for South Africa - Sustainable solutions for water security & sanitation. WRC and Toilet Board Coalition, pp. 1–40.

Ali, M., Emch, M., Park, J.K., Yunus, M., Clemens, J. (2011) Natural cholera infection–derived immunity in an endemic setting. *Journal of Infectious Diseases* 204, 912–918.

Austin, A. (2001) Health aspects of ecological sanitation, Abstract Volume, First International Conference on Ecological Sanitation. Citeseer, pp. 104–111.

Bai, X., Li, Z., Zhang, Y., Ni, J., Wang, X., Zhou, X. (2018) Recovery of ammonium in urine by biochar derived from faecal sludge and its application as soil conditioner. *Waste and Biomass Valorization* 9, 1619–1628.

Bame, I.B. (2012) A laboratory and glasshouse evaluation of an anaerobic baffled reactor effluent as a nutrient and irrigation source for maize in soils of KwaZulu-Natal, South Africa, Soil Science. University of KwaZulu-Natal, Pietermaritzburg, South Africa, p. 156.

Bhagwan, J., Pillay, S., Koné, D. (2019) Sanitation game changing: Paradigm shift from end-of-pipe to off-grid solutions. *Water Practice and Technology* 14, 497–506.

Bischel, H.N., Schindelholz, S., Schoger, M., Decrey, L., Buckley, C.A., Udert, K.M., Kohn, T. (2016) Bacteria inactivation during the drying of struvite fertilizers produced from stored urine. *Environmental Science & Technology* 50, 13013–13023.

Bishoge, O.K. (2021) Challenges facing sustainable water supply, sanitation and hygiene achievement in urban areas in sub-Saharan Africa. *Local Environment* 26, 893–907.

Bonvin, C., Etter, B., Udert, K.M., Frossard, E., Nanzer, S., Tamburini, F., Oberson, A. (2015) Plant uptake of phosphorus and nitrogen recycled from synthetic source-separated urine. *Ambio* 44, S217–S227.

Bright-Davies, L., Lüthi, C., Jachnow, A. (2015) DEWATS for urban Nepal: A comparative assessment for community wastewater management. *Waterlines* 34(2), 119–138.

Bumb, B.L., Baanante, C.A. (1996) World trends in fertilizer use and projections to 2020. International Food Policy Research Institute (IFPRI), Brief 38. Washington DC, p. 4.

Burchard-Dziubinska, M. (2017) Cradle to cradle approach in development of resource-efficient economy. Ekonomia i Środowisko.

Busari, T.I., Senzanje, A., Odindo, A.O., Buckley, C.A. (2020) Effect of intercropping madumbe (Colocasia esculenta) and rice (Oryza sativa L.) on yield and land productivity under different irrigation water management techniques with effluent water. *Water SA* 46, 205–212.

Camilleri, M.A. (2020) European environment policy for the circular economy: Implications for business and industry stakeholders. *Sustainable Development* 28, 1804–1812.

Chen, D., Li, Y., Cen, K., Luo, M., Li, H., Lu, B. (2016) Pyrolysis polygeneration of poplar wood: Effect of heating rate and pyrolysis temperature. *Bioresource Technology* 218, 780–788.

Chen, W., Bai, Y., Zhang, W., Lyu, S., Jiao, W. (2015) Perceptions of different stakeholders on reclaimed water reuse: The case of Beijing, China. *Sustainability* 7, 9696–9710.

Chitaka, T.Y., Schenck, C. (2022) Transitioning towards a circular bioeconomy in South Africa: Who are the key players? *South African Journal of Science* 118, 1–8.

Cofie, O., Nikiema, J., Impraim, R., Adamtey, N., Paul, J., Koné, D. (2016) *Co-composting of solid waste and fecal sludge for nutrient and organic matter recovery*. IWMI.

Cui, B.-J., Cui, E.-P., Hu, C., Fan, X.-Y., Gao, F. (2020a) Effects of selected biochars application on the microbial community structures and diversities in the rhizosphere of water spinach (Ipomoea aquatica Forssk.) irrigated with reclaimed water. *Huan Jing ke Xue= Huanjing Kexue* 41, 5636–5647.

Cui, C., Liu, Y., Xia, B., Jiang, X., Skitmore, M. (2020b) Overview of public-private partnerships in the waste-to-energy incineration industry in China: Status, opportunities, and challenges. *Energy Strategy Reviews* 32, 100584.

Dariva, M.A., Araujo, A. (2021) The potential of BIM in disseminating knowledge about decentralized wastewater treatment systems: Learning through the design process. *Water* 13, 1504.

Das, M., Uppal, H.S., Singh, R., Beri, S., Mohan, K.S., Gupta, V.C., Adholeya, A. (2011) Co-composting of physic nut (Jatropha curcas) deoiled cake with rice straw and different animal dung. *Bioresource Technology* 102, 6541–6546.

Demirbas, A., Arin, G. (2002) An overview of biomass pyrolysis. *Energy Sources* 24, 471–482.

Deng, P., Liu, C., Wang, M., Lan, G., Zhong, Y., Wu, Y., Fu, C., Shi, H., Zhu, R., Zhou, L. (2022) Effect of dewatering conditioners on phosphorus removal efficiency of sludge biochar. *Environmental Technology*, 44(20), 1–9.

Department of Environment Forestry and Fisheries (2021) National Waste Management Strategy 2020. Government Gazette No 44116, 28, Pretoria, South Africa, p. 71. (Accessed January 2021).

Department of Environmental Affairs (2017) Maximising the circular waste economy in South Africa PCEA colloquium on waste management industry waste plans presentation. Department of Environmental Affairs, Pretoria, South Africa.

Department of Environmental Affairs (2018) South Africa state of waste. A report on the state of the environment. Department of Environmental Affairs, Pretoria, South Africa.

Department of Science and Innovation (2021) Science, technology and innovation decadal plan 2021–2031. Department of Science and Innovation (DSI), Pretoria, South Africa, p. 39.

Department of Science and Technology (2014) A waste research, development and innovation roadmap for South Africa (2015–2025). Summary report. Department of Science and Technology, Pretoria, South Africa.

Department of Science and Technology (2019) White paper on science, technology and innovation. Department of Science and Technology, Pretoria, South Africa.

Department of Water and Sanitation (2016) The National Sanitation Policy 2016. Department of Water and Sanitation (DWS), Pretoria, South Africa, p. 72.

Department of Water and Sanitation (2022) Green Drop National Report 2022. Department of Water and Sanitation, Pretoria, South Africa.

Desmond, P., Asamba, M. (2019) Accelerating the transition to a circular economy in Africa: Case studies from Kenya and South Africa, in: Schröder, P., Anantharaman, M., Anggraeni, K., Foxon, T.J. (Eds.), *The Circular Economy and the Global South*. Routledge, London, pp. 152–172.

Devkota, G., Pandey, M., Maharjan, S. (2020) Urine diversion dry toilet: A narrative review on gaps and problems and its transformation. *European Journal of Behavioral Sciences* 2, 10–19.

Diener, S., Semiyaga, S., Niwagaba, C.B., Muspratt, A.M., Gning, J.B., Mbéguéré, M., Ennin, J.E., Zurbrugg, C., Strande, L. (2014) A value proposition: Resource recovery from faecal sludge—Can it be the driver for improved sanitation? *Resources, Conservation and Recycling* 88, 32–38.

Diener, S., Zurbrügg, C., Gutiérrez, F.R., Nguyen, D.H., Morel, A., Koottatep, T., Tockner, K. (2011) Black soldier fly larvae for organic waste treatment-prospects and constraints. *Proceedings of the WasteSafe* 2, 13–15.

Drabe, V., Herstatt, C. (2016) Why and how companies implement circular economy concepts–The case of Cradle to Cradle innovations, R&D Management Conference.

Drangert, J.-O., Tonderski, K., McConville, J. (2018) Extending the European Union waste hierarchy to guide nutrient-effective urban sanitation toward global food security—Opportunities for phosphorus recovery. *Frontiers in Sustainable Food Systems* 2, 3.

Dumontet, S., Dinel, H., Baloda, S. (1999) Pathogen reduction in sewage sludge by composting and other biological treatments: A review. *Biological Agriculture & Horticulture* 16, 409–430.

Eastman, B.R. (1999) Achieving pathogen stabilization using vermicomposting. *BioCycle* 40, 62–64.

Ellen Macarthur Foundation (2013) Towards the circular economy. *Journal of Industrial Ecology* 2, 23–44.

Etter, B., Udert, K.M., Gounden, T. (2015a) Valorisation of Urine Nutrients: Promoting Sanitation & Nutrient Recovery through Urine Separation, in: Etter, B., Udert, K.M., Gouden, T. (Eds.), VUNA final report. VUNA, Dübendorf, Switzerland.

Etter, B., Udert, K.M., Gounden, T. (2015b) VUNA: Valorisation of urine nutrients. Promoting sanitation & nutrient recovery through urine separation. Final Project Report 2015. ETH Zurich.

European Commission (2017) Prevention of waste in the circular economy: Analysis of strategies and identification of sustainable targets: the food waste example. Publications Office.

European Commission (2018) Circular economy: Closing the loop: The circular economy and chemicals, products and waste legislation. Publications Office.

European Environment Agency (2019) The European environment—State and outlook 2020: Knowledge for transition to a sustainable europe. Publications Office of the European Union Luxembourg.

European Union (2020) Circular economy action plan for a cleaner and more competitive Europe. Publications Office of the European Union Luxembourg.

Fuhrmann, A., Wilde, B., Conz, R.F., Kantengwa, S., Konlambigue, M., Masengesho, B., Kintche, K., Kassa, K., Musazura, W., Späth, L., Gold, M., Mathys, A., Six, J., Hartmann, M. (2022) Residues from black soldier fly (Hermetia illucens) larvae rearing influence the plant-associated soil microbiome in the short term. *Frontiers in Microbiology* 13, 994091.

Godfrey, L., Roman, H., Smout, S., Maserumule, R., Mpofu, A., Ryan, G., Mokoena, K. (2021) Unlocking the opportunities of a circular economy in South Africa, in: Sadhan, K.G., Sannidhya K.G. (Eds.), *Circular economy: Recent trends in global perspective.* Springer, Singapore, pp. 145–180.

Gounden, T., Pfaff, B., Macleod, N., Buckley, C. (2006) Provision of free sustainable basic sanitation: The Durban experience. 32nd WEDC International Conference, Colombo, Sri Lanka, p. 5.

Grantina-Ievina, L., Rodze, I. (2020) Survival of pathogenic and antibiotic-resistant bacteria in vermicompost, sewage sludge, and other types of composts in temperate climate conditions. *Biology of Composts*, 58, 107–124.

GRID-Arendal (2021) Circular economy on the African Continent: Perspectives and potential. GRID-Arendal, Arendal, Norway. 48 pages.

Guo, S., Zhou, X., Simha, P., Mercado, L.F.P., Lv, Y., Li, Z. (2021) Poor awareness and attitudes to sanitation servicing can impede China's rural toilet revolution: Evidence from Western China. *Science of The Total Environment* 794, 148660.

Gutterer, B., Sasse, L., Panzerbieter, T., Reckerzügel, T. (2009) *Decentralised wastewater treatment systems (DEWATS) and sanitation in developing countries.* BORDA, Bremen, Germany.

Gwara, S., Wale, E., Odindo, A., Buckley, C. (2021) Attitudes and perceptions on the agricultural use of human excreta and human excreta derived materials: A scoping review. *Agriculture* 11, 153.

H2O Sanitation Services (2022) Transforming lives through sanitation solutions. H2O Sanitation Services. Durban, South Africa.

Harder, R., Wielemaker, R., Molander, S., Öberg, G. (2020) Reframing human excreta management as part of food and farming systems. *Water Research* 175, 115601.

Harrison, J., Wilson, D. (2012) Towards sustainable pit latrine management through LaDePa. *Sustainable Sanitation Practice* 13, 25–32.

Hatfield-Dodds, S., Schandl, H., Newth, D., Obersteiner, M., Cai, Y., Baynes, T., West, J., Havlik, P. (2017) Assessing global resource use and greenhouse emissions to 2050, with ambitious resource efficiency and climate mitigation policies. *Journal of Cleaner Production* 144, 403–414.

Heffer, P., Prud'homme, M. (2016) Fertilizer Outlook 2016–2020, 84th IFA Annual Conference, Moscow, Russia, pp. 1–5.

Herselman, J. (2013) Guidelines for the utilisation and disposal of water treatment residues. Water Research Commission, Pretoria, South Africa.

Hove, J., D'Ambruoso, L., Twine, R., Mabetha, D., Van Der Merwe, M., Mtungwa, I., Khoza, S., Kahn, K., Witter, S. (2021) Developing stakeholder participation to address lack of safe water as a community health concern in a rural province in South Africa. *Global Health Action* 14, 1973715.

Innovus (2022) Sustainable through Entomology (SUSENTO). Innovus, Stellenbosch, South Africa.

Inyang, M., Dickenson, E. (2015) The potential role of biochar in the removal of organic and microbial contaminants from potable and reuse water: A review. *Chemosphere* 134, 232–240.

Iqbal, M.K., Shafiq, T., Hussain, A., Ahmed, K. (2010) Effect of enrichment on chemical properties of MSW compost. *Bioresource Technology* 101, 5969–5977.

Ismail, H., Smith, A.M., Tau, N.P., Sooka, A., Keddy, K.H., for the Group for Enteric, Respiratory and Meningeal Disease Surveillance in South Africa (2013) Cholera outbreak in South Africa, 2008–2009: Laboratory analysis of Vibrio cholerae O1 strains. *The Journal of Infectious Diseases* 208, S39–S45.

Jensen, L. (2022) The sustainable development goals report 2022. United Nations, New York.

Junior, I.V., de Almeida, R., Cammarota, M.C. (2021) A review of sludge pretreatment methods and co-digestion to boost biogas production and energy self-sufficiency in wastewater treatment plants. *Journal of Water Process Engineering* 40, 101857.

Kabundu, E., Mbanga, S., Makasa, P., Ngema, N. (2022) User-acceptance of sanitation technologies in South Africa and Malawi. *Town and Regional Planning* 80, 88–104.

Kalulu, K., Thole, B., Mkandawire, T., Kululanga, G. (2020) Application of process intensification in the treatment of pit latrine sludge from informal settlements in Blantyre City, Malawi. *International Journal of Environmental Research and Public Health* 17(9), 3296.

Kerstens, S., Legowo, H., Hendra Gupta, I. (2012) Evaluation of DEWATS in Java, Indonesia. *Journal of Water, Sanitation and Hygiene for Development* 2, 254–265.

Khajuria, A., Rudra, C. (2016) Role of public-private-partnerships in circular economy ~ experience from IPLA ~A Rio+20 Partnership.

Kirchherr, J., Reike, D., Hekkert, M. (2017) Conceptualizing the circular economy: An analysis of 114 definitions. *Resources, Conservation and Recycling* 127, 221–232.

Korhonen, J., Nuur, C., Feldmann, A., Birkie, S.E. (2018) Circular economy as an essentially contested concept. *Journal of Cleaner Production* 175, 544–552.

Körner, I., Braukmeier, J., Herrenklage, J., Leikam, K., Ritzkowski, M., Schlegelmilch, M., Stegmann, R. (2003) Investigation and optimization of composting processes—Test systems and practical examples. *Waste Management* 23, 17–26.

Krueger, B.C., Fowler, G.D., Templeton, M.R., Moya, B. (2020) Resource recovery and bio-char characteristics from full-scale faecal sludge treatment and co-treatment with agricultural waste. *Water Research* 169, 115253.

Kvarnström, E., Emilsson, K., Stintzing, A.R., Johansson, M., Jönsson, H., af Petersens, E., Schönning, C., Christensen, J., Hellström, D., Qvarnström, L. (2006) Urine diversion: One step towards sustainable sanitation. EcoSanRes Programme.

Lai, F.-Y., Chang, Y.-C., Huang, H.-J., Wu, G.-Q., Xiong, J.-B., Pan, Z.-Q., Zhou, C.-F. (2018) Liquefaction of sewage sludge in ethanol-water mixed solvents for bio-oil and biochar products. *Energy* 148, 629–641.

Lehmann, J. (2009) Biochar for environmental management: An introduction, in: Lehmann, J., Joseph, S. (Eds.) *Biochar for environmental management: Science and technology.* Earthscan Publications Ltd., London, p. 13.

Leng, L., Yuan, X., Huang, H., Shao, J., Wang, H., Chen, X., Zeng, G. (2015) Bio-char derived from sewage sludge by liquefaction: Characterization and application for dye adsorption. *Applied Surface Science* 346, 223–231.

Lim, M.K., Lai, M., Wang, C., Lee, S.Y. (2022) Circular economy to ensure production operational sustainability: A green-lean approach. *Sustainable Production and Consumption* 30, 130–144.

Lopez, A.B., Martin, A., Killeen, B., Iversen, C., Russo, G., Andersen, H.K., Daniell, J., Galea, L., Giannini, M., Jol, A. (2020) The European Environment State and Outlook 2020. European Environment 2021.

Magwaza, S.T., Magwaza, L.S., Odindo, A.O., Mashilo, J., Mditshwa, A., Buckley, C. (2020) Evaluating the feasibility of human excreta-derived material for the production of hydroponically grown tomato plants - Part I: Photosynthetic efficiency, leaf gas exchange and tissue mineral content. *Agricultural Water Management* 234, 12.

Maleba, V., Barnard, P., Rodda, N. (2016) Using black soldier fly larvae to treat faecal sludge from urine diversion toilets. University of KwaZulu-Natal, Durban, South Africa.

Mallory, A., Akrofi, D., Dizon, J., Mohanty, S., Parker, A., Rey Vicario, D., Prasad, S., Welivita, I., Brewer, T., Mekala, S., Bundhoo, D., Lynch, K., Mishra, P., Willcock, S., Hutchings, P. (2020a) Evaluating the circular economy for sanitation: Findings from a multi-case approach. *Science of the Total Environment* 744, 140871.

Mallory, A., Holm, R., Parker, A. (2020b) A review of the financial value of faecal sludge reuse in low-income countries. *Sustainability* 12, 8334.

Mamera, M., van Tol, J.J., Aghoghovwia, M.P. (2022) Treatment of faecal sludge and sewage effluent by pinewood biochar to reduce wastewater bacteria and inorganic contaminants leaching. *Water Research* 221, 118775.

Mamera, M., van Tol, J.J., Aghoghovwia, M.P., Mapetere, G.T. (2020) Community faecal management strategies and perceptions on sludge use in agriculture. *International Journal of Environmental Research and Public Health* 17, 1–21.

Mamera, M., van Tol, J.J., Aghoghovwia, M.P., Nhantumbo, A.B., Chabala, L.M., Cambule, A., Chalwe, H., Mufume, J.C., Rafael, R.B. (2021) Potential use of biochar in pit latrines as a faecal sludge management strategy to reduce water resource contaminations: A review. *Applied Sciences* 11, 11772.

Mathews, J.A., Tan, H. (2011) Progress toward a circular economy in China. *Journal of Industrial Ecology* 15, 435–457.

Merli, R., Preziosi, M., Acampora, A. (2018) How do scholars approach the circular economy? A systematic literature review. *Journal of Cleaner Production* 178, 703–722.

Mhatre, P., Panchal, R., Singh, A., Bibyan, S. (2021) A systematic literature review on the circular economy initiatives in the European Union. *Sustainable Production and Consumption* 26, 187–202.

Mirara, S., Septien, S., Singh, A., Velkushanova, K., Buckley, C. (2020) Drying of faecal sludge from VIP latrines through a medium infrared radiation process. *Water SA* 46, 540–546.

Mishra, J.L., Chiwenga, K.D., Ali, K. (2021) Collaboration as an enabler for circular economy: A case study of a developing country. *Management Decision* 59, 1784–1800.

Mjoli, N. (2010) Review of sanitation policy and practice in South Africa from 2001–2008. Water Research Commission, Pretoria, South Africa.

Mkhize, N., Taylor, M., Udert, K.M., Gounden, T.G., Buckley, C.A. (2017) Urine diversion dry toilets in eThekwini Municipality, South Africa: Acceptance, use and maintenance through users' eyes. *Journal of Water, Sanitation and Hygiene for Development* 7, 111–120.

Mnkeni, P., Austin, L. (2009) Fertiliser value of human manure from pilot urine-diversion toilets. *Water SA* 35, pp. 133–138.

Mogollón, J., Beusen, A., Van Grinsven, H., Westhoek, H., Bouwman, A. (2018) Future agricultural phosphorus demand according to the shared socioeconomic pathways. *Global Environmental Change* 50, 149–163.

Montwedi, M., Munyaradzi, M., Pinoy, L., Dutta, A., Ikumi, D.S., Motoasca, E., Van der Bruggen, B. (2021) Resource recovery from and management of wastewater in rural South Africa: Possibilities and practices. *Journal of Water Process Engineering* 40, 101978.

Msaki, G.L., Njau, K.N., Treydte, A.C., Lyimo, T. (2022) Social knowledge, attitudes, and perceptions on wastewater treatment, technologies, and reuse in Tanzania. *Water Reuse* 12, 223–241.

Murray, A., Skene, K., Haynes, K. (2017) The circular economy: An interdisciplinary exploration of the concept and application in a global context. *Journal of Business Ethics* 140, 369–380.

Musazura, W., Odindo, A., Tesfamariam, E., Hughes, J.C., Buckley, C. (2018) Decentralised wastewater treatment effluent fertigation: Preliminary technical assessment. *Water SA* 44, 250–257.

Musazura, W., Odindo, A.O. (2021) Suitability of the decentralised wastewater treatment effluent for agricultural use: Decision support system approach. *Water* 13(18), 2454.

Mutsakatira, E., Buckley, C., Mercer, S.J. (2018) Potential use of the black soldier fly larvae in faecal sludge management: A study in Durban, South Africa, 41st WEDC International Conference, Egerton University, Nakuru, Kenya, 2018, Paper 2994, p. 7.

Nahman, A., Godfrey, L., Oelofse, S., Trotter, D. (2021) Driving economic growth in South Africa through a low carbon, sustainable and inclusive circular economy. *The Circular Economy as Development Opportunity*, CSIR Opinion Piece and Briefing Note, Pretoria, South Africa, p. 4.

Nakakawa, A., Bommel, P.V., Proper, H.A. (2010) Challenges of involving stakeholders when creating enterprise architecture, Volume 662 of the 5th SIKS/BENAIS Conference on Enterprise Information Systems, Eindhoven, The Netherlands, pp. 43–54.

Nancarrow, B.E., Leviston, Z., Po, M., Porter, N.B., Tucker, D.I. (2008) What drives communities' decisions and behaviours in the reuse of wastewater. *Water Science and Technology* 57, 485–491.

National Institute for Communicable Disease (2009) Bulletin. Cholera outbreak in South Africa: Preliminary descriptive epidemiology on laboratory-confirmed cases, 15 November 2008 to April 2009. National Institute for Communicable Disease.

Naustdalslid, J. (2014) Circular economy in China – The environmental dimension of the harmonious society. *International Journal of Sustainable Development & World Ecology* 21, 303–313.

Ndegwa, P., Thompson, S. (2001) Integrating composting and vermicomposting in the treatment and bioconversion of biosolids. *Bioresource Technology* 76, 107–112.

Negi, S., Hu, A., Kumar, S. (2021) 24- Circular bioeconomy: Countries' case studies, in: Pandey, A., Tyagi, R.D., Varjani, S. (Eds.), *Biomass, biofuels, biochemicals*. Elsevier, Amsterdam, Netherlands, pp. 721–748.

Nikolaou, I.E., Tsagarakis, K.P. (2021) An introduction to circular economy and sustainability: Some existing lessons and future directions. *Sustainable Production and Consumption* 28, 600–609.

Nkomo, N., Odindo, A.O., Musazura, W., Missengue, R. (2021) Optimising pyrolysis conditions for high-quality biochar production using black soldier fly larvae faecal-derived residue as feedstock. *Heliyon* 7, e07025.

Oakdene, H. (2018) *Towards a circular economy: Waste management in the EU*. European Union, Brussels, Belgium, IP/G/STOA/FWC/2013-001/LOT 3/C3, p. 140.

Odindo, A.O., Musazura, W., Migeri, S., Hughes, J.C., Buckley, C.B. (2022) Intergrating sustainable agricultural production in the design of low-cost sanitation technologies by using plant nutrients and wastewater recovered from human excreta derived materials. Guideline for sustainable agricultural use of human excreta derived materials. Water Research Commission, Pretoria, South Africa, p. 96.

OECD (2019) *Global Material Resources Outlook to 2060*. Economic drivers and environmental consequences. OECD, Paris, Francis, p. 214.

Oliveira, F., Doelle, K., List, R., O'Reilly, J.R. (2015) Assessment of Diptera: Stratiomyidae, genus Hermetia illucens (L., 1758) using electron microscopy. *Journal of Entomology and Zoology Studies* 3, 147–152.

Otoo, M., Drechsel, P., Hanjra, M.A. (2015) Business models and economic approaches for nutrient recovery from wastewater and fecal sludge, Wastewater, in: Drechsel, P., Qadir, M., Wichelns, D. (Eds.), Springer, Dordrecht, pp. 247–268.

Ouma, J., Septien, S., Velkushanova, K., Pocock, J., Buckley, C. (2016) Characterization of ultrafiltration of undiluted and diluted stored urine. *Water Science and Technology* 74, 2105–2114.

Owojori, O.M., Mulaudzi, R., Edokpayi, J.N. (2022) Student's Knowledge, Attitude, and Perception (KAP) to solid waste management: A survey towards a more circular economy from a rural-based tertiary institution in South Africa. *Sustainability* 14, 1310.

Owojori, O.M., Okoro, C. (2022) The private sector role as a key supporting stakeholder towards circular economy in the built environment: A scientometric and content analysis. *Buildings* 12, 695.

Özkan, P., Yücel, E.K. (2020) Linear economy to circular economy: Planned obsolescence to cradle-to-cradle product perspective, in: *Handbook of research on entrepreneurship development and opportunities in circular economy*. IGI Global, Pennsylvania, USA, pp. 61–86.

Paredes, C., Bernal, M., Cegarra, J., Roig, A., Navarro, A. (1996) Nitrogen transformation during the composting of different organic wastes, in: *Progress in nitrogen cycling studies,* Van Cleemput, O., Hofman, G., Vermoesen, A. (Eds). Springer, Dordrecht, pp. 121–125.

Pearce, D.W., Turner, R.K., Turner, R.K. (1990) *Economics of natural resources and the environment*. Johns Hopkins University Press, Baltimore, USA, p. 392.

Peguero, D.A., Mutsakatira, E.T., Buckley, C.A., Foutch, G.L., Bischel, H.N. (2021) Evaluating the microbial safety of heat-treated fecal sludge for black soldier fly larvae production in South Africa. *Environmental Engineering Science* 38, 331–339.

Petric, I., Helić, A., Avdić, E.A. (2012) Evolution of process parameters and determination of kinetics for co-composting of organic fraction of municipal solid waste with poultry manure. *Bioresource Technology* 117, 107–116.

Pillay, S., Bhagwan, J. (2021) SaNiTi–A WRC research strategy and response to transforming sanitation into the future. Water Research Commission (WRC) Research Report, Pretoria, South Africa, p. 31.

Principato, L., Ruini, L., Guidi, M., Secondi, L. (2019) Adopting the circular economy approach on food loss and waste: The case of Italian pasta production. *Resources, Conservation and Recycling* 144, 82–89.

Reynaud, N., Buckley, C. (2015) Field-data on parameters relevant for design, operation and monitoring of communal decentralized wastewater treatment systems (DEWATS). *Water Practice and Technology* 10, 787–798.

Roma, E., Philp, K., Buckley, C., Xulu, S., Scott, D. (2013) User perceptions of urine diversion dehydration toilets: Experiences from a cross-sectional study in eThekwini Municipality. *Water SA* 39, 305–311.

Samal, K., Moulick, S., Mohapatra, B.G., Samanta, S., Sasidharan, S., Prakash, B., Sarangi, S. (2022) Design of faecal sludge treatment plant (FSTP) and availability of its treatment technologies. *Energy Nexus* 7, 100091.

Sánchez, Ó.J., Ospina, D.A., Montoya, S. (2017) Compost supplementation with nutrients and microorganisms in composting process. *Waste Management* 69, 136–153.

Scheutz, C., Pedicone, A., Pedersen, G.B., Kjeldsen, P. (2011) Evaluation of respiration in compost landfill biocovers intended for methane oxidation. *Waste Management* 31, 895–902.

Sehnem, S., Vazquez-Brust, D., Pereira, S.C.F., Campos, L.M. (2019) Circular economy: Benefits, impacts and overlapping. *Supply Chain Management: An International Journal* 24(6), 784–804

Septien, S., Singh, A., Mirara, S., Teba, L., Velkushanova, K., Buckley, C. (2018) 'LaDePa' process for the drying and pasteurization of faecal sludge from VIP latrines using infrared radiation. *South African Journal of Chemical Engineering* 25, 147–158.

Sharma, H.B., Vanapalli, K.R., Samal, B., Cheela, V.S., Dubey, B.K., Bhattacharya, J. (2021) Circular economy approach in solid waste management system to achieve UN-SDGs: Solutions for post-COVID recovery. *Science of the Total Environment* 800, 149605.

Sharma, M., Yadav, A., Mandal, M.K., Pandey, S., Pal, S., Chaudhuri, H., Chakrabarti, S., Dubey, K.K. (2022) Chapter 7- Wastewater treatment and sludge management strategies for environmental sustainability, in: Stefanakis, A., Nikolaou, I. (Eds.), *Circular economy and sustainability*. Elsevier, Eindhoven, Netherlands, pp. 97–112.

Shevelov, A. (2020) The transition towards circular economy: Circular supply chain management and digital technologies as the key enablers towards circular economy. Master of Science, Institute of Corporate Management & Economics, Zeppelin University, Friedrichshafen, Germany, p. 62.

Shumo, M., Osuga, I.M., Khamis, F.M., Tanga, C.M., Fiaboe, K.K., Subramanian, S., Ekesi, S., van Huis, A., Borgemeister, C. (2019) The nutritive value of black soldier fly larvae reared on common organic waste streams in Kenya. *Scientific Reports* 9, 1–13.

Sidley, P. (2001) Cholera sweeps through South African province. *BMJ* 322, 71.

Simha, P., Barton, M., Perez Mercado, L., McConville, J., Lalander, C., Magri, M., Dutta, S., Kabir, H., Selvakumar, A., Zhou, X., Martin, T., Kizos, T., Kataki, R., Gerchman, Y., Herscu-Kluska, R., Al-Rousan, D., Goh, E.G., Elenciuc, D., Głowacka, A., Vinnerås, B. (2020) Willingness among food consumers to recycle human urine as crop fertiliser: Evidence from a multinational survey. *Science of The Total Environment* 765, 144438.

Simha, P., Lalander, C., Vinnerås, B., Ganesapillai, M. (2017) Farmer attitudes and perceptions to the re–use of fertiliser products from resource–oriented sanitation systems – The case of Vellore, South India. *Science of The Total Environment* 581–582, 885–896.

Singh, A., Sawant, M., Kamble, S.J., Herlekar, M., Starkl, M., Aymerich, E., Kazmi, A. (2019) Performance evaluation of a decentralized wastewater treatment system in India. *Environmental Science and Pollution Research* 26, 21172–21188.

Singh, R., Embrandiri, A., Ibrahim, M., Esa, N. (2011) Management of biomass residues generated from palm oil mill: Vermicomposting a sustainable option. *Resources, Conservation and Recycling* 55, 423–434.

Snyman, H.G., Herselman, J. (2006) Guidelines for the utilisation and disposal of wastewater sludge: Requirements for the on-site and off-site disposal of sludge. Water Research Commission (WRC), Report No. TT 349/09, Pretoria, South Africa, p. 117.

Still, D., Foxon, K., O'Riordan, M. (2012) Tackling the challenges of full pit latrines. Water Research Commission, South Africa.

Suthar, S. (2009) Vermicomposting of vegetable-market solid waste using Eisenia fetida: Impact of bulking material on earthworm growth and decomposition rate. *Ecological Engineering* 35, 914–920.

Tahulela, A.C., Ballard, H.H. (2020) Developing the circular economy in South Africa: Challenges and opportunities. Conference proceedings, *Sustainable Waste Management: Policies and Case Studies*, 7th IconSWM—ISWMAW 2017, Telangana State Agricultural University, Hyderabad, India, 1, 125–133.

ten Berge, H.F.M., Hijbeek, R., van Loon, M.P., Rurinda, J., Tesfaye, K., Zingore, S., Craufurd, P., van Heerwaarden, J., Brentrup, F., Schröder, J.J., Boogaard, H.L., de Groot, H.L.E., van Ittersum, M.K. (2019) Maize crop nutrient input requirements for food security in sub-Saharan Africa. *Global Food Security* 23, 9–21.

Tuyens, C., Wilson, D., Schmidt, A., Buckley, C. (2018) Bridging the gap between onsite and conventional sanitation: Decentralised Wastewater Treatment Solutions (DEWATS). Institute of Municipal Engineering of Southern Africa (IMESA). 31 October–02 November 2018, Broadwalk hotel, East London, South Africa.

Udert, K.M., Buckley, C.A., Wächter, M., McArdell, C.S., Kohn, T., Strande, L., Zöllig, H., Fumasoli, A., Oberson, A., Etter, B. (2015) Technologies for the treatment of source-separated urine in the eThekwini Municipality. *Water SA* 41, 212–221.

Udert, K.M., Buckley, C.A., Wächter, M., McArdell, C.S., Kohn, T., Strande, L., Zöllig, H., Fumasoli, A., Oberson, A., Etter, B. (2016) Technologies for the treatment of source-separated urine in the eThekwini Municipality. *Water SA* 41, 212–221.

UNEP, I. (2016) Global material flows and resource productivity. Assessment report for the UNEP international resource panel. United Nations Environment Programme, Nairobi.

UNGA (2015) Transforming our world: The 2030 agenda for sustainable development. United Nations General Assembly (UNGA), New York, USA, p. 35.

Varma, V.G., Jha, S., Raju, L.H.K., Kishore, R.L., Ranjith, V. (2022) A review on decentralized wastewater treatment systems in India. *Chemosphere*, 300, 134462.

Velenturf, A.P., Purnell, P. (2021) Principles for a sustainable circular economy. *Sustainable Production and Consumption* 27, 1437–1457.

Velenturf, A.P.M., Archer, S.A., Gomes, H.I., Christgen, B., Lag-Brotons, A.J., Purnell, P. (2019) Circular economy and the matter of integrated resources. *Science of The Total Environment* 689, 963–969.

Vinnerås, B. (2001) Faecal separation and urine diversion for nutrient management of household biodegradable waste and wastewater. Master of Science, Swedish University of Agricultural Sciences, Department of Agricultural Engineering, Uppsala, Sweden, Report 244, p. 79.

Vinnerås, B., Jönsson, H. (2002) The performance and potential of faecal separation and urine diversion to recycle plant nutrients in household wastewater. *Bioresource Technology* 84, 275–282.

Halbert-Howard, A., Häfner, F., Karlowsky, S., Schwarz, D., Krause, A. (2021) Evaluating recycling fertilizers for tomato cultivation in hydroponics, and their impact on greenhouse gas emissions. *Environmental Science and Pollution Research* 28, 59284–59303.

Wand, H., Vacca, G., Kuschk, P., Krüger, M., Kästner, M. (2007) Removal of bacteria by filtration in planted and non-planted sand columns. *Water Research* 41, 159–167.

Wang, S., Dai, G., Yang, H., Luo, Z. (2017) Lignocellulosic biomass pyrolysis mechanism: A state-of-the-art review. *Progress in Energy and Combustion Science* 62, 33–86.

Water and Sanitation Program (2013) Water and sanitation program end of year report, Fiscal Year 2013.

Wijkman, A., Skånberg, K. (2015) The circular economy and benefits for society: Jobs and Climate Clear Winners in an Economy Based on Renewable Energy and Resource Efficiency. Club of Rome, Zurich, Switzerland, p. 62.

World Economic Forum (2019) Harnessing the fourth industrial revolution for the circular economy: Consumer electronics and plastics packaging. World Economic Forum (WEF), Geneva, Switzerland, p. 29.

Yan, S., Zhang, S., Yan, P., Aurangzeib, M. (2022) Effect of biochar application method and amount on the soil quality and maize yield in Mollisols of Northeast China. *Biochar* 4, 1–15.

Yescombe, E.R. (2011) *Public-private partnerships: principles of policy and finance: Principles of Policy and Finance.* Elsevier, Eindhoven, Netherlands, p. 531.

Zuma, L., Velkushanova, K., Buckley, C. (2015) Chemical and thermal properties of VIP latrine sludge. *Water SA* 41(4), 534–540.

7 The circular economy as a catalyst for environmental and human health

Nonhlanhla Kalebaila, Mpho Kapari,
Luxon Nhamo, and Sylvester Mpandeli

7.1 INTRODUCTION

The world's population is expected to reach 10 billion people by 2050, and this growth is accompanied by increasing demand for natural resources (https://www.un.org/en/global-issues/population). The population increase is taking place at a time when natural resources are degrading and depleting, giving an insecure resource future outlook (Nhamo et al., 2019). According to a 2019 report by the Organization for Economic Co-operation and Development, the annual material consumption increased from 37 billion tonnes in 1990 to 88 billion tonnes in 2017, while the average daily materials used per capita went from 22 kg in 1990 to 33 kg in the same period (Wiedmann et al., 2015; Yamaguchi, 2018). Most of these products are integral to all sectors of society, with tangible benefits on sustainable development due to their use in improving health and the quality of life, agriculture and food production, consumer goods, clean technologies and their related industries contributing to poverty alleviation. However, when improperly used or disposed of unsafely, some of these products and their wastes pose significant environmental and human health risks (De and Debnath, 2016; Nhamo and Ndlela, 2021).

Furthermore, global material consumption currently contributes to about half of global CO_2 emissions worldwide (Yamaguchi, 2018). It is estimated that about 62% of these greenhouse gas (GHG) emissions (excluding those from land use and forestry) are released during the extraction, processing, and manufacturing of goods to serve society's needs, while the remaining 38% is released in the delivery and use of related products and services (Naidoo et al., 2021; Yamaguchi, 2018). Climate change poses many environmental and human health risks (Nhamo and Ndlela, 2021). Climate change–related risks to environmental health include the increased risk of extreme heat-related diseases and environmental degradation due to flooding (Nhamo and Ndlela, 2021). It is now increasingly recognized that linear models have reached their limits and are longer capable of addressing today's interlinked challenges as they have been promoting the "take-make-dispose" concept of production and consumption. Transformative and circular models are envisaged to drive

DOI: 10.1201/9781003327615-7 **139**

towards the reuse and regeneration of materials or products and sustainable development (Naidoo et al., 2021).

The CE concept entails redefining economic growth, prioritizing sustainability, reducing waste and repurposing and recycling materials and products already in use (Kirchherr et al., 2017). Therefore, the CE approach is a systems solution framework that can effectively be used to tackle global challenges like climate change, biodiversity loss, waste disposal, and pollution control and, in the process, facilitate the achievement of several SDGs (Hartley et al., 2020; Kirchherr et al., 2017). Today, it is estimated that only 8.6% of the global economy is circular, as the rest remains linear, posing a huge risk to human and environmental health (Hamam et al., 2021). Therefore, there is an urgent need to speed up the production and consumption trends towards a circular, sustainable, and regenerative bioeconomy, which takes into consideration the immediate, medium-term, and long-term environmental sustainability (Hamam et al., 2021; Naidoo et al., 2021).

The implication of linear economy approaches to environmental and human health is multifaceted and not fully understood (Iacovidou et al., 2021). To gain more understanding of the impact of the "take-make-dispose" production and consumption approach on environmental and human health, this chapter re-examines the impacts of linear and circular economic processes on natural resource degradation and the emergence of infectious diseases. Specifically, the impact of natural resource degradation as a driver for biodiversity loss, climate change and the spread of diseases. The chapter provides successful examples of the application of the CE and its role in transitioning towards a regenerative production and consumption of goods and materials.

7.1.1 THE NEED TO TRANSITION FROM A LINEAR TO A CIRCULAR ECONOMY

Economies are currently supported by intertwined product value chains, sustained by more than 100 billion tons of raw materials that are produced through the extraction and use of key natural resources (Wiedmann et al., 2015). Since the first industrial revolution, production has followed two approaches: circular and linear systems, but generally inclined towards the linear economy (Scheel and Bello, 2022). The traditional linear system maximizes the use of raw materials, while the circular system minimizes the use of raw materials and maximizes the lifecycle of products (Naidoo et al., 2021). An example of how the linear economy operates is shown in Box 7.1. The increasing demand for products drives the irrational use of the available resources, the outcome of which is mass production and consumption (Scheel and Bello, 2022). In the linear economy, once the products are no longer useful, they are disposed of, resulting in negative impacts on the environment (Naidoo et al., 2021; Scheel and Bello, 2022). Ninety percent of land use–related biodiversity loss in today's linear economy is caused by the way we extract and process natural resources to make the things we want. The environmental impacts associated with raw materials extraction and use range from land degradation to the release of toxic pollutants that affect both humans and the ecosystem to the emission of GHGs into the atmosphere, thus contributing to climate change (Wiedmann et al., 2020).

The relationship between environmental quality and health becomes more complex due to climate change (Mpandeli et al., 2018). The frequency and intensity of storms, flooding, heat waves, rising ground-level ozone concentrations, food short-ages as crop production and aquaculture are negatively impacted, and forced migration as a result of drought, habitat modification, and sea-level rise are just a few of the health threats that climate change exposes humans to (Mpandeli et al., 2018). Also, commodity-based land degradation resulting from the linear economy processes is contributing to deforestation contributing to climate change (Nobre et al., 2016). Climate change is the most pressing issue in the world today, having a negative effect on both natural and social systems. This is especially true for the human livelihoods of a nature-based economy. In turn, natural events resulting from climate change affect the ecosystem's productivity and, therefore, affect the availability and distribution of goods and services (Nhamo et al., 2021a).

According to a 2019 Global Resources Outlook Report, over 80% of the global land use–related rapid biodiversity and ecosystem loss results from the extraction and processing of biomass using the linear system (Marques et al., 2019). With the extraction of raw materials for product development projected to double by 2060, the negative environmental impacts posed by mass material extraction and use are also expected to more than double, with adverse consequences for human health, ecosystems, and the economy (Wiedmann et al., 2020). The nexus between materials used under a linear economy and the degradation of natural resources, such as land, water and biodiversity are extremely critical, such that increasing pressures on one medium is likely to intensify pressures on others, resulting in dire environmental health consequences (Nhamo and Ndlela, 2021). Specifically, pollution develops when wastes from socioeconomic activities are released at levels that are higher than what the ecosystem can safely handle. Therefore, due to the ecosystem's finite capacity to meet the ever-increasing demands from socioeconomic activities, natural resource shortages develop. These socioeconomic activities demand is mainly driven by the ever-increasing.

Population increase is leading to the intensification and extension of agriculture, contributing to the degradation and depletion of resources and contributing to disastrous consequences for the environment (Reynolds et al., 2015). Agriculture and urban expansion have been the primary cause of encroachment on critical biodiversity areas, resulting in habitat loss and emergency of novel infectious diseases (Nhamo and Ndlela, 2021). Therefore, the demand for more agricultural products for food drives the repurposing of rangelands and forests into large croplands and livestock farms (Nhamo et al., 2021b). Increased food production is a major driver of biodiversity loss and air and water pollution, deforestation, soil degradation, antimicrobial resistance, and water scarcity (Nhamo et al., 2021b; Sena and Ebi, 2021). The challenge is compounded by other agricultural activities, including the use of plant biomass as a source of liquid fuel (Ketov et al., 2022). These are referred to as biofuels and are resulting in land-use change patterns in most regions around the world, including sensitive and most diverse regions (Ketov et al., 2022). In this instance, the first indication of biodiversity loss is habitat loss, which is followed by land conversion for crop production. With many biofuel crops located in tropical areas, an increase in biofuel production would mean potentially converting natural ecosystems to feedstock plantations, thus the loss of wild biodiversity. However, agriculture is not the only sector causing environmental pollution, as the construction sector accounts for about 30% of natural resource extraction and 25% of solid waste generation globally (Benachio et al., 2020).

BOX 7.1   PAPER MANUFACTURING COMPANY LOCATED IN SOUTH INDIA'S LINEAR ECONOMY (ABC MANUFACTURERS)

ABC manufactures writing paper and newsprint. ABC is an ISO 9001-2000 organization that manufactures newsprint and authoring paper; it differentiates itself as an organization that is keen on applying sustainable supply chain practices. Before then, the organization's supply chain was following the linear economic model, meaning raw materials used to produce final products were under-utilized. Bagasse is one of the raw materials used to manufacture paper by the organization. The company followed the linear economic procedures in the sense that the raw materials would be collected from the sugar factory, then loaded in the truck to be transported to the manufacturing plant, and from there, the materials would be processed to generate pulp. Thereafter, the process of fermentation, boiling, and bleaching follows, then the calendaring process. The final finished product is the outcome of the calendaring process, which can be newsprint or authoring paper, depending on the raw materials used. Thereafter, the finished product is taken to the distribution centre, to retailers, and then to consumers.

Waste generated during the production process is not accounted for or considered useful and, therefore, not utilized. This waster can be reused for other products, including pulp residue (can be used for spirit manufacturing) and fly ash from coal burning during the boiling process (can be used in cement manufacturing).

Source: Manavalan and Jayakrishna (2019).

All the aforementioned factors show that the use of raw materials in the linear economy is not only environmentally damaging but also unsustainable. It is increasingly becoming less effective with the increasing population because more resources are in demand to meet people's needs. However, credit must be given to the linear economy's ability to improve economic gains and how it has reached its limits and is no longer sustainable if ever humankind is to achieve the SDGs (Nhamo and Ndlela, 2021). Its shortcoming in mitigating environmental impacts and little consideration of social impacts allows adopting a better economic model to rectify and account for these in the quest for sustainable development. Unfortunately, this means that the newly adopted economic model does not only have to account for what was neglected by the linear economy but also work towards environmental rehabilitation. Biodiversity loss, pollution resulting from waste disposals, and climate change are some of the challenges resulting from linear economy practices, and the adoption of the CE is anticipated to address these socioeconomic and ecological challenges.

7.2   LINEAR ECONOMY AND THE EMERGENCE OF INFECTIOUS DISEASES

The range of negative environmental impacts associated with materials extraction and use range from land degradation to the release of toxic pollutants that affect human health to the emission of GHG into the atmosphere, thus contributing to the effects of climate change. Climate change has led to altered rainfall patterns with evidence of extensive, intensive rainfall that is causing floods, resulting in increased climate change–related diseases (Mpandeli et al., 2018). This has witnessed the emergence of novel infectious diseases and other widespread long-standing disorders such as diarrheal diseases, lower respiratory infections, and unintentional injuries (Nhamo and Ndlela, 2021). The lack of sufficient data to assess climate change impacts (Box 7.2) and the degradation of natural resources are causing the emergence and transmission of novel diseases. For instance, water quality has been affected by land degradation, soil erosion, and waste disposal into surface water resources. This has resulted in waterborne diseases, which are a major cause of child mortality. Unfortunately, the burden of environmental degradation and contamination is felt by some of the poorest populations, which form the majority, especially in developing countries (Remoundou and Koundouri, 2009). For example, in the Northern Cape Province, lead poisoning in children is more pronounced in nearby mining areas than in non-mining towns (Orton, 2019). In addition, land degradation, soil erosion, droughts, and floods all have direct and indirect effects on child mortality, maternal health, and other diseases such as malaria and bilharzia (Remoundou and Koundouri, 2009).

BOX 7.2   THE SCIENTIFIC CONSIDERATION OF CLIMATE CHANGE IMPACTS ON INFECTIOUS DISEASES

Since climate change is a gradual process with effects that are difficult to differentiate from larger natural disparities occurring over various seasons, it has been challenging to predict the influence of weather and climate itself on the transmission of diseases. Also, this has been made more challenging by the ability of non-climatic factors to moderate climate change impacts on infectious diseases. For instance, better socioeconomic conditions, behavioural changes, and improved treatment may reduce the severity of climate-induced pathogen transmission to clinical illness.

For a successful determination of climate change's direct impact on the increase of infectious diseases, there should be a standard monitoring of exposure, in this case, climate, and the result, which is the occurrence of a particular disease and other factors of diseases (treatment, immunity, etc.) over numerous years. Unfortunately, such datasets are rare to none, especially in developing countries where climate change impacts are experienced the most. This lack of data leads to the 'absence of evidence' of climate change effects on vector-borne diseases. Therefore, improved disease surveillance will lead to direct evidence of obvious climate change impacts on infectious diseases.

Best estimations in the meantime, the current and future impacts of climate change. These are based on theoretical consideration of known climate impacts on diseases and assessment of reported climate impacts on infectious diseases.

Source: Saker et al. (2004).

Furthermore, food production practices produce around a quarter of global GHG emissions (Lynch et al., 2021). Large animal production farms can also serve as a source for the spill-over of infections from animals to people and proliferate the dissemination of antimicrobial resistance within the environment (Economou and Gousia, 2015). Infectious diseases and their causative agents have occurred regularly throughout history, with others resulting in pandemics (Nhamo and Ndlela, 2021). However, there is evidence of many infectious diseases leading to pandemics having been transmitted to humans due to increased contact with animals due to habitat loss (land degradation), climate change impacts and water pollution (WHO, 2020). Examples of major pandemics and epidemics that have severely impacted humanity include the plague, cholera, flu, severe acute respiratory syndrome coronavirus (SARS-CoV), Middle East respiratory syndrome coronavirus (MERS-CoV) and the recent COVID-19 (Nhamo and Ndlela, 2021; WHO, 2020).

On the other hand, poor pesticide management techniques on farms and the possibility of long-term health effects, suicide, and unintentional poisoning in agricultural settings have resulted in disease outbreaks (WHO, 2020). Although the understanding of the mechanisms of transmission of pathogens to humans allowed the establishment of methods to prevent and control infections, the emergence of antimicrobial-resistant agents has been a major setback, presenting a new environmental and human health crisis (WHO, 2020). Poor water and sanitation hygiene, land use and climate changes are expected to further compound the impact of these infectious diseases (Mpandeli et al., 2018).

As aforementioned, the linear economy promotes the disposal of waste (Box 7.1) and contributes to pollution, including water pollution. Therefore, leading to waterborne diseases can be divided into two categories based on their mode of transmission: waterborne (ingested) diseases and water-washed illnesses (caused by lack of hygiene). More than 2 billion people live in the world's dry regions, where they are disproportionately affected by malnutrition, infant mortality, and diseases caused by contaminated or insufficient water (Mpandeli et al., 2018). Climate change impacts make conditions more favourable to the spread of some infectious diseases, such as Lyme disease, waterborne diseases such as cholera, and mosquito-borne diseases such as malaria and dengue fever (Mpandeli et al., 2018). This is the case with the Highland Malaria in East African highlands (Himeidan and Kweka, 2012).

7.2.1 Malaria in the Eastern African highlands

The highlands are characterized by an altitude above 1,500 m elevation above sea level together with mean daily temperatures of below 20°C (Himeidan and Kweka, 2012). Rwanda, Ethiopia, Uganda, Burundi, Madagascar, Kenya, and Tanzania make up about 82.4% of the highlands. Moderate temperature, enough rainfall and productive soils found in the highlands make the area suitable for agricultural development. As such, large populations of humans and livestock have occupied this area to benefit from the high agricultural production potential. As a result, agriculture is the main source of livelihood for the populations occupying this area (Himeidan and Kweka, 2012). Over 75%, 80%, and 85% of the labour force are focused on agriculture in the highlands of Kenya, Uganda, and Ethiopia, respectively (Himeidan and Kweka, 2012).

The highlands documented the first malaria case post the influenza pandemic during resettlement and troop demobilization post World War 1 in 1918 and 1919. Thereafter, between the 1920s and 1950s, there were infrequent malaria epidemics with no reports between the 1960s and the early 1980s (Himeidan and Kweka, 2012). Malaria, however, re-emerged in the 1980s in the Kericho district of Western Kenya highlands. Since then, several malaria epidemics were reported between the 1980s and 1990s in countries including Kenya, Uganda, Tanzania, Ethiopia, Rwanda, and Madagascar. In their *The Africa Malaria Report 2003*, the WHO revealed that malaria epidemics are estimated to kill 1 million people per year, with children under 5 being the highest number (Himeidan and Kweka, 2012). Hypotheses on what influenced malaria include but are not limited to climate change, land use and land cover changes. The controversy behind the cause of malaria in the East African highlands also included antimalarial drug resistance, healthcare infrastructure degradation, and global warming.

Furthermore, land use and land cover changes also lay the ground for the spread of infectious diseases. Studies have found that land cover change is the main driver for African highlands' rising temperatures and, therefore, increases the rate of malaria vectors *Anopheles gambiae* ssp., *An. funestus*, and *An. arabiensis* (Himeidan and Kweka, 2012). Anthropogenic activities have led to the loss of forest areas in the region, with a record range between 8,000 ha in Rwanda and 2,838,000 ha in Ethiopia (Himeidan and Kweka, 2012). This is driven by agricultural expansion, land degradation, overpopulation, and deforestation. The loss of forestland to agriculture is one of the greatest environmental changes leading to disequilibrium in the local natural balance as well as global biodiversity loss in the African highlands, one of the most fragile ecologies in the world (Himeidan and Kweka, 2012). Agricultural production systems, including farming practices, farm location, and farming technologies, may result in land use changes that create favourable ecological and climatic conditions for the breeding and survival of Anopheline mosquitos, which transmit malaria (Janko et al., 2018). This increases malaria transmission, considering deforestation leads to changes in the microclimate of adult and larval habitats, therefore, increasing larval survival, population density, and gametocyte development in adult mosquitos (Janko et al., 2018). For instance, the re-emergence of malaria in Western Kenya highlands has been reported to be the consequence of clearing forest land for developing tea estates (Himeidan and Kweka, 2012). As such, changes in land cover and use, particularly for agricultural use, have been identified as one of the drivers of increased malaria transmission.

Increasing resource demand can increase pollution and spread of diseases under the linear approach. As it stands, linear economy practices are contributing to climate change, which then favours the spread of diseases and thus becoming a double tragedy of having to mitigate the direct impacts of the practice, such as water pollution and adapt and mitigate climate change impacts like water scarcity due to uneven rainfall patterns and extremely high temperatures. More than ever, the urgency to adopt sustainable, circular, and transformative solutions as a backbone to production and consumption activities has become a priority in many policy initiatives across the globe. To safeguard environmental and human health, it is imperative that governments, businesses, and civil society put in place measures to increase resource efficiency, close material loops, and improve overall environmental management. Otherwise, continued mass production and consumption using the linear approach will continue to exert negative environmental pressures, including land degradation, GHG emissions, and the dispersion of toxic substances in the environment.

7.3   THE CIRCULAR ECONOMY MODEL

Figure 7.1 depicts a circular concept that describes a loop that includes production, consumption, and reuse/repair/recycling (WHO, 2018). The Ellen MacArthur Foundation (EMF) developed a complex representation that outlines the principles of (Ellen MacArthur, 2015):

1. preserving and enhancing natural capital by controlling finite stocks and balancing renewable resource flows,
2. optimizing resource yields by circulating products, components, and materials at the highest utility, and
3. fostering system effectiveness by revealing and designing out negative externalities.

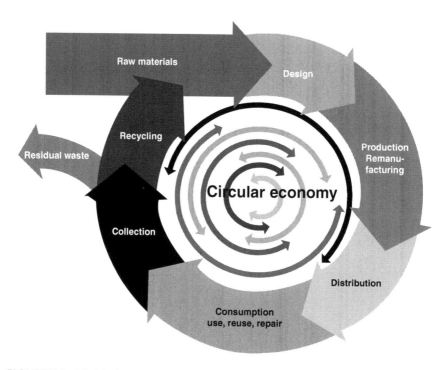

FIGURE 7.1   Model of the CE.

Source: WHO (2018).

There has been an increase in the uptake of CE by businesses. One such organization promoting the CE concept is the EMF, which has promoted the CE benefits to its "CE100 companies" (Dell, Coca-Cola, IKEA and others). However, it is unclear what these companies have achieved in consideration of this. EMF employs the Regenerate, Share, Optimize, Loop, Virtualize, and Exchange (ReSOLVE) framework, which identifies six types of CE actions that businesses and governments can take. Such frameworks represent a transition that necessitates a collaborative effort from various stakeholders. These include the state's role in strategy, regulatory and fiscal frameworks, and funding for some measures such as research and business support. Businesses play a critical role in implementing CE principles, including through innovation, while NGOs and business associations help by promoting and sharing knowledge. Furthermore, the European Union (EU) has promoted the CE concept by investing €650 million towards the transition to CE. In addition, China has become the first country to adopt a law for CE.

7.4   ECOLOGICAL BENEFITS OF THE CIRCULAR ECONOMY

Various strategies and methods have been used over the years to address the issues of natural resource scarcity and environmental pollution. For instance, economic restructuring, technical advancement, resource/energy conservation, institutional reform, and economic restructuring are all utilized to increase economic efficiency (Naidoo et al., 2021; Zvimba et al., 2021). To reduce pollution, abatement facilities are also put in place, as well as prevention strategies, including environmental impact assessments and cleaner production (WHO, 2018). The reality, however, reveals that these strategies and actions are insufficient to fully address the issues of natural resource scarcity and environmental deterioration. With the aforementioned knowledge, the CE defines its goal as solving issues from the standpoint of lowering material waste and achieving a balance in material flow between the ecosystem and the socioeconomic system (WHO, 2018). The CE is defined as a production and consumption paradigm that minimizes waste generation by renting, sharing, reusing, repairing, refurbishing, and recycling existing resources and products (Hamam et al., 2021). The strategy involves (1) changing the material flow from a linear to a circular one, that is, from resources to products to wastes, and then further converting the wastes into new resources; and (2) improving resource utilization efficiency and lowering the intensity of emissions (Hamam et al., 2021).

This distinguishing feature makes the CE a popular paradigm for transforming conventional production and consumption patterns into sustainable ones. For this reason and others, the CE has been adopted in light of its ability to build a strong foundation for innovation and investment. For instance, the European Commission adopted an action plan in 2015 to quicken Europe's shift to a CE, boost global competitiveness, encourage sustainable economic growth, and add new jobs (Naidoo et al., 2021). The action plan includes 54 steps to "close the loop" on the product life cycle, from manufacturing and consumption to trash disposal and the market for recycled raw materials. The plan also designated five priority industries (plastics, food waste, crucial raw resources, construction and demolition, biomass, and bio-based materials) to speed up the transformation along the value chain.

7.4.1 The 3R's principle

The CE undoubtedly has special principles, techniques, and indicators thanks to its distinct theoretical foundation in ecology and economics. Reduce, reuse, and recycle, or the "3Rs" philosophy, is an excellent guide for how the CE has been put into action. Reducing the flow of resources into the production and consumption processes is the goal of the **reduce** input technique. **Reuse** is a technical method used to increase the time-intensiveness of a product or service. **Recycle** is an output method that calls for materials to be returned to renewable resources after usage (Zhao et al., 2012).

The manufacturing and consumption processes are the focus of efforts to reduce the usage of resources and energy (D'Amato, 2021). The reuse fully utilizes items that have been used together with any residual utilization function. This also applies to pieces of products that have been used up and any leftover materials from the manufacturing process.

Recycling is a crucial step in the development of the CE because it connects the production and consumption sectors by converting trash into new resources. The socioeconomic system's material flow is sometimes compared to the human body's blood circulation system, and the recycling industry in Japan is frequently contrasted with the "arterial industry" of production (Zhao et al., 2012). This means that resources like energy and materials shouldn't be released into the environment before being used for less valuable purposes (D'Amato, 2021).

7.4.1.1 The concept of mining CE and the 3Rs

The traditional mining industry relies on mineral resources to operate in a one-way fashion, including "mineral exploration exploitation-primary product processing-fine product manufacturing-product consumption-waste disposal." The term "mining CE" refers to an economic system that prioritizes the highly effective exploitation and complete utilization of mineral resources while adhering to the features and inherent ecological laws of mineral resources and mineral products. According to mineral exploration, exploitation, processing, melting, deep processing, consumption, and other activities, it constitutes a closed-loop material flow as "mineral resources—mineral products—renewable mineral resources" (Zhao et al., 2012). And to achieve the harmonious growth of the global environment and social progress, the material flow to the inner overlaps with the energy and information flow.

7.4.1.2 3R principles

The mining CE adheres to the 3R principle, which has the following specific meaning:

a. Reduce: The Reduce is primarily demonstrated during the process of exploitation, processing, and utilization of mineral resources (Zhao et al., 2012):

- realizing the efficient exploitation of resources through mechanization, automation, and exploit optimization.
- reducing mining dilution and ore loss ratios and enhancing the recovery rate of mineral-processing and smelting to improve total resource recovery by studying mining processing and melting technology of complex, difficult mining and refractive materials.

- increasing the overall benefit of resource development by lowering pollution emissions from tailings, gangue, and wastewater.

b. Reuse: Here, mine wastewater is primarily produced by discharged ore pit water and wastewater from concentration plants or coal preparation plants. Regarding wastewater treatment, there are three options: physical, chemical, and biological. The basic idea behind each method is to separate or convert harmful substances into harmless substances. An increasing number of concentration or coal preparation plants are utilizing closed-cycle technology. They do not discharge wastewater but instead dispose of it within the system, after which the water is reused. According to the study, the cost of mine pit water treatment is roughly half that of tap water (Zhao et al., 2012). Compared to groundwater supply projects, mine pit water saves groundwater resource fees and hoisting costs and discharge over standard will, which has significant environmental and economic benefits. Mine wastewater isn't the only reusable remnants of the mining process, as tailing processing can also lead to the production of other products, particularly building materials such as ceramics and cement.

c. Recycle: Recycling entails reducing garbage production as much as possible by processing mineral resource products that have completed their functions so that they become available resources again and can enter the secondary market or production process. The total value of renewable resource recovery in major developed countries has reached $250 billion annually and is increasing at a 10%–20% annualized rate (Zhao et al., 2012). Renewable resource recovery accounts for 45% of global steel output, 62% of copper output, 22% of aluminium output, 40% of lead output, 30% of zinc output, and 30% of paper product output. The rapid development of metal secondary use technology and markets is beneficial to relieve the pressure on mineral resource supply, energy consumption, and the environment. Recovering aluminium from beverage cans, for example, recovering manganese, zinc, and hydrargyrum from waste batteries, converting waste plastic into petrol and diesel, and so on. Reusing aluminium scrap collected in the social recovery network, for example, by mixing with primary aluminium, which formed a closed loop and, as a result, developed the secondary aluminium industry. Using aluminium scrap reduced the production of primary aluminium and aluminium product waste and saved a lot of electricity. It is estimated that recycling 1 kg of aluminium scrap can save approximately 46 kWh of electricity (Zhao et al., 2012).

7.4.1.3   The practice of the mining circular economy

A coal mining enterprise is used as an example to illustrate a CE model at the enterprise level (Figure 7.2) (Zhao et al., 2012). According to this illustration, following coal mine extraction, a portion of the raw coal enters the coal preparation plant. After coal preparation, the remainder blends with commercial coal to supply other industries and enterprises with fuel. Coal gangue and slime are used after coal preparation,

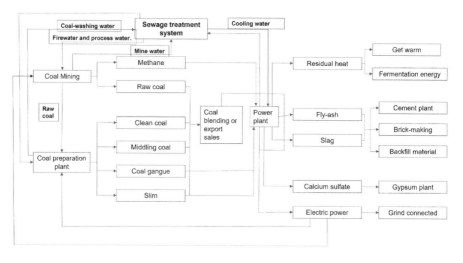

FIGURE 7.2   The CE model of coal mine enterprise.

Source: Zhao et al. (2012).

and methane is extracted from a fiery mine as the main fuel in a power plant built near a coal enterprise.

The electricity generated by power plants is primarily used for productive power in coal mining and preparation and household electricity in mining areas. The remaining power was paralleled in the grid. Slag and fly ash as wastes can be used as raw materials for building materials, cement and brick manufacturing, and green mining backfill. Residual heat can be used to warm up and as fermentation energy in the brewing industry. The raw material for a gypsum plant can be calcium sulphate. After being disposed of, coal-washing wastewater and mine water can be used as power plant cooling water, as well as firewater and process water of mine and coal preparation plants, which not only saves water resources but also solves the problem of environmental pollution caused by direct discharge.

7.4.1.4   Plastics alternatives in support of CE

Plastics are used in various products, ranging from consumer durables such as clothes, televisions, and toys to construction materials, vehicles and packaging for beverages and food. Furthermore, it is used in health as clothing protecting against viruses, while in the environmental industry, it is used for leaching chemicals from waste sites and preventing soil erosion. Because of their robust use, they have had a large negative environmental impact, leading to plastic pollution, especially in the oceans. Therefore, there have been requests and efforts to reduce plastic pollution and 'end-of-life' disposal previously promoted by the linear economy.

BOX 7.3   CASE EXAMPLES OF PRODUCTION AND USE OF NATURAL MATERIALS AND BIOBASED POLYMERS TO REPLACE CONVENTIONAL POLYMERS

CASE EXAMPLE 1: BIO4PACK

"Bio4Pack is a German company that has been a specialist in the field of compostable, sustainable packaging and has reportedly developed the 'first meat tray in the entire world which is completely compostable in accordance with the strict EN-13432 norm.' The tray, transparent film, label, and absorption pad will all be bio-based and compostable and indistinguishable, with the product being produced at only a fraction higher than the cost of a traditional plastic tray. Production of the tray has been a challenge. Given the fragility of PLA relative to other types of plastic, the use of approved additives has been necessary. The package must also have 'good barrier properties and be able to be mechanically processed with ease.' Retailers also benefit by being exempt from packaging tax. The company also manufactures paddy-straw trays that can be used for packing fruits and vegetables made from paddy straw waste generated in the paddy fields of Malaysia, thus providing farmers there a new source of income and avoiding other negative environmental externalities such as the air-pollution and groundwater pollution in the region caused by burning of paddy-waste. In addition to complying with the EN13432 composting standard, the Paddy Straw Trays may also be disposed of with the wastepaper after use" (UNCTAD, 2022).

CASE EXAMPLE 2: ENVIGREEN

"Envigreen is an Indian company that produces 100% organic, biodegradable, and eco-friendly bags to replace conventional single end-use plastic bags. The bags are made out of twelve ingredients, including potato, tapioca, corn, natural starch, vegetable oil, banana, and flower oil. The raw materials are converted into liquid form and then taken through a six-step procedure before the end product is ready. According to the company, no chemicals are used, and the paint used for printing on the bags is also natural and organic. The bags are water-soluble and do not melt or release any toxic fumes when burnt, unlike conventional plastic bags and have undergone numerous tests by various government agencies. The ingredients are also edible and do not harm animals that consume it. In addition to India, the company's bags are available in 13 countries, including Qatar, the United Arab Emirates, the United States, the United Kingdom and Kenya" (UNCTAD, 2022).

Source: (UNCTAD, 2022).

The 2017 Declaration of the United Nations Ocean Conference *Our Ocean, Our Future: Call for Action (UNGA, 2015)* refers that it is necessary to address consumption patterns and how they influence marine pollution while mentioning plastics. It called to countries, among other things,

> (i) promote waste prevention and minimization, develop sustainable consumption and production patterns, adopt the 3Rs – reduce, reuse and recycle – including through incentivizing market-based solutions to reduce waste and its generation, improving mechanisms for environmentally-sound waste management, disposal and recycling, and developing substitutes such as reusable or recyclable products, or products biodegradable under natural conditions; and (ii) Implement long-term and robust strategies to reduce the use of plastics and micro-plastics, particularly plastic bags and single-use plastic.
>
> *(Hopewell et al., 2009)*

Accounting for and spotting the negative environmental impacts of plastic production, use, and disposal. Therefore, plastic substitutes where its use is convenient and useful but inappropriate have been developed. As such, companies have invested in the shift from conventional plastic (Box 7.3) use to biobased plastics, for instance, as an alternative (UNCTAD, 2022).

7.4.2 SUSTAINABLE DEVELOPMENT

It has been widely believed that the CE idea originated with the work of Boulding in 1966, who proposed that the Earth was a closed system with "limited assimilative capacity, and as such, the economy and environment must coexist in equilibrium" (Boulding, 1966). When the CE concept was introduced in 1966 by Boulding, it was mainly rooted in ecological and environmental issues: "a man should find his place in a circular environmental system" (Boulding, 1966). Even though the CE concept is not an analogue of the green economy, its pursuit of achieving sustainable development is an integral part of it (Gureva and Deviatkova, 2021). However, it appears that in the CE, the economic system has been prioritized with principal benefits for the environment (Figure 7.3) and only implicit gains for social aspects (D'Amato, 2021; Kledyński et al., 2020).

However, the social component of sustainability is not usually openly addressed. The social issues typically include job creation or fairer taxation, while other societal problems go unmentioned. This shows the demerit in the whole concept since social equality achievement is not clearly articulated in terms of gender, race, religion and other diversity, and financial equality, and thus needs to be improved on. This has resulted in concerns about CE's ability to reach sustainable development.

Nonetheless, all types of waste, such as clothing, scrap metal, and obsolete electronics, are recycled or reused in such an economy. This can be used to protect the environment and use natural resources, develop new sectors, create jobs, and develop new capabilities more wisely. Perhaps the way the CE contributes to climate change adaption and mitigation is its drive for sustainable development. The CE is envisaged to address the adverse impacts of climate change and resource insecurity (Zvimba et al., 2021). As such, realizing the CE's ability to correct the shortcomings of the linear economy and possibly contribute to sustainable

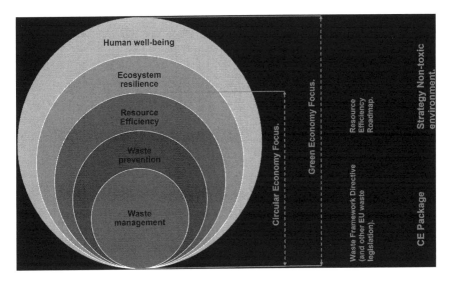

FIGURE 7.3   Environmental context of CE.

Source: Kledyński et al. (2020).

development is leading to its popularity and adoption (Naidoo et al., 2021). The concept has been prominently promoted by the EU, which invested €650 million in their package to move to a CE, and China has become the first country to pass a CE law (Millar et al., 2019).

The displacement of linear production by CE will still stimulate economic growth. Some argue that economic growth under a CE model will be greater in the long run than growth forecasts under current linear models (Millar et al., 2019). This demonstrates the current narrative of the CE as a model that stimulates growth while causing minimal environmental damage and thus explains the popular "win-win" catchphrase that is increasingly associated with the CE. The full implementation of CE, with not just mass but also power flow throughout the entire life cycle and the issue of harmful emissions, positions CE as a green economy pioneer (Figure 7.3) (Kledyński et al., 2020). As far as environmental quality is concerned, CE can promote low-carbon energy and advocates that ecological processes in natural and semi-natural systems can be leveraged for the benefit of human beings without endangering these ecosystems (D'Amato, 2021). The CE encompasses elements of a green economy, including recycling and reuse, greener supply chains, and reduction of energy and material inputs in the production process (D'Amato, 2021). The CE objective is to extract materials, wastes, and energy for the advantage of the industry and to achieve low-carbon development and pollution abatement for the power industry using the 3R principle (reduce, reuse, and recycle) (Kledyński et al., 2020).

The United Nations' 17 SDGs, established in 2015, have renewed a global vision for addressing sustainability challenges, emphasizing the importance of coordinated efforts by multiple societal actors (D'Amato, 2021). The significant benefits of the CE in terms of boosting the achievement of the SDGs, such as SDG 12 on responsible consumption and production), SDG 11 on (Sustainable Cities and Communities), SDG 6, and SDG 3 (Good Health and Well-Being) are well recognized. The recognition of SDG 3 as critical in achieving all SDGs places the CE approach at the core of sustainable development (WHO, 2018). The significant benefits of the CE in terms of boosting the achievement of the SDGs, such as SDG 12 on responsible consumption and production), SDG 11 on sustainable cities and communities), SDG 6 and SDG 3 on good health and well-being are well recognized (UNGA, 2015). The recognition of SDG 3 as critical in achieving all SDGs places the CE approach at the core of sustainable development (WHO, 2018). Therefore, in the quest to achieve SDGs, incorporating circularity principles would address many global challenges, such as the climate crisis, pollution, biodiversity loss, land/water degradation, etc.

Overall, a system thinking for circular products and a business model redesign economy can provide a major opportunity to avoid natural resource degradation and increase resource efficiency and environmental gains (WHO, 2018). Specifically, implementing CE principles can support the transition to a low-carbon economy, resource efficiency, value retention, waste management, and many of the root causes of climate change. Implementation of sustainable development and circular principles into value chain of commodity production and consumption, with a focus on minimizing the need for new inputs of materials and energy and reducing the environmental pressures related to resource extraction, emissions and waste, could reduce the level of global GHG emissions by up to 45% (WHO, 2018).

Innovative solutions brought by science and technology, practices and systems have been identified as key enablers for sustainable resource utilization, waste management, environmental surveillance, and overall advancing the CE and sustainable development agendas. While research and development are fundamental to facilitating the adoption of circularity and sustainable development measures, digital transformation, especially now in the 4th Industrial Revolution, provides an opportunity to accelerate the transition to a CE by using disruptive digital technologies or smart tools. In turn, limiting impacts on the environment is a major step towards the fight against biodiversity loss and climate change. For the transition to succeed, coordination and collaboration amongst all stakeholders, including citizens, is also necessary for promoting and implementing the CE and sustainable agenda globally.

7.5 USING THE CIRCULAR ECONOMY TO BUILD RESILIENCE AGAINST FUTURE INFECTIOUS DISEASES

If ecosystems continue to degrade and fail to provide the necessary services that sustain the environment and human life, it will become unbearable for every form of life (Laininen, 2019). The CE approach can potentially reduce environmental waste, and its adoption and implementation should be sooner rather than later. The key is to reduce waste and pollution, keep products and materials in use for longer, and regenerate natural systems to contribute positively to achieving SDGs.

At the macroeconomic level, perhaps the most significant trend influencing CE initiatives is globalization: the increased interdependence of countries and world regions for financial, human, and material resources as transportation and communication costs have decreased (WHO, 2018). One likely outcome of this trend is that technological innovations resulting from resource reuse and reduction strategies in

CE initiatives in one country are more likely to be exported to others. As a result, manufacturing economies of scale can be realized, increasing the competitiveness of technologies even further (Naidoo et al., 2021). The annual benefit of adopting advanced CE technologies rather than current technologies could be €1.8 trillion by 2030 (Rizvi et al., 2021). This technological diffusion will have far-reaching health benefits that otherwise would not have been possible. The global adoption of digitization in communication and other technologies will further amplify these trends (Rizvi et al., 2021). Other things being equal, increased global trade, including technological advancements that support the CE, would improve employment opportunities and reduce poverty (Naidoo et al., 2021). Higher employment is thought to have both direct positive psychological and physical health benefits and indirect benefits from higher income, which allows for healthier foods (Mpandeli et al., 2018). Additional health benefits associated with GDP growth due to globalization stem from the possibility of increased expenditures on public and private health care. The same benefits are likely to apply to the diffusion of technological innovations in pollution reduction, which, when implemented, reduce health impacts.

In contrast to this optimistic view of globalization and its relationship to the CE initiatives and their health implications, one tangible disadvantage of this trend is that comparative advantage encourages higher-income countries to export their waste—as well as polluting production—to lower-income countries (WHO, 2018). Furthermore, the increasing movement of chemical production to low- and middle-income countries, where public health and environmental protections are often lacking, is likely to exacerbate the health impacts of emerging chemicals of concern (Suk et al., 2016).

Recycling and reusing products, components, and materials have many positive implications, such as cost savings in the healthcare sector and indirect health benefits from reducing environmental impacts (air, water, and soil pollution, as well as GHG emissions) from manufacturing and extraction processes (Box 7.3). The health implications identified in the other broad categories of CE processes—maintaining the highest value of materials and products and changing utilization patterns—are also overwhelmingly positive. Performance models of utilization, in particular, show the potential for significant direct health benefits for the hospital/health care sector, as well as a wide range of indirect health benefits from resource-efficient agricultural practices, increased use of renewable energy and energy efficiency, building with circular principles, and shifts to new product-sharing and product-as-service models (WHO, 2018). These processes are expected to reduce waste generation and improve resource efficiency, reducing environmental impacts from economic activity across various sectors and the morbidity and mortality endpoint impacts.

7.6 CONCLUSION

In a global economic system bound by finite natural resources, faced with a growing population and increasing global demand for products and services, considering its negative impacts on the environment and human health, humankind cannot depend on the linear economic process to drive the production and consumption processes if ever the growing gap between demand and supply of resources is to be reduced and achieve the SDGs. As such, moving towards a CE approach is one of the best ways for the world to build resilience against the impacts of resource degradation, climate change and future pandemics. Particularly because the CE promotes reusing, reducing, and recycling waste materials instead of waste disposal. The CE approach presents a shift away from the current linear take-make-waste extractive systems, thereby helping achieve several SDGs, including enhancing environmental and human health and conserving the planet's finite resources.

REFERENCES

Benachio, G.L.F., Freitas, M.D.C.D., Tavares, S.F. (2020) Circular economy in the construction industry: A systematic literature review. *Journal of Cleaner Production* 260, 121046.

Boulding, K.E. (1966) *The economics of the coming spaceship earth*. Earthscan, New York.

D'Amato, D. (2021) Sustainability narratives as transformative solution pathways: Zooming in on the circular economy. *Circular Economy and Sustainability* 1, 231–242.

De, S., Debnath, B. (2016) Prevalence of health hazards associated with solid waste disposal-A case study of kolkata, India. *Procedia Environmental Sciences* 35, 201–208.

Economou, V., Gousia, P. (2015) Agriculture and food animals as a source of antimicrobial-resistant bacteria. *Infection and Drug Resistance* 8, 49.

Ellen MacArthur, F. (2015) *Delivering the circular economy: A toolkit for policymakers*. Ellen MacArthur Foundation: Isle of Wight, United Kingdom, p. 177.

Gureva, M., Deviatkova, Y. (2021) Formation of the concept of a circular economy. *Revista Inclusiones*, 5(2), 156–169.

Hamam, M., Chinnici, G., Di Vita, G., Pappalardo, G., Pecorino, B., Maesano, G., D'Amico, M. (2021) Circular economy models in agro-food systems: A review. *Sustainability* 13, 3453.

Hartley, K., van Santen, R., Kirchhrr, J. (2020) Policies for transitioning towards a circular economy: Expectations from the European Union (EU). *Resources, Conservation and Recycling* 155, 104634.

Himeidan, Y.E., Kweka, E.J. (2012) Malaria in East African highlands during the past 30 years: Impact of environmental changes. *Frontiers in Physiology* 3, 315.

Hopewell, J., Dvorak, R., Kosior, E. (2009) Plastics recycling: Challenges and opportunities. *Philosophical Transactions of the Royal Society B: Biological Sciences* 364, 2115–2126.

Iacovidou, E., Hahladakis, J.N., Purnell, P. (2021) A systems thinking approach to understanding the challenges of achieving the circular economy. *Environmental Science and Pollution Research* 28, 24785–24806.

Janko, M.M., Irish, S.R., Reich, B.J., Peterson, M., Doctor, S.M., Mwandagalirwa, M.K., Likwela, J.L., Tshefu, A.K., Meshnick, S.R., Emch, M.E. (2018) The links between agriculture, Anopheles mosquitoes, and malaria risk in children younger than 5 years in the Democratic Republic of the Congo: A population-based, cross-sectional, spatial study. *The Lancet Planetary Health* 2, e74–e82.

Ketov, A., Sliusar, N., Tsybina, A., Ketov, I., Chudinov, S., Krasnovskikh, M., Bosnic, V. (2022) Plant biomass conversion to vehicle liquid fuel as a path to sustainability. *Resources* 11, 75.

Kirchherr, J., Reike, D., Hekkert, M. (2017) Conceptualizing the circular economy: An analysis of 114 definitions. *Resources, Conservation and Recycling* 127, 221–232.

Kledyński, Z., Bogdan, A., Jackiewicz-Rek, W., Lelicińska-Serafin, K., Machowska, A., Manczarski, P., Masłowska, D., Rolewicz-Kalińska, A., Rucińska, J., Szczygielski, T. (2020) Condition of circular economy in Poland. *Archives of Civil Engineering* 66(3), 37–80.

Laininen, E. (2019) *Transforming our worldview towards a sustainable future, sustainability, human well-being, and the future of education*. Palgrave Macmillan, Cham, pp. 161–200.

Lynch, J., Cain, M., Frame, D., Pierrehumbert, R. (2021) Agriculture's contribution to climate change and role in mitigation is distinct from predominantly fossil CO_2-emitting sectors. *Frontiers in Sustainable Food Systems*, 4, 518039.

Manavalan, E., Jayakrishna, K. (2019) An analysis on sustainable supply chain for circular economy. *Procedia Manufacturing* 33, 477–484.

Marques,A.,Martins,I.S.,Kastner,T.,Plutzar,C.,Theurl,M.C.,Eisenmenger,N.,Huijbregts,M.A., Wood, R., Stadler, K., Bruckner, M. (2019) Increasing impacts of land use on biodiversity and carbon sequestration driven by population and economic growth. *Nature Ecology & Evolution* 3, 628–637.

Millar, N., McLaughlin, E., Börger, T. (2019) The circular economy: Swings and roundabouts? *Ecological Economics* 158, 11–19.

Mpandeli, S., Naidoo, D., Mabhaudhi, T., Nhemachena, C., Nhamo, L., Liphadzi, S., Hlahla, S., Modi, A. (2018) Climate change adaptation through the water-energy-food nexus in southern Africa. *International Journal of Environmental Research and Public Health* 15, 2306.

Naidoo, D., Nhamo, L., Lottering, S., Mpandeli, S., Liphadzi, S., Modi, A.T., Trois, C., Mabhaudhi, T. (2021) Transitional pathways towards achieving a circular economy in the water, energy, and food sectors. *Sustainability* 13, 9978.

Nhamo, L., Matchaya, G., Mabhaudhi, T., Nhlengethwa, S., Nhemachena, C., Mpandeli, S. (2019) Cereal production trends under climate change: Impacts and adaptation strategies in southern Africa. *Agriculture* 9, 30.

Nhamo, L., Mpandeli, S., Senzanje, A., Liphadzi, S., Naidoo, D., Modi, A.T., Mabhaudhi, T. (2021a) Transitioning toward sustainable development through the water–energy–food nexus, in: Ting, D., Carriveau, R. (Eds.), *Sustaining tomorrow via innovative engineering*. World Scientific, Singapore, pp. 311–332.

Nhamo, L., Ndlela, B. (2021) Nexus planning as a pathway towards sustainable environmental and human health post Covid-19. *Environment Research* 192, 110376.

Nhamo, L., Rwizi, L., Mpandeli, S., Botai, J., Magidi, J., Tazvinga, H., Sobratee, N., Liphadzi, S., Naidoo, D., Modi, A., Slotow, R., Mabhaudhi, T. (2021b) Urban nexus and transformative pathways towards a resilient Gauteng City-Region, South Africa. *Cities* 116, 103266.

Nobre, C.A., Sampaio, G., Borma, L.S., Castilla-Rubio, J.C., Silva, J.S., Cardoso, M. (2016) Land-use and climate change risks in the Amazon and the need of a novel sustainable development paradigm. *Proceedings of the National Academy of Sciences* 113, 10759–10768.

Orton, J. (2019) Heritage Impact Assessment: Proposed Aggeneys 1–100MW Solar PV Facility and associated infrastructure near Aggeneys, Namakwaland Magisterial District, Northern Cape. Savannah Environmental (Pty) Ltd, Capt Town, South Africa, p. 43.

Remoundou, K., Koundouri, P. (2009) Environmental effects on public health: An economic perspective. *International Journal of Environmental Research and Public Health* 6, 2160–2178.

Reynolds, T.W., Waddington, S.R., Anderson, C.L., Chew, A., True, Z., Cullen, A. (2015) Environmental impacts and constraints associated with the production of major food crops in Sub-Saharan Africa and South Asia. *Food Security* 7, 795–822.

Rizvi, S.W.H., Agrawal, S., Murtaza, Q. (2021) Circular economy under the impact of IT tools: A content-based review. *International Journal of Sustainable Engineering* 14, 87–97.

Saker, L., Lee, K., Cannito, B., Gilmore, A., Campbell-Lendrum, D.H. (2004) *Globalization and infectious diseases: A review of the linkages*. World Health Organization (WHO), Geneva, Switzerland, p. 67.

Scheel, C., Bello, B. (2022) Transforming linear production chains into circular value extended systems. *Sustainability* 14, 3726.

Sena, A., Ebi, K. (2021) When land is under pressure health is under stress. *International Journal of Environmental Research and Public Health* 18, 136.

Suk, W.A., Ahanchian, H., Asante, K.A., Carpenter, D.O., Diaz-Barriga, F., Ha, E.-H., Huo, X., King, M., Ruchirawat, M., da Silva, E.R. (2016) Environmental pollution: An under-recognized threat to children's health, especially in low-and middle-income countries. *Environmental Health Perspectives* 124, A41–A45.

UNCTAD (2022) Substitutes for single-use plastics in sub-Saharan Africa and south Asia: Case studies from Bangladesh, Kenya and Nigeria. United Nations Conference on Trade and Development (UNCTAD), Geneva, Switzerland, p. 121.

UNGA (2015) Transforming our world: The 2030 Agenda for Sustainable Development, Resolution adopted by the General Assembly (UNGA). United Nations General Assembly, New York, p. 35.

WHO (2018) Circular economy and health: Opportunities and risks. World Health Organization. Regional Office for Europe, Brussels, Belgium.

WHO (2020) Coronavirus disease 2019 (COVID-19): Situation report. World Health Organization (WHO), Geneva, Switzerland, p. 9.

Wiedmann, T., Lenzen, M., Keyßer, L.T., Steinberger, J.K. (2020) Scientists' warning on affluence. *Nature Communications* 11, 1–10.

Wiedmann, T.O., Schandl, H., Lenzen, M., Moran, D., Suh, S., West, J., Kanemoto, K. (2015) The material footprint of nations. *Proceedings of the National Academy of Sciences* 112, 6271–6276.

Yamaguchi, S. (2018) International trade and the transition to a more resource efficient and circular economy: A Concept paper. Organisation for Economic Co-operation and Development (OECD), Paris, France, p. 23.

Zhao, Y., Zang, L., Li, Z., Qin, J. (2012) Discussion on the model of mining circular economy. *Energy Procedia* 16, 438–443.

Zvimba, J.N., Musvoto, E.V., Nhamo, L., Mabhaudhi, T., Nyambiya, I., Chapungu, L., Sawunyama, L. (2021) Energy pathway for transitioning to a circular economy within wastewater services. *Case Studies in Chemical and Environmental Engineering* 4, 100144.

8 Gender norms and social transformation of agriculture in Sub-Saharan Africa

Everisto Mapedza

8.1 INTRODUCTION

This chapter problematizes gender equality, whose role is often ignored despite playing a pivotal role in societal transformation. Scholars have analysed discourses on gender as a continuum from gender-blind, practical approaches to Gender Transformative Approaches (GTAs) (Cole, 2014). The chapter posits that gender equality, which ensures equal access to and control over resources and services by men and women within the family and society, has not yet been achieved in African Food Systems (Quisumbing, 2019). The chapter further argues that understanding gender norms is a key pillar in the social transformation of agricultural commons within Sub-Saharan Africa (Badstue, 2020; Grashuis, 2021; Ostrom, 1990; Rose, 2020). Not much attention has been paid to understanding better norms and their enabling and disabling roles in agriculture, as revealed by the study conducted by several Consultative Group on Agricultural Research centres (Aregu, 2018; Badstue, 2020; Mudege, 2017; Petesch, 2018; van den Bold, 2015). Gender norms limit or act as a glass ceiling on what women may contribute towards agricultural transformation. Whilst many researchers have emphasized restrictions on women, there is a strand of research that looks at how gender norms interact with gender relations, thereby impacting women's innovation, adoption, and benefit from new technologies (Aregu, 2018). The COVID-19 pandemic, which began as a health pandemic, has significantly led to the exposure of gendered inequalities and fault lines in Sub-Saharan Africa and beyond (Altieri, 2021; Clapp, 2020; Levine, 2021; Liegeois, 2020; McKinsey, 2020; Mooi-Reci, 2021; Saba, 2020; UN Women, 2020). The mantra of building back better has often highlighted the need to be inclusive and transformative on gender. The mantra, however, needs to be further grounded in terms of what it means for gender inequalities, especially in Sub-Saharan Africa, where patriarchy has resulted in unequal access to and control over the means of agricultural production, such as land. The following section situates gender within Africa.

A better understanding of norms will likely result in an inclusive social change within the African agricultural sector. This will ensure the achievement of the vision of a better Africa that will leave no one behind. Agricultural-driven transformation

DOI: 10.1201/9781003327615-8

requires normative and structural change for equality through agricultural outcomes. The need for a just society has been further illustrated by the COVID-19 pandemic, which has exposed the deep-seated fissures of inequality that are highly gendered and intersectional. This chapter argues that the mantra of building back better will not yield positive outcomes if the underlying causes of gendered and intersectional inequalities towards common access are not strategically and meaningfully addressed. The chapter recommends that gender norms are central to accessing the means of agricultural production in Sub-Saharan Africa, such as land and water. It is, therefore, important that the region's gender norms are seriously considered in all local and global initiatives aiming to transform agriculture within the Sub-Saharan Africa region.

8.2   SITUATING GENDER WITHIN AFRICA

Agenda 2063 aims to transform a 'prosperous Africa based on inclusive growth' and sustainable development. Agriculture, which is one of the key pillars, highlights that such a transformation must ensure that the 'full potential of women and youth, boys and girls are realized' (Union, 2020: 1). The African Union's Comprehensive Africa Agricultural Development Program and the 2014 Malabo Declaration on Accelerated Agricultural Growth and Transformation for Shared Prosperity and Improved Livelihoods place emphasis on agriculture as a key driver for livelihood improvements. This is the vehicle for agricultural development across Africa and supports the first aspiration in the A.U. Agenda 2063. Point 45 in the Agenda 2063 highlights the leading role of women and youths. The Agenda 2063 does not address the broader patriarchy and power, 'Africa's women and youth shall play an important role as drivers of change. An inter-generational dialogue will ensure that Africa is a continent that adapts to social and cultural change.' Whilst this seems to imply a change in norms, the report also mentions the importance of maintaining culture, which in gender power relationships could be contradictory. Maintaining culture could entail re-enforcing oppressive and unequal gender practices. This then requires the use of a GTA.

8.2.1   DEFINING NORMS AND GENDER

Norms are defined as "collective definitions of socially approved conduct, stating rules, or ideals; and gender norms are such definitions applied to groups constituted in the gender order – mainly, to distinctions between women and men" (Pearse, 2016:30). Norms can be further defined as the 'invisible barriers' undermining women's engagement in agriculture (Quisumbing, 2019). Norms are further embedded within society and its institutions. Gender norms govern access and control over the means of agricultural production, such as land and labour. Norms are part of the socialization process, and feminist researchers argue for campaigning on changes in the social norms, which are often labelled as 'culture' by those defending them (Agarwal, 1997; Gray, 1999; Narayan, 1998; Seguino, 2007; Tavenner, 2018; Zibani, 2016). It is also important to note such transformational approaches call for a GTA (Cole, 2014). A GTA approach aims at ensuring that gender relationships are changed by changing the structure (norms, beliefs, culture) which are producing gender inequalities. The GTA approach is opposed to the Practical Gender Approach, which aims to introduce changes to lessen the women's burden without challenging the underlying power dynamics producing inequality (Cole, 2014).

Building back better in Sub-Saharan Africa will entail changing gender norms and norms and building towards a future society. COVID-19 has further highlighted the following gender issues: the absence of comprehensive sex-disaggregated data highlighted pre-existing social inequalities and intersectionalities, the increased burden of childcare on women, especially during lockdowns, women being disproportionally affected by unemployment, the unequal access to social protection, equality in access to health care is important for both the poor and the rich and the lack of access to inputs and markets for some of the agricultural activities which were gendered (Altieri, 2021; Anthony, 2020; Cardwell, 2020; Clapp, 2020; Connell, 2020; Corburn, 2020; de Wit, 2021; Franco, 2020; Hupkau, 2020; Leigh, 2020; Levine, 2021; Liegeois, 2020; McKinsey, 2020; Meine, 2020; Mooi-Reci, 2021; Ryan Cardwell, 2020; UN Women, 2020). The norm-driven transformation will need the co-production of knowledge. Such co-production must be inclusive, bringing together women, men, the old, the young, the youth, the better-off and the not-so better-off and other intersectionalities in crafting inclusive solutions. "The natural world, in particular, played an important part in defining gender norms, such as notions of appropriate femininity and masculinity" (Sundberg, 2017:2).

Gender in this chapter will be defined as the roles and responsibilities that society ascribes to an individual based on their sex, age, ethnicity, religion, caste, or any other social criteria. Gender is a social construct referring to relations between and among sexes. This chapter understands that gender is not equal to sex, but it portrays the differential power dynamics in an intersectional manner. Gender is dynamic; it is contested and reconfigured over time as the norms influencing gender change over time (Chant, 2002; Cleaver, 2002; Doss, 2020; FAO, 2011; Lawless, 2019; Mapedza, 2013, 2019; Peters, 2002; Quisumbing, 2019). Intersectionality is understood as how the various dimensions of inequality further intersect to compound the inequalities and disadvantages men and women face (Crenshaw, 1989; Nightingale, 2011; Sundberg, 2017; Viruell-Fuentes, 2012).

This chapter also understands the importance of agency by women (Farnworth, 2010; Leder, 2017; Meinzen-Dick, 1999; O'Hara, 2018; Petesch, 2018). The chapter, however, argues that gender norms provide barriers that will make it more difficult for women to exercise their agency (Adams, 2018; Holdo, 2020; Lawless, 2019; Leder, 2017; O'Hara, 2018; Petesch, 2018; Suhardiman, 2013, 2016; Victor, 2013; Waldman, 2005). The chapter does not downplay the role of agency by women in Sub-Saharan Africa; the point being made is that restrictive norms undermine women's agency in contributing towards, influencing, accessing and benefiting from agricultural-driven social transformation (Lawless, 2019). The following section justifies why gender norms are important for the social transformation of the agricultural commons.

8.2.2 GENDER NORMS AND AGRICULTURAL TRANSFORMATION

Research in agriculture has not focused on the changing gender norms and their implications for women's engagement within agriculture (Quisumbing, 2015). Understanding the nature and evolution of gender norms will better inform how women could meaningfully engage in the agricultural transformation which Africa has been advocating for under Agenda 2063. Norms, jointly with other gender barriers, hinder women from accumulating assets through agriculture and further reduce women's ability to control the accumulated assets (Njuki, 2021; Quisumbing, 2015). Incorporating positive norms in agriculture is one of the key aspects of inclusive and resilient food systems (Njuki, 2021). The following section briefly introduces gender norms and agriculture in Africa.

Agricultural resources are important common resources (Ostrom, 1990, 2002) for agricultural production. Gender norms further negatively impact how women could be key players in the transformation of agriculture in Africa. This section is not meant to be exhaustive. Still, it aims to highlight how norms are a major barrier to women's livelihood transformation, negatively impacting everyone. According to the Malabo Montpellier Panel Report (2021), in Ethiopia, Malawi, Rwanda, Uganda, and Tanzania, agricultural productivity would go up by 19% if women had the same access to means of agricultural production and support as men. Changing the norms contributing towards persistent gender inequalities will result in poverty reduction, food and nutrition security, economic growth and a positive impact on the food systems in Africa. The following sections will look at gender norms regarding the access and control of land, water resources, agricultural extension and livestock.

8.2.2.1   Access and control of land

'The Sustainable Development Goals (SDGs) 1.4.2 and 5.A.1 refer to the strengthening of women's land and property rights as a fundamental pathway towards poverty reduction and women's empowerment' (Prindex, 2020:5). Land ownership and security of tenure are key tenets for increased agricultural production in Sub-Saharan Africa (Doss, 2020). The Rights Resources Institute notes that land ownership gives agency towards power, authority and governance (Danson, 2021; RRI, 2015). The SDGs indicators 5.A.1 and 1.4.2 specifically focus on women's land rights (Doss, 2020: 1). According to Prindex (2020), land inequality is a global phenomenon that is more pronounced in regions such as Sub-Saharan Africa. It is important to note that there are intra-regional variations, with Benin having the most tenure-insecure women. Whilst women comprise about 51% of the Sub-Saharan African population, the land is largely patriarchal, so it is inherited through the male lines. Women comprise 47% of the agricultural labour force in Sub-Saharan Africa (FAO, 2017). Patriarchy, therefore, shapes the norms which set barriers for women to inherit land in their name.

Islam allows the inheritance of land by both sons and daughters. However, most daughters pass on their inheritance to their brothers or other male relatives (Jones-Casey, 2011). When discussing the issue of land rights, it is important to understand better the bundle of rights that women have over a piece of land (Doss, 2020). According to Schlager (1992), there are bundles of rights that include the following: access, withdrawal, management, regulation, exclusion and transfer (Schlager, 1992). Women tend to have very limited rights as opposed to men. Such insecure rights are negative for women and have negative agricultural, economic, environmental and social outcomes for society (Prindex, 2020).

Some countries, such as Ethiopia, have begun to change the patriarchal norms of passing land through male lines by beginning the land registration process, which has seen both husband and wife and all their children registered as land co-owners. In the event of the husband passing away, the wife will remain the owner of the land. Whilst this is a progressive step, some research has noted that this does not automatically translate into equality, land tenure security, or even increased land-based investments by women (Prindex, 2020; Quisumbing, 2014).

Ownership and rights to land matter not only in terms of allowing women to cultivate the land but also the limited security of tenure and rights that will not enable women to make long-term investments, especially if the husband or male relative dies. Agricultural support through financing is usually linked to ownership of land. Women will get less access to credit and finance if they do not have secure rights to the land they are cultivating, as land is usually used as collateral (FAO, 2011; Panel, 2021). A better understanding of women's land tenure status will inform pathways towards inclusive poverty reduction (Meinzen-Dick, 2017).

8.2.2.2 Access to water resources

The African Union passed the Framework for Irrigation Development and Agricultural Water Management in Africa (Union, 2020). Improved agricultural water productivity is central to transforming livelihoods (de Jong, 2020). Approximately 18.6 million hectares of land are irrigated in Africa (Union, 2020). Africa is estimated to irrigate only 36% of the estimated 42.5 million hectares potential (Molden, 2007). However, this tends to be underestimated as Farmer Led Irrigation Development (FLID) is often not documented in national irrigated statistics (de Bont, 2020; Muturi, 2019; Union, 2020; Woodhouse, 2017; World Bank, 2018).

Access to either supplementary or full irrigation water resources is an important mechanism for building resilience against shocks such as drought, climate variability and climate change. Water resources for agricultural water production are often closely related to land ownership. Whilst legislation across the continent varies with water rights being separate or linked to land, norms governing land ownership make it difficult for women to access water for irrigated agriculture (Schreiner, 2004; Sokile, 2004; van Koppen, 2007). For the FLID, which is also referred to as informal irrigation, even for this type of irrigation, usually 0.5–2 hectares (can be as small as 100 square metres), it is more difficult for women to access this type of irrigation (de Bont, 2020; Osewe, 2020; World Bank, 2018). Whilst Peters (2013) noted that customary land tenure did not stop agricultural intensification through irrigation in West Africa, norms and customary restrictions tend to form major barriers for women to access irrigated agriculture. Based on patriarchal norms, most formal irrigation schemes assume that men are the landowners and tend to develop irrigation schemes targeting men. Research has also demonstrated that even in West Africa, where crops such as rice were normally considered female crops when formal rice irrigation schemes were established, men were the main beneficiaries as they were perceived to be the household heads and owners of the land (Zwarteveen, 1996b). This further brings to the fore the intersectionality lens, which goes beyond the binary of men and women by highlighting the multiple and interlinked systems of oppression that women, who fall into differentiated categories, face in their bid to access water and other productive resources (Carastathis, 2014; Crenshaw, 1989; Leder, 2019; Sundberg, 2017; Tavenner, 2019).

8.2.2.3   Access to agricultural extension

According to FAO (2011), increasing women's access to agricultural extension services in Sub-Saharan Africa will result in a 20%–30% increase in productivity. An intersectional analysis would further interrogate FAO's findings regarding which women, when and under what farming circumstances. Such a political feminist lens will help deepen our understanding of agricultural extension solutions. Farnworth (2010) points out that agricultural extension is not a technical and gender-neutral intervention. Extension plays out in a complex way structured by gender and power relationships. The intersectionality of gender, class, culture and place was seen as creating women's subjectivities in Egypt. Women struggled to access land and irrigated agriculture even in instances where drainage water (wastewater) was to be used (Rap, 2019).

In Sub-Saharan Africa, the norms which see men as the household heads and the farmers have resulted in most extension services being directed towards men rather than women. In Malawi, one of the female extension officers interviewed re-enforced patriarchal norms that men are the heads of households and the farmers (Mapedza, 2017). In some instances, agricultural extension staff negatively stereotyped women by viewing them as unknowledgeable helpers and carers (Mudege, 2017). In Ethiopia, despite large investments in agricultural extension, women were not getting the same quality of extension services as men (Ragasa, 2013). This means that agricultural extension must also be tailored to the specific needs of female farmers. This must be grounded in the changing of gender norms, which do not see women as farmers.

8.2.2.4   Access to and control of livestock

In mixed farming and pastoral communities' women have less access to and control over livestock. The gendered inequalities in ownership, labour and benefits continue (Archambault, 2016). For most of Africa, women tend to focus on small livestock (Aredo, 2006; Mapedza, 2008; Pica-Ciamarra, 2007; Scoones, 1990, 2020; Van Hoeve, 2006; van Koppen, 2005). Norms governing livestock units that are bigger and more valuable, such as cattle, are usually labelled as men's assets. In Africa, according to the Malabo Report (2021), there are 249 million women livestock keepers, with the majority keeping the livestock around the homestead (Njuki, 2013; Panel, 2021). In a study conducted by Quisumbing et al. in Mozambique, men's mean cattle ownership was 3.08 in 2009 and rose to 3.46 in 2011 (Quisumbing, 2015). Women owned a mean of 0.23 cattle in 2009, which fell to 0.20 in 2011. The joint ownership of cattle mean was 1.47 and rose to a mean of 1.53 in 2011. Whilst this clearly shows the differences in ownership of cattle, which resonates with most of Sub-Saharan Africa, a feminist political ecology lens would be important to further interrogate which men or women own livestock and why that is the case. Some literature points to women's agencies, for instance, in eastern Africa (Parsons, 2017) and among the Masaai women as managers of milk, especially in the dairy commercialization context (Allegretti, 2018; Bischot, 2017).

8.3   METHODOLOGY

This research is informed by a simple literature review. The literature review focused on the literature on gender, norms, and agriculture transformation in Africa. A search for literature on gender and COVID-19 was also conducted. There was also a focus on the commons literature and the International Association's Africa Commons webinar series held in 2020. The main research question that the literature addressed was: How can gender norms contribute towards social transformation in Africa? The literature review was based on peer-reviewed publications up to 2021 looking at gender and agricultural transformation and change in Sub-Saharan Africa, which was generic and not based on specific databases. Over the past year, there has been an increase in peer-reviewed articles looking at COVID-19 and how it impacts many things, including food systems in Africa and developing countries. This literature has also been made freely available by several journal publications. The literature review was also conducted for regional and international policy documents and webinar summaries, especially by the African Union and other development agencies looking at agricultural transformation in Sub-Saharan Africa. There have also been several preparatory dialogues on agricultural transformation for the United Nations Food Systems Summit. Lessons from past agricultural water management research, feminization of agriculture and agricultural extension by the International Water Management Institute (IWMI) in Africa were also drawn. Lastly, as part of the Deutsche Gesellschaft für Internationale Zusammenarbeit Gender Baseline in Ghana, a review was made on the role of gender and access to land on increasing agricultural productivity and hence transformation in Ghana and within Sub-Saharan Africa at large. The following section briefly overviews emerging themes from the literature review.

8.4   EMERGING THEMES FROM THE LITERATURE REVIEW

8.4.1   The landed commons are not common for both men and women

Land as a key means of production is difficult for women to access and, further, leverage it for other means of production such as finance, which usually demands collateral such as land. If women are part of the agricultural transformation narrative in Africa, improved access to and control over land will be necessary for women. In Tanzania, game-changing norm adjustments are taking place with the investment in gender and land champions as part of the Women's Land Tenure Security (Daley, 2021). In most of Sub-Saharan Africa, efforts are being made to ensure that traditional leaders, who are custodians of a patriarchal culture, are also part of the harmful norm change dialogue. Even when women have access to a plot, patriarchy still prioritizes male-owned land regarding labour required. Some authors noted that women's land would be allocated labour after completing the labour requirements in men's fields (Gray, 1999). The increase in investments in land in Africa, which accelerated after the 2008 global food crisis, further marginalized women (Behrman, 2012; Peters, 2020).

8.4.2 Water tenure access is gendered

In the context of climate change and climate variability, access to water for full or supplementary irrigation would entail increased and more resilient production. In most developing countries, access to water is closely linked to access to land. If women have norms and cultural barriers making it difficult to access land, it will also be challenging to access land for agricultural production. Irrigation increases land value and is often associated with masculinity. Men then tend to take over irrigated production. Studies in West Africa showed that women were traditional cultivators of lowlands and valleys for rice production. However, once irrigation was introduced, men were members of the irrigation scheme, undermining productivity (Dembele, 2012; van Koppen, 2007; Zwarteveen, 1996).

8.4.3 Norms on who is a farmer

Despite more recent changes, the perception of a farmer is usually of men, with women seen as belonging to the men's household. Most development partners have also re-enforced that view despite attempts to incorporate gender in agriculture (Manfre, 2013). However, several researchers are now looking at understanding the intra-household dynamics to capture the differentiated interests within the house-hold (Adimassu, 2015; Agarwal, 1997; Alderman, 1995; Bastidas, 1999; Doss, 2009; Kamo, 2000; Udry, 1996). Such norms on who a farmer is are important in that they also have implications on access to services and knowledge targeting. Access to agricultural extension and modern information communication technology depends on different education levels. According to the UNESCO, 9 million girls drop out of school in Africa compared to 6 million boys (UNESCO, 2021). Less educated women will have a disadvantage in accessing, understanding and using modern technologies and accessing agricultural extension more broadly.

8.4.4 Norms on what type of livestock women should own

It is usually the norm that men own bigger livestock, with women owning smaller livestock. However, once there are better commercialization opportunities for the smaller livestock, men will also start taking over access and control over such livestock. The findings in Burkina Faso clearly show that men owned more large livestock than smaller livestock, especially when the financial gain from small live-stock increased. The underlying assumption is that women can be responsible for livestock with lower returns. This is linked to the norms of toxic masculinity, which makes boys providers who should control and decide upon the most productive assets (Harrington, 2021).

8.5 DISCUSSION

The African Union and the member countries are advocating for a path of agricul-tural transformation to improve agricultural production, improve available food for consumption and for the market to support the much-needed economic development.

Whilst this is a noble approach, the provided solutions must not leave women behind. Emphasizing the technical solutions of increasing agricultural productivity without addressing the underlying norms that disadvantage women will not be as successful as the planners envisioned. Social transformation needs to accompany agricultural transformation. It is also important to understand that the call for building back better after COVID-19 must be grounded on the co-production of knowledge for resilient and sustainable agricultural transformation. Such co-production will address the norms on, for instance, access to land and water. Lack of access to land and water also has implications on access to agricultural extension services, which, in some instances, is largely based on patriarchy, assuming that a farmer is a man (Peters, 2013). Despite the feminization of agriculture due to migration by men, most extension departments still view men as the 'farmers.' Field observation in Malawi also showed that even female agricultural extension workers needed to be educated on why gender and intersectionality matter (Mapedza, 2017).

In the socially inclusive building back better, a GTA needs to form the basis for such an agricultural-driven transformation. A Gender Transformative Approach will go beyond the binary of men and women towards an intersectionality approach, which will help focus on women's multiple disadvantages in agriculture in Sub-Saharan Africa (IWMI, 2020). This will help inform new inclusive landscapes. A GTA will contribute by removing three types of barriers. Firstly, a GTA approach will address the challenge of entry barriers. These include socially defined roles, access to assets and complex intersectional inequalities such as gender, age, class, ethnicity, caste, and disability. The second type of barrier is structural barriers, including masculinity and cultures of privilege and hierarchy. The third systemic barriers include climate challenge, knowledge, technology distance and language (IWMI, 2020).

The chapter also highlights the importance of collecting nuanced and feminist political ecology sex-disaggregated data. To understand the transformation process, sex-disaggregated gender data will also help measure changes in norms over time. Whilst there is a better understanding of internal household dynamics, land and water ownership rights and decision-making processes (Meinzen-Dick, 2017; Peters, 2019), these are still case study specific. To better understand internal household dynamics, sex-disaggregated data will need to be collected to better inform the agricultural social transformation agenda (Doss, 2013; Peters, 2019). According to Peters (2019), such an understanding will shine a light on kinship, descent and marriage, which interact with norms, culture, economy, and politics. This chapter further reinforces the importance of understanding norms and their relevance for making structural changes to incorporate gender in the agricultural transformation process. This calls, as Quisumbing et al. (2019) argue, for transforming the agricultural system to serve women farmers better. This goes beyond the past calls which put the burden on women to change themselves to be amenable to a patriarchal-based food system.

8.6   CONCLUSION

This chapter highlights the importance of understanding norms in agricultural transformation. An intersectionality approach grounded within a GTA will impact agricultural transformation more. Whilst this chapter notes some initiatives to include gender, a much more GTA needs to be part of the solution as it questions the norms and cultural practices which deny women access and opportunities within the agricultural sector. For some women who might have access to agricultural resources, control over the benefits and decisions is still dominated by men. Sex-disaggregated data at the intra-household level, collected by researchers, national statistics offices, and development partners, would be an important resource for agricultural development. Such sex-disaggregated data will be a key resource for monitoring how gender inequalities are being addressed in agriculture, beginning from the intra-household level to community, sub-national, national and global levels. The best way forward advocates for the change of norms that produce less favourable outcomes for women and is a setback towards gender equality. The chapter acknowledges that gender norms alone will not transform the agricultural landscape. The policy implication is that agricultural-driven transformation in Africa has to be socially inclusive, especially if lessons are to be drawn from the COVID-19 inequality fault lines and the feminist political ecology insights. Going beyond technical solutions by addressing the norms, underlying beliefs, and cultural barriers that undermine women's efforts to access agricultural resources such as land and water will lead to gender-inclusive agricultural development.

REFERENCES

Adams, E.A., Juran, L., Ajibade, I. (2018) Spaces of exclusion' in community water governance: A feminist political ecology of gender and participation in Malawi's Urban Water User Associations. *Geoforum* 95, 133–142.

Adimassu, Z., Langan, S., Johnston, R. (2015) Understanding determinants of farmers' investments in sustainable land management practices in Ethiopia: Review and synthesis. *Environment, Development and Sustainability* 18, 1005–1023.

Agarwal, B. (1997) "Bargaining" and gender relations: In and beyond the household. *Feminist Economics* 3, 1–51.

Alderman, H., Hoddinott, J., Haddad, L., Udry, C. (1995) Gender differentials in farm productivity: Implications for household efficiency and agricultural policy IFPRI, Washington DC.

Allegretti, A. (2018) Respatializing culture, recasting gender in peri-urban sub-Saharan Africa: Maasai ethnicity and the 'cash economy' at the rural-urban interface, Tanzania. *Journal of Rural Studies* 60, 122–129.

Altieri, M.A., Nicholls, C.I. (2021) Agroecology and the reconstruction of a post-COVID-19 agriculture. *Journal of Peasant Studies* 47, 881–898.

Anthony, K.M. (2020) Making COVID-19 prevention etiquette of social distancing a reality for the homeless and slum dwellers in Ghana: Lessons for consideration. *Local Environment* 25, 536–539.

Archambault, C.S. (2016) Re-creating the commons and re-configuring Maasai women's roles on the rangelands in the face of fragmentation. *International Journal of the Commons* 10, 728–746.

Aredo, D., Peden, D., Taddese, G. (2006) Gender, irrigation and livestock: Exploring the nexus. ILRI, Nairobi.

Aregu, L., Choudury, A. Rajaratnam, S., Locke, C., McDougall, C. (2018) Gender norms and agricultural innovation: Insights from six villages in Bangladesh. *Journal of Sustainable Development* 11, 270–287.

Badstue, L., Elias, M., Kommerell, V., Petesch, P., Prain, G., Pyburn, R., Umantseva, A. (2020) Making room for manoeuvre: Addressing gender norms to strengthen the enabling environment for agricultural innovation. *Development in Practice* 30, 541–547.

Bastidas, E.P. (1999) *Gender issues and women's participation in irrigated agriculture: the case of two private irrigation canals in Carchi, Ecuador.* International Water Management Institute (IWMI). Colombo, Sri Lanka, pp. v, 21p.

Behrman, J.A., Meinzen-Dick, R.S., Quisumbing, A.R. (2012) The gender implications of large-scale land deals. *Journal of Peasant Studies* 39, 49–79.

Bischot, K. (2017) *Patriarchy, colonialism and gender: A long shadow of British colonial institutions amongst the Kenyan Maasai.* Leiden University College, The Hague Universiteit Leiden, The Hague.

Carastathis, A. (2014) The concept of intersectionality in feminist theory. *Philosophy Compass* 9, 304–314.

Cardwell, R.G., Pascal, L. (2020) COVID-19 and International Food Assistance: Policy proposals to keep food flowing. *World Development* 135, 105059.

Chant, S., Gutmann, M.C. (2002) Men-streaming' gender? Questions for gender and development policy in the twenty-first century. *Progress in Development Studies* 2, 269–282.

Clapp, J., Moseley, W.G. (2020) This food crisis is different: COVID-19 and the fragility of the neoliberal food security order. *The Journal of Peasant Studies* 47, 1393–1417.

Cleaver, F. (2002) *Masculinities matter! Men, gender, and development.* ZED Books, London.

Cole, S.M., van Koppen, B., Puskur, R., Estrada, N., DeClerck, F., Baidu-Forson, J.J., Remans, R., Mapedza, E., Longley, C., Muyaule, C., Zulu, F. (2014) *Collaborative effort to operationalize the gender transformative approach in the Barotse Floodplain.* WorldFish, Penang, Malaysia.

Connell, R. (2020) COVID-19 sociology. *Journal of Sociology* 00, 1–7.

Corburn, J., Vlahov, D., Mberu, B., Riley, L., Caiaffa, W.T., Rashid, S.F., Ko, A., Patel, S., Jukur, S., Martínez-Herrera, E., Jayasinghe, S., Agarwal, S., Nguendo-Yongsi, B., Weru, J., Ouma, S., Edmundo, K., Oni, T., Ayad, H. (2020) Slum health: Arresting COVID-19 and improving well-being in urban informal settlements. *Journal of Urban Health* 97, 348–357.

Crenshaw, K. (1989) Demarginalizing the intersection of race and sex: A black feminist critique of antidiscrimination doctrine, feminist theory and antiracist politics. *University of Chicago Legal Forum* 1989(1), 139–167.

Daley, E., Grabham, J., Narangerel, Y., Ndakaru, J. (2021) *Women and community land rights: Investing in local champions.* Mokoro Ltd, with HakiMadini (Tanzania) and PCC (Mongolia), Oxford, UK.

Danson, M., Burnett, K.A. (2021) Current Scottish land reform and reclaiming the commons: Building community resilience. *Progress in Development Studies*, 21(3), 280–297.

de Bont, C., Veldwisch, G.J. (2020) State engagement with farmer-led irrigation development: Symbolic irrigation modernisation and disturbed development trajectories in Tanzania. *The Journal of Development Studies* 56, 2154–2168.

de Jong, I.H., Arif, S.S., Gollapalli, P.K.R., Neelam, P., Nofal, E.R., Reddy, K.Y., Röttcher, K., Zohrabi, N. (2020) Improving agricultural water productivity with a focus on rural transformation. Irrigation and Drainage Special Issue: Development for water, food and nutrition security in a competitive environment. Selected papers of the 3rd World Irrigation Forum, Bali, Indonesia, 458–469.

Dembele, Y., Yacouba, H., Keita, A., Sally, H. (2012) Assessment of irrigation system performance in south-western Burkina Faso. *Irrigation and Drainage* 61(63), 306–315.

de Wit, M.M. (2021) What grows from a pandemic? Toward an abolitionist agroecology. *The Journal of Peasant Studies* 48, 99–136.

Doss, C. (2013) *Data needs for gender analysis in agriculture.* IFPRI, Washington DC.

Doss, C., Meinzen-Dick, R. (2009) Collective action within the household. Feminist Economics Annual Coneference.

Doss, C., Meinzen-Dick, R. (2020) Land tenure security for women: A conceptual framework. *Land Use Policy* 99, 105080.

FAO (2011) *Women in Agriculture: Closing the gender gap for development.* FAO, Rome.

FAO (2017) *The state of food and agriculture: Leveraging food systems for inclusive rural transformation.* Food and Agriculture Organisation of the United Nations, Rome.

Farnworth, C.R. (2010) Gender aware approaches in agricultural programmes: A study of Sida-supported agricultural programmes. Swedish International Development Cooperation Agency (Sida) Evaluation Sundbyberg, Sweden, p. 104.

Franco, I.D., Ortiz, C., Samper, J., Millan, G. (2020) Mapping repertoires of collective action facing the COVID-19 pandemic in informal settlements in Latin American cities. *Environment & Urbanization* 1, 1–24.

Grashuis, J., Stanley, K.D. (2021) Design principles of common property institutions: The case of farmer cooperatives in the Upper West Region of Ghana. *International Journal of the Commons* 15, 50–62.

Gray, L.C., Kevane, M. (1999) A woman's field is made at night: Gendered land rights and norms in Burkina Faso. *Feminist Economics* 5, 1–26.

Harrington, C. (2021) What is "toxic masculinity" and why does it matter? *Men and Masculinities* 24, 345–352.

Holdo, M. (2020) Contestation in participatory budgeting: Spaces, boundaries, and agency. *American Behavioral Scientist* 64, 1348–1365.

Hupkau, C., Petrongolo, B. (2020) COVID-19 and gender gaps: Latest evidence and lessons from the UK. https://cepr.org/voxeu/columns/covid-19-and-gender-gaps-latest-evidence-and-lessons-uk.

IWMI (2020) *IWMI Gender and Inclusion Strategy 2020–2023: New landscapes of water equality and inclusion.* International Water Management Institute (IWMI), Colombo, Sri Lanka, p. 16.

Jones-Casey, K., Knox, A., Chenitz, Z. (2011) *Women, inheritance and Islam in Mali, focus on land in Africa.* WRI, Washington D.C, USA, Gates Open Research, 3(1372), 1372.

Kamo, Y. (2000) "He said, she said": Assessing discrepancies in husbands' and wives' reports on the division of household labor. *Social Science Research* 29, 459–476.

Lawless, S., Cohen, P.J., McDougall, C.L., Orirana, G., Siota, F., Doyle, K. (2019) Gender norms and relations: Implications for agency in coastal livelihoods. *Maritime Studies* 18, 347–358.

Leder, S., Clement, F., Karki, E. (2017) Reframing women's empowerment in water security programmes in western Nepal. *Gender and Development* 25(22), 235–251.

Leder, S., Sachs, C.E. (2019) *Gender, agriculture and agrarian transformations.* Routledge, London.

Leigh, J. (2020) 'It was the best of times; it was the worst of times': The impact of Covid-19 on families in the child protection process. *Qualitative Social Work* 19(5–6), 1–5.

Levine, M., Meriggi, N., Mobarak, A.M., Ramakrishna, V., Voors, M. (2021) *How COVID-19 is making gender inequality worse in low-income countries—and what to do about it.* Yale Insights.

Liegeois, F., Baudron, F. (2020) Fixing our global agricultural system to prevent the next COVID-19. *Outlook on Agriculture* 49, 111–118.

Manfre, C., Rubin, D., Allen, A., Summerfield, G., Colverson, K., Akeredolu, M. (2013) *Reducing the gender gap in agricultural extension and advisory services: How to find the best fit for men and women farmers modernising extension and agricultural sevices (MEAS).* USAID, Washington DC.

Mapedza, E. (2013) Why gender matters: For farming within the Limpopo River Basin. *AGRIDEAL Magazine*, 2(September), 83–85.

Mapedza, E., Amarnath, G., Matheswaran, K., Nhamo, L. (2019) Drought and the gendered livelihoods implications for smallholder farmers in the Southern Africa Development Community [SADC] region, in: Mapedza, E., Tsegai, D., Bruntrup, M., McLeman, R. (Eds.), *Drought challenges: Policy options for developing countries.* Elsevier, Amsterdam, Netherlands, pp. 87–99.

Mapedza, E., Amede, T., Geheb, K., Peden, D., Boelee, E., Demissie, S., van Hoeve, E., Van Koppen, B. (2008) Why gender matters: Reflections from the livestock-water productivity research project. *Fighting Poverty through Sustainable Water Use*, CGIAR Research Project No. 614-2016-40926, pp. 10–14.

Mapedza, E., Tagutanazvo, E.M., van Koppen, B., Manyamba, C. (2017) Agricultural water management in matrilineal societies in Malawi: Landownership and implications for collective action, in: Suhardiman, D., Nicol, A., Mapedza, E. (Eds.), *Water governance and collective action – Multi scale challenges*. Routledge, Taylor and Francis Group, London and New York, pp. 82–95.

McKinsey (2020) *COVID-19 and gender equality: Countering the regressive effects*. McKinsey Global Institute, New York, USA.

Meine, C. (2020) Peering through the portal: COVID-19 and the future of agriculture. *Agriculture and Human Values* 37, 563–564.

Meinzen-Dick, R., Bakker, M. (1999) Irrigation systems as multiple-use commons: Water use in Kirindi Oya, Sri Lanka. *Agriculture and Human Values* 16(13), 281–293.

Meinzen-Dick, R.S., Quisumbing, A., Doss, C.R., Theis, S. (2017) Women's land rights as a pathway to poverty reduction a framework and review of available evidence, IFPRI Discussion Paper 01663. IFPRI, Washington DC.

Molden, D.J. (2007) Summary for Decision-makers, in: Molden, D. (Ed.), *Water for food, water for life: A comprehensive assessment of water management in agriculture*. Earthscan & Colombo: International Water Management Institute, London, pp. 1–37.

Mooi-Reci, I., Risman, B.J. (2021) The gendered impacts of COVID-19: Lessons and reflections: Editorial. *Gender & Society* 35, 161–167.

Mudege, N.N., Mdege, N., Abidin, P.E., Bhatasara, S. (2017) The role of gender norms in access to agricultural training in Chikwawa and Phalombe, Malawi. *Gender, Place & Culture* 24, 1689–1710.

Muturi, J., Oggema, S., van Veldhuizen, L., Mehari, A., Veldwisch, G.J. (2019) *Accelerating farmer-led irrigation development: Theory and practice of the smart water for agriculture in Kenya project*. SNV, SWA Kenya, SNV Netherlands Development Organisation, Nairobi.

Narayan, U. (1998) Essence of culture and a sense of history: A feminist critique of cultural essentialism. *Hypatia* 13, 86–106.

Nightingale, A.J. (2011) Bounding difference: Intersectionality and the material production of gender, caste, class and environment in Nepal. *Geoforum* 42, 153–162.

Njuki, J., Eissler, S., Malapit, H., Meinzen-Dick, R., Bryan, E., Quisumbing, A. (2021) *A review of evidence on gender equality, women's empowerment, and food systems*. IFPRI, Washington DC, p. 28.

Njuki, J., Sanginga, P.C. (2013) *Women, livestock ownership, and markets: Bridging the gender gap in Eastern and Southern Africa*. Routledge, Abingdon, Oxon and New York; International Development Research Centre, Ottawa.

O'Hara, C., Clement, F. (2018) Power as agency: A critical reflection on the measurement of women's empowerment in the development sector. *World Development* 106, 111–123.

Osewe, M., Liu, A., Njagi, T. (2020) Farmer-led irrigation and its impacts on smallholder farmers' crop income: Evidence from Southern Tanzania. *International Journal of Environmental Research and Public Health* 17, 1–13.

Ostrom, E. (1990) *Governing the commons: The evolution of institutions for collective action*. Cambridge University Press, Cambridge.

Ostrom, E., Dietz, T., Dolasak, N., Stern, P., Stonich, S., Weber, E. (2002) *The drama of the commons*. National Academy Press, Washington DC.

Panel, M.M. (2021) Connecting the dots: Policy innovations for food systems transformation in Africa, A Malabo Montpellier Panel Report 2021. Akademiya 2063.

Parsons, I., Lombard, M. (2017) The power of women in dairying communities of eastern and southern Africa. *Azania: Archaeological Research in Africa* 52, 33–48.

Pearse, R., Connell, R. (2016) Gender norms and the economy: Insights from social research. *Feminist Economics* 22, 30–53.

Peters, P. (2013) Land appropriation, surplus people and a battle over visions of agrarian futures in Africa. *Journal of Peasant Studies* 40, 537–562.

Peters, P. (2019) Revisiting the social bedrock of kinship and descent in the anthropology of Africa, in: Richard Grinker, S.C.L., Steiner, C.B., Gonçalves, E. (Eds.), *A companion to the anthropology of Africa*, First Edition. John Wiley & Sons, Inc., New York, USA, pp. 33–62.

Peters, P.E. (2002) Gender and broadening access to land and waer in Southern Africa. Broadening Access and Strengthening Input Market Systems.

Peters, P.E. (2020) Land grabs: The politics of the land rush in Africa, in: Cheeseman, N. (Ed.), *The encyclopedia of African politics*, Published ahead of print, 2018, oxfordre.com/politics. ed. Oxford University Press, New York, pp. 855–879.

Petesch, P., Badstue, L., Prain, G. (2018) *Gender norms, agency, and innovation in agriculture and natural resource management: The GENNOVATE methodology*. CIMMYT, Mexico.

Pica-Ciamarra, U., Otte, J., Chilonda, P. (2007) Livestock policies, land and rural conflicts in Sub-Saharan Africa. *Land Reform* 1, 19–33.

Prindex (2020) Women's perceptions of tenure security: Evidence from 140 countries. ODI, UK.

Quisumbing, A., Meinzen-Dick, R., Njuki, J. (2019) Gender equality in rural Africa, from commitments to outcomes. Annual Trends and Outlook Report (ATOR) IFPRI ReSAKSS, Washington DC.

Quisumbing, A.R., Kumar, N. (2014) Land rights knowledge and conservation in Rural Ethiopia: Mind the gender gap. International Food Policy Research Institute (IFPRI) Discussion Paper No. 01386. IFPRI, Washington, DC.

Quisumbing, A.R., Rubin, D., Manfre, C., Waithanji, E., van den Bold, M., Olney, D., Johnson, N., Meinzen-Dick, R. (2015) Gender, assets, and market-oriented agriculture: Learning from high-value crop and livestock projects in Africa and Asia. *Agriculture and Human Values* 32, 705–725.

Ragasa, C., Berhane, G., Tadesse, F., Taffesse, A.S. (2013) Gender differences in access to extension services and agricultural productivity. *The Journal of Agricultural Education and Extension* 19, 437–468.

Rap, E., Jaskolski, M.S. (2019) The lives of women in a land reclamation project: Gender, class, culture and place in Egyptian land and water management. *International Journal of the Commons* 13, 84–104.

Rose, C.M. (2020) Thinking about the commons. *International Journal of the Commons* 14, 557–566.

RRI (2015) Who owns the world's land? A global baseline of formally recognized indigenous and community land rights Google Scholar. Rights Resources Institute, Washington DC.

Ryan Cardwell, P.L.G. (2020) COVID-19 and international food assistance: Policy proposals to keep food flowing. *World Development* 135, 105059.

Saba, C.K.S. (2020) COVID-19: Implications for food, water, hygiene, sanitation, and environmental safety in Africa: A case study in Ghana. University of Development Studies, Tamale, Ghana, Department of Biotechnology, Faculty of Agriculture, University for Development Studies.

Schlager, E., Ostrom, E. (1992) Property rights regimes and natural resources: A conceptual analysis. *Land Economics* 68, 249–262.

Schreiner, B., Mohapi, N., van Koppen, B. (2004) Washing away poverty: Water, democracy and gendered poverty eradication in South Africa. *Natural Resources Forum* 28(23), 171–178.

Scoones, I. (2020) Pastoralists and peasants: Perspectives on agrarian change. *The Journal of Peasant Studies*, 48, 1–47.

Scoones, I.C. (1990) *Livestock populations and the household economy: A case study from southern Zimbabwe*. Imperial College London (University of London).

Seguino, S. (2007) PlusCa change? Evidence on global trends in gender norms and stereotypes. *Feminist Economics* 13, 1–28.

Sokile, C.S., van Koppen, B. (2004) Local water rights and local water user entities: The unsung heroines of water resource management in Tanzania. *Physics and Chemistry of the Earth, Parts A/B/C* 29(15–18), 1349–1356.

Suhardiman, D. (2013) The power to resist: Irrigation management transfer in Indonesia. *Water Alternatives* 6(1), 25–41.

Suhardiman, D. (2016) Irrigation management transfer and the shaping of irrigation realities in Indonesia: From means to empower farmers to a tool to transfer rent seeking? *Human Organization* 75(74), 326–335.

Sundberg, J. (2017) Feminist political ecology, in: Richardson, D., Castree, N., Goodchild, M. F., Kobayashi, A., Liu, W., Marston, R.A. (Eds.), *The international encyclopedia of geography*. John Wiley & Sons, Ltd, New York, p. 9.

Tavenner, K., Crane, T.A. (2018) Gender power in Kenyan dairy: Cows, commodities, and commercialization. *Agriculture and Human Values* 35, 701–715.

Tavenner, K., Crane, T.A. (2019) Beyond "women and youth": Applying intersectionality in agricultural research for development. *Outlook on Agriculture* 48, 316–325.

Udry, C. (1996) Gender, agricultural production, and the theory of the household. *Journal of Political Economy* 104(5), 010–1046.

UNESCO (2021) *Education in Africa*. UNESCO Institute for Statistics, Montreal, Quebec, Canada.

Union, A. (2020) *Framework for irrigation development and agricultural water management in Africa*. African Union, Addis Ababa.

UN Women (2020) *From insight to action: Gender equality in the wake of COVID-19*. UN Women, New York, USA, p. 17.

van den Bold, M., Dillon, A., Olney, D., Ouedraogo, M., Pedehombga, A., Quisumbing, A. (2015) Can integrated agriculture-nutrition programmes change gender norms on land and asset ownership? Evidence from Burkina Faso. *The Journal of Development Studies* 51, 1155–1174.

Van Hoeve, E., van Koppen, B. (2006) Beyond fetching water for livestock: A gendered sustainable livelihood framework to assess livestock-water productivity, ILRI working paper number 1 ILRI, Nairobi.

van Koppen, B., Boelee, E., Bekele, S. (2005) Integrating water needs for livestock-based livelihoods in multiple-use water development approaches: State of the art and knowledge gaps. International Water Management Institute, 2 September 2005.

van Koppen, B., Hussain, I. (2007) Gender and irrigation: Overview of issues and options. *Irrigation and Drainage* 56, 289–298.

Victor, B., Fischer, E.F., Cooil, B., Vergara, A., Mukolo, A., Blevins, M. (2013) Frustrated freedom: The effects of agency and wealth on wellbeing in rural Mozambique. *World Development* 47, 30–41.

Viruell-Fuentes, E.A., Miranda, P.Y., Abdulrahim, S. (2012) More than culture: Structural racism, intersectionality theory, and immigrant health. *Social Science & Medicine* 75, 2099–2106.

Waldman, L., with contributions from Barrance, A., Ramos, R.F.B., Gadzekpo, A., Mugyenyi, O., Nguyen, Q., Tumushabe, G., Stewart, H. (2005) Environment, politics, and poverty: Lessons from a review of PRSP stakeholder perspectives. Synthesis review. Institute of Development Studies, Study initiated under the Poverty Environment Partnership (PEP), and jointly funded and managed by the Canadian International Development Agency (CIDA), Department for International Development (DFID) of the United Kingdom, and German Technical Cooperation Agency (GTZ), pp. ix, 34.

Woodhouse, P., Jan Veldwisch, G., Venot, J.-P., Brockington, D., Komakech, H., Manjichi, Â. (2017) African farmer-led irrigation development: Re-framing agricultural policy and investment? *The Journal of Peasant Studies* 44, 213–233.

World Bank (2018) Innovation, entrepreneurship, positive change. Join the farmer-led irrigation revolution. World Bank, Washington DC.

Zibani, T. (2016) The Triple burden and triple role of women. Empowerwomen.org, https://www.empowerwomen.org/en/community/discussions/2016/11/the-triple-burden-and-triple-role-of-women.

Zwarteveen, M.Z. (1996) A plot of one's own: Gender relations and irrigated land allocation policies in Burkina Faso. International Irrigation Management Institute (IIMI). Colombo, Sri Lanka.

9 Sustainable water management
Does gender matter?

Dalia Saad

9.1 INTRODUCTION

Women have historical and traditional ties with water by virtue of their domestic functions. Playing the role of a provider, women suffer more than men from water scarcity and pollution, particularly in developing countries (Bhattacharya, 2016; Richman Gambe, 2019). As a result of the time spent finding and carrying water, women and girls have less time to improve their lives through education and productive activities (Silva et al., 2020), and that is how the cycle of poverty sustains itself in developing countries (Richman Gambe, 2019; van Houweling et al., 2012). However, it is important to expand the discussion of water and gender beyond the household level and not to limit our objectives to reducing the time and burdens associated with water collection. The ultimate goal should be women's inclusion "along with men" in every aspect of water management.

In addition to their role in providing water for domestic use, women also have notable "but often invisible" roles in productive uses of water, such as agriculture. According to the FAO report (2011), women are responsible for half of the world's food production. In most developing countries, they account for 43% of the agricultural labour and produce 60%–80% of the food. Yet, they do not have adequate access to resources, information, and credit and own only 1% of the world's land (Doss et al., 2018; Khandker et al., 2020; Raney et al., 2011).

Water projects that overlook the central role of women in water management and exclude them bypass not only half the population but also reduce the projects' efficiency, effectiveness, and sustainability (Doss, 2018). As a result of their dependence on water resources and their role as water providers, women have accumulated rich knowledge and skills about water resources; however, because of their limited participation in decision-making, this valuable knowledge is often lost (Brewster et al., 2006). Societal and cultural values that determine men as heads of households and main decision-makers in the public sphere marginalize women's views and preferences. Despite this, it is important to avoid addressing gender in water management as a woman's agenda. Rather, it should be about the involvement of both genders to achieve equal rights, opportunities, and access to water and decision-making positions (Najjingo Mangheni et al., 2021).

DOI: 10.1201/9781003327615-9

The lack of men's awareness of women's specific needs with regards to, e.g., menstrual hygiene management, as well as the underrepresentation of women in projects, often lead to women's specific needs being ignored or badly translated into technology/policies (Bhattacharya, 2016; Dickin and Caretta, 2022; Doss, 2018; Nigussie et al., 2017). Gender-sensitive strategies are therefore critical in creating a framework of cooperation between men and women by adopting the insights and abilities of both in shaping programmes and meeting project objectives. This optimizes social and economic development and reduces competition and conflicts over water resources (Sülün, 2018).

9.2   UNDERSTANDING GENDER IN THE CONTEXT OF WATER

The relationship between people and water is not gender-neutral; people have different needs, interests and access and use of water resources based on several factors, including gender. Both women and men are natural and essential agents for sustainable water management; thus, considering the potential is crucial for socio-economic and sustainable water management (Najjingo Mangheni et al., 2021). However, due to different roles in water supply and food production, women and men have different priorities, demands and knowledge of water management (Saad et al., 2017). For example: (i) women prefer to have domestic water supply and irrigation structures close to their households to effectively manage both their productive and domestic responsibilities, whereas men are more mobile and flexible (Saad, 2019), and (ii) women and men have different productive uses, for instance, women are mostly responsible for subsistence agricultural production (low-value crops that are harvested in small quantities over long periods for low regular income), while men are most likely engaged in commercial agricultural production (high-value crops that are harvested in bulk and sold in larger quantities) (Bjornlund et al., 2019; Drechsel et al., 2017; Njuki et al., 2014). This disparity in roles and interests creates differences in their respective needs; the best practice is, therefore, to acknowledge these differences in water resource management strategies (Elias, 2015). However, it is important to note that gender-based roles can also be context-specific and may differ from region to region and/or country to country (Drechsel et al., 2006). Applying a gender lens in analysing, designing and implementing a gender-sensitive approach in water management programmes is imperative if effective outcomes are to be achieved (Doss, 2018). Involving women and men in water management planning often makes for fewer oversights in technical planning, improves resource and financial management, and allows for greater transparency (Doss, 2018; Khandker et al., 2020; Macarthur et al., 2020).

Gender is not yet mainstreamed into water research and water resource management. Current approaches to research, urban planning, and development projects concerning water are highly segregated, focusing on technical improvements and infrastructure with limited attention to the social aspects of water issues regarding users (Macarthur et al., 2020). In addition to environmental and economic sustainability elements, various sociological factors should be considered, including sociocultural respect, community participation, public acceptance, and, of most importance, gender roles. Gender, among others, is influential in achieving sustainable water management (Saad et al., 2017). Water management is always perceived as not having a gender dimension. Still, the fact is that the whole process, including technology choices, decision-making, implementation, benefits and risks, are all gendered. For instance, men always develop technologies with male users' interests in mind "by default". As a result, most developed technologies don't cater to differences in usage patterns and, therefore, are not women-friendly (Bhattacharya, 2016; Dickin and Caretta, 2022; Doss, 2018; Nigussie et al., 2017). This is an obvious outcome when infrastructures are designed and developed without consideration of their gender impact. In a sense, facilities are more likely to be technically appropriate, well-used, and maintained when women and men are considered potential stakeholders.

9.3   WOMEN AND WATER "THE UNTAPPED CONNECTION"

Women are prime domestic water users; they fulfil important roles in providing drinking water and water for other household purposes; the burden of fetching water from outdoor sources historically falls on them. They collect, draw, transport, store and use water for preparing food, washing, cleaning, and hygiene. This is particularly true in developing countries where water is often not pipped directly into houses (Garcia, 2019; Richman Gambe, 2019). Studies from developing countries reported that in two-thirds of households without a water source on the premises, women and girls collect water, walk many hours, and spend four to six hours a day. When children are involved in collecting water, girls are found to be twice as likely as boys to be responsible (Dickin and Caretta, 2022; Richman Gambe, 2019).

Besides domestic uses, women also need water for productive uses alongside men. They play key roles in agriculture, fisheries, and livestock. Women are often responsible for specific crops, and they take responsibility for specific farming tasks such as weeding and transplanting (Drechsel et al., 2017; Obuobie et al., 2004). As part of their agricultural work, they contribute to retaining plant and animal species, conserving genetic resources, and retaining indigenous knowledge. These multiple/divergent uses of water exert additional pressure on women and enrich them with special knowledge, experience and skills (Adoukonou-Sagbadja et al., 2006; Olango et al., 2014). For instance, women are more knowledgeable than men about the location, reliability, quality, and seasonal variation of local water resources and soil conditions. They gain these knowledge/skills through personal experiences and interpersonal and intergenerational contacts with other women (see Box 9.1).

**BOX 9.1   WOMEN'S VALUABLE KNOWLEDGE
ABOUT WATER LOCATION**

In Burkina Faso, the participation of women added a special value to the success of water projects. They found to have information on the year-round reliability of traditional water sources, whereas village chiefs, men and elders lacked such knowledge.

Source: Narayan (1995).

Traditionally, women are responsible for managing and maintaining the communal water supply. This is quite common in many African communities where women regulate and control the use and maintenance of water resources. For example, women restrict cattle watering to particular sites and washing to specific downstream sites on the river (Yanıkkaya and Nairn, 2021). However, these tasks are performed informally, and thus, women are not involved in strategic planning and decision-making. For example, in India, most of the work women perform is informal, and more than 50% is unpaid (Taron et al., 2021).

Additionally, women are often responsible for finding alternatives and solving problems related to water and food supply. A few examples include (i) modifying farming practices when crop yield is low due to soil exhaustion, (ii) developing alternative strategies in response to soil deterioration and erosion, (iii) evaluating water sources and analysing supply patterns, and (iv) lobbying relevant authorities and organize protests when water availability reaches dire levels (Saad, 2019). As key actors in water management, their full participation in all spheres of the water management cycle is fundamental for achieving sustainable solutions (Caruso et al., 2015; Caruso and Sinharoy, 2019). In a sense, involving the people directly engaged in collecting, using, managing and developing water resources from the household level and up makes a significant difference in terms of short-term effectiveness and long-term sustainability.

9.4   WATER AND WOMEN'S WELFARE

Because of the intertwined relationship between women and water, water resources directly affect the welfare of women, and thus, any improvements in water resource management practices will ultimately reflect on women's welfare (Caruso et al., 2018). Water availability in quantity and quality directly impacts the life and health of women and girls in developing countries. For instance, regarding quality, women are at higher risk and most vulnerable to water-related diseases. Many infectious diseases associated with water pollution are reported among the fifth biggest killer of women worldwide (Pouramin et al., 2020). Water pollution is also directly linked to maternal and child mortality and sexual violence. Additionally, many women in developing countries give birth at home without access to clean water, exposing themselves and their babies to infections (Svetanoff and Ilobodo, 2022).

In terms of quantity, due to the time spent collecting water and handling other problems related to water scarcity, women cannot work or have an income, which is also a major reason girls drop their education (Silva et al., 2020). Simple improvements like providing water closer to the households allow them to improve their lives through education and work to generate their income. For example, it will increase girls' free time and boost their school attendance (see Box 9.2).

Women suffer more than men from water scarcity and pollution. Because they are the household caretakers, they are more concerned about health issues and other problems associated with water scarcity (Stevenson et al., 2016). Studies revealed that women participate more effectively and diligently in water management programmes (Box 9.3). It is simply a self-interest; nevertheless, it benefits the entire community.

BOX 9.2 EFFECT OF WATER SUPPLY ON GIRLS' EDUCATION

In Morocco, the Rural Water Supply and Sanitation Project of the World Bank aimed to reduce the burden of girls traditionally involved in fetching water to improve their school attendance. In the six provinces where the project is based, **it was found that girls' school attendance increased by 20%, which was attributed to the fact that girls spent less time fetching water. At the same time, convenient access to safe water reduced the time spent collecting water by women and young girls by 50%–90%.**

In the Ejura-Sekyedumasi, Ghana, World Vision Ghana (WVG) initiated the Ghana Rural Water Project (GRWP) to address a serious infestation of guinea worms and poor access to potable drinking water. **The project has shifted from a strictly technology-driven approach to a community-based, people-oriented, demand-driven focus, including gender mainstreaming, poverty alleviation and the well-being of children.** Through the GRWP initiative, WVG supplied the village with two boreholes fitted with hand pumps, two public Ventilated Improved Pit (VIP) latrines and a urinal. The community has since identified this water project as having had a high level of community participation and gender integration. **It has improved the education of girls, who accounted for 53% of primary school students in 2005, compared to 43% in 1995.**

A study in Tanzania showed a 12% increase in school attendance when water was available within 15 min compared to more than half an hour away.

Sources: Serwah and Sam (2006); World Bank (2003).

BOX 9.3 EFFECTIVE PARTICIPATION AND BENEFITS

In **Nigeria**, constructing a tourist resort on the Obudu plateau led to deforestation and exacerbated preexisting pressures on water resources and the environment, such as overgrazing and unsustainable agricultural practices. The local Becheve women complained about wasted time collecting water, poor water quality and low-income family health. Consequently, the Nigerian Conservation Foundation started a Watershed Management Project on the Obudu plateau and encouraged women to get involved in the project's decision-making process. Women leaders were elected to the management committee, became involved in constructing and maintaining a water reservoir and showed commitment and dedication. **The reduced time spent collecting water allowed women more time to generate income through farming and marketing.** A conflict between the Becheve women and the Fulani tribesmen over access to water was resolved through negotiation, and the women were ensured timely access to water. **Moreover, the women's healthcare burden was reduced, with a 45% reduction in cases of diarrhoea in 2004.**

Source: "Gender, Water and Sanitation: A Policy Brief" (2006).

9.5 GENDER AND DECISION-MAKING POWER

Although women are defined as essential providers and users of water, their participation in developing strategies and decision-making is very limited, if any (Saad, 2019). Social and cultural values and stereotypes sustain traditional gendered norms and roles and create a lot of inequalities between women and men regarding the potential for having their voices heard. Men are considered community leaders by both men and women (Eagly et al., 2019; Stewart et al., 2021). Cultural beliefs regarding men's superiority made men uncomfortable when women were allowed to participate in meetings and decision-making (Kilsby, 2012) (see Box 9.4). This cultural bias against public participation, even when women have more experience and expertise than men, is one of the most serious impediments to women's involvement in water management.

Other problems women face include a lack of confidence, family commitment, time constraints, and, most importantly, limited control and access to productive resources, often the main determinant for decision-making at the household level and publicly (Kilsby, 2012).

BOX 9.4   BREAKING CULTURAL BARRIERS

Meetings of local councils and development committees are restricted mainly to men. A study of 18 communities in Tanzania shows that the average number of women councillors is **2 out of 25 members**. In two provinces in Colombia, the representation of women on **3,500 community development committees is between 8% and 17%**. The absence of women in decision-making organizations is also reported in **Kenya, Thailand, South Korea, and Guatemala**. According to Indian law, one-third of the members elected to the local council must now be women. A study in three villages in Madhya Pradesh reported: "Often these women did not know they were elected". Women in the lake zone in Tanzania said they often were not informed of elections and were not given voting cards.

In Hoto village, Pakistan, where women follow a strict form of purdah, a participatory action research team went to help the village improve its water management in 1994. **The men would not permit the action team to meet the village women for a year**. Eventually, the women were able to participate in a joint meeting. They proposed building a new water tank on unused land, providing water to the non-functioning public standpipes. The women's solution, which was far more cost-effective, was adopted over the men's proposal.

Moreover, after this initial success, women became active participants in decision-making, and significant changes have been made in their lives through hygiene education. Most significant has been the demand for education for their daughters. In 1998, a new girls' school was opened in Hoto. Traditional leaders have been impressed by the result of the project. The same approach is now taken in other villages.

Source: "Gender, Water and Sanitation: A Policy Brief" (2006).

> **BOX 9.5   CULTURE ADOPTION**
>
> The water-wise women's initiative in Jordan exemplifies how adopting cultural values improves water management. More than 300 women were trained in plumbing and water management skills in 15 communities in Jordan. This was particularly important because of the traditional and social values in some communities in Jordan where a male plumber cannot enter the house to fix leakages or repair broken taps and pipes in the absence of a male family member. This initiative has resulted in a 30% reduction in water losses in households. Additionally, women were provided with the opportunity to generate income and reduce expenses.
>
> *Source: Sayyed (2018).*

Thus, women are rarely represented in committees and meetings where men are known to represent the family. Even when women manage domestic water, sanitation and hygiene, they are still excluded from public infrastructure planning and public consultative processes. However, even if women can participate and engage in water management activities and have valuable knowledge and skills to share, they often lack the capacity to enable their voices and formulate their needs. Training women to gain technical skills and involve them in different capacity-building programmes is an entry point for breaking down some traditional values, cultural barriers and stereotypes (Foster, 2013; Kilsby, 2012; Sayyed, 2018; Yanıkkaya and Nairn, 2021). On the other hand, water agencies operating at the community level must be aware of cultural dynamics within the specific community. It is important to gradually disempower culturally constructed barriers; however, in some cases, it is also important to adopt them if project goals are to be met (see Box 9.5). Project managers may opt for special measures in programme design to ensure that women's demands and priorities are addressed. For instance, timing and allocation of committee meetings to consider women's other responsibilities (e.g., domestic work). Otherwise, women may choose not to participate to avoid conflict in their responsibilities. Where necessary, they can also arrange separate male and female sessions "before reaching a consensus for the group as a whole". Also, project information should be availed in a user-friendly manner and provided to both men and women.

9.6   GENDER AND CAPACITY BUILDING

Effective capacity-building programmes could bridge the gap between water managers and users. Training both men and women to gain different water management skills is crucial in supporting the success and sustainability of water services. Besides acquiring skills (both technical and other skills), capacity building can build women's self-confidence and encourage them to participate in technical and managerial roles and in the decision-making process (Caruso Id et al., 2022). It enables them to be seen as valuable stakeholders and potential decision-makers (Box 9.6).

BOX 9.6   EMPOWERING WOMEN
THROUGH CAPACITY BUILDING

The Watersheds and Gender project in El Salvador exemplifies how women learned new skills through participation and involvement. The project has promoted women as leaders and trained them as community promoters and managers of small-scale companies. As a result, women have acquired technical agricultural knowledge and are now performing tasks previously considered suitable only for men.

In India, a project to train women hand pump mechanics was successful when modified to accommodate women's specific needs. Village women also found the women mechanics more accessible and responsive than male mechanics, which enhanced the rate of preventive maintenance, with a much lower hand pump breakdown rate than male mechanics.

Source: Agua Project Report (2002); Narayan (1995).

9.7   BENEFITS OF GENDER-BALANCED WATER MANAGEMENT

Incorporating gender issues in water resources management contributes to effectiveness and sustainability (Sülün, 2018). When women and men are equally involved in decision-making, decisions and solutions represent the entire community's needs. It allows for fair water sharing and maximises social and economic benefits for the entire community. It also strengthens the contributions of women to a healthier environment and develops the community through empowering women (Garcia, 2019; Sülün, 2018). In a sense, improving water services gives women more time for production, education and empowerment activities. Thus, a gender-sensitive approach in water resources management leads to greater benefits in many terms, including:

9.7.1   ECONOMIC BENEFITS

Women's effective participation and involvement in water management boost economic production in agriculture and small industries (van Houweling et al., 2012). Irrigation methods can be improved and tailored to women's needs, thus contributing to food security and cash crop production (Doss et al., 2018; Nigussie et al., 2017; van Houweling et al., 2012). This would then contribute to the women's welfare across the four dimensions of gender-sensitive poverty alleviation factors that include (a) increase women's opportunities to access resources and gain employment, (b) increase women's capabilities and skills to perform more efficiently and gain from them, (c) strengthen women's security (their risk-bearing capacities), and (d) empower women at the household and community levels.

9.7.2   TECHNICAL SUSTAINABILITY

Infrastructure and facilities are more widely used, well maintained, and sustained when both genders are consulted (see Box 9.7). Improving water services through

BOX 9.7   GENDER-ORIENTED FACILITIES

In the Est-Mono region of Togo, where only 10% of the population has access to potable water, a project aimed at improving access to water and sanitation facilities in schools did not adequately take a gender perspective into account. Thus, the facilities did not meet everyone's needs and fell into disuse. A new project design encouraged the participation of all villagers, boy and girl students, men and women teachers and administrators. The schools and the villages approved an action plan for hygiene promotion. The project provided separate water and sanitation facilities for boys and girls and educational resources to each village school. Addressing gender imbalances among students and ensuring the entire community's participation has led to impacts far beyond the immediate results. Girls have taken a leadership role and increased their self-esteem. Gender-balanced School Health Committees are responsible for the equipment and overseeing hygiene.

Source: Gender, Water and Sanitation Case
Studies on Best Practices (2006).

social processes provides water and increases consumption, production, income generation, environmental security, and health (Theis et al., 2018).

9.7.3   SOCIAL BENEFITS AND SOCIAL SUSTAINABILITY

A larger share of community responsibility by sharing burdens and benefits between men and women increases mutual respect in the community and within the families. This unlocks enormous potential imprisoned by traditional stereotyping, challenges men's perspectives on women's roles, and ultimately changes negative perceptions about women's capabilities (Fauconnier et al., 2018). As such, natural skills will develop and flow to the surface, increasing incomes and national development. Improving social women's position also greatly affects the entire community. The contribution of half the world's population is more effectively mobilized towards other sustainable development goals (Khandker et al., 2020). For instance, because women are more concerned about health, nutrition and hygiene, their control over water use will ultimately boost the health and well-being of the entire community. It will also spread concern for nutrition, childcare and health among men (Caruso et al., 2018, 2015; Mitra and Rao, 2019; Pouramin et al., 2020).

9.7.4   WOMEN EMPOWERMENT

Putting women at the centre of water resource management empowers them by being recognised as having skills and knowledge that are outside the scope of their traditional roles (Mthiyane et al., 2016). This strengthens their voice within the family and the community to negotiate their needs. Contributing to the success and improvement of water services, women become more confident to take public leadership roles, and their relationship with men becomes more equitable (O'Reilly, 2006). This leads to changes in attitudes among both women and men and pushes women to have greater autonomy and independence (Kayser et al., 2019).

9.7.5 ENVIRONMENTAL SUSTAINABILITY

Broader social participation results in the more effective use of water resources through rehabilitation activities, pollution protection, waste reduction and water conservation. Greater women's involvement in water management facilitates freshwater ecosystem maintenance and protection. An overall improvement in water conservation, management, and supply strategies is an imperative outcome (Nigussie et al., 2017; Sülün, 2018).

9.8 CONCLUSIONS

Evidence has shown that involving women leads to greater improvements in water resources management. Services are more efficient, user-focused, financially viable, and environmentally sustainable. However, showing that water projects work better with women's involvement has a greater impact on mobilizing finance towards gender-balanced projects than highlighting the impact of water access on gender equality.

Incorporating the gender dimension enables water professionals to make informed choices/decisions during water management projects' planning, design, and implementation. It is thus very important to train professionals in the water sector to appreciate the significance of the gender dimension of water resource management. However, gender mainstreaming should be adopted as a holistic approach covering all projects' cycles from planning and design to implementation and relevant policies. This includes governmental level, management programming, research, and policies. For instance, women-specific uses and priorities should be protected through government regulations. At a project level, projects should be planned for and designed to consider both productive and domestic uses of water resources. Also, when employment opportunities are planned within a project, the recruitment process should include measures to ensure that women are informed of the opportunities and equally paid for. Further, projects should offer women equal access to technical, financial and administrative training. At the institutional level, technical staff working in research and development should be trained to integrate the gender dimension into the socio-economic aspects of research work to address the differential impacts of structural interventions and the appropriation of new technologies.

REFERENCES

Adoukonou-Sagbadja, H., Dansi, A., Vodouhe'3, R., Vodouhe'3, V., Akpagana, K., 2006. Indigenous knowledge and traditional conservation of fonio millet (Digitaria exilis, Digitaria iburua) in Togo. *Biodiversity and Conservation* 15, 2379–2395. https://doi.org/10.1007/s10531-004-2938-3.

Bhattacharya, S., 2016. Changing dimensions and interactions of water crisis and human rights in developing countries. *World Scientific News* 34, 86–97.

Bjornlund, H., Zuo, A., Wheeler, S.A., Parry, K., Pittock, J., Mdemu, M., Moyo, M., 2019. The dynamics of the relationship between household decision-making and farm household income in small-scale irrigation schemes in southern Africa. *Agricultural Water Management* 213, 135–145. https://doi.org/10.1016/J.AGWAT.2018.10.002.

Brewster, M.M., Herrmann, T.M., Bleisch, B., Pearl, R., 2006. A gender perspective on water resources and sanitation. *Wagadu* 3, ISSN: 1545–6196.

Caruso, B.A., Cooper, H.L.F., Haardörfer, R., Yount, K.M., Routray, P., Torondel, B., Clasen, T., 2018. The association between women's sanitation experiences and mental health: A cross-sectional study in Rural, Odisha India. *SSM - Population Health* 5, 257–266. https://doi.org/10.1016/J.SSMPH.2018.06.005.

Caruso, B.A., Sevilimedu, V., Chun-Hai Fung, I., Patkar, A., Baker, K.K., 2015. Gender disparities in water, sanitation, and global health. *The Lancet Global Health* 386. https://doi.org/10.1016/S0140-6736(15)61497-0.

Caruso, B.A., Sinharoy, S.S., 2019. Gender data gaps represent missed opportunities in WASH. *The Lancet Global Health* 7, e1617. https://doi.org/10.1016/S2214-109X(19)30449-8.

Caruso Id, B.A., Conrad, A., Patrick Id, M., Owens, A., Kviten, K., Zarella, O., Rogers, H., Sinharoy, S.S., 2022. Water, sanitation, and women's empowerment: A systematic review and qualitative metasynthesis. *PLOS Water* 1. https://doi.org/10.1371/journal.pwat.0000026.

Dickin, S., Caretta, M.A., 2022. Examining water and gender narratives and realities. *WIREs Water*. https://doi.org/10.1002/wat2.1602.

Doss, C., Meinzen-Dick, R., Quisumbing, A., Theis, S., 2018. Women in agriculture: Four myths. *Global Food Security* 16, 69–74. https://doi.org/10.1016/J.GFS.2017.10.001.

Doss, C.R., 2018. Women and agricultural productivity: Reframing the Issues. *Dev Policy Rev* 36, 35–50. https://doi.org/10.1111/dpr.12243.

Drechsel, P., Graefe, S., Cofie, O.O., 2006. *Informal irrigation in Urban West Africa: An overview*. International Water Management Institute, p. 40 (IWMI Research Report 102).

Drechsel, P., Hope, L., Cofie, O., 2017. Gender mainstreaming: Who wins? Gender & irrigated Urban vegetable production in West Africa. *The Journal of Gender and Water* 2, 14–17.

Eagly, A.H., Abele, A.E., Haines, E., Hentschel, T., Heilman, M.E., Peus, C.V., 2019. The multiple dimensions of gender stereotypes: A current look at men's and women's characterizations of others and themselves. *Frontiers in Psychology* 10. https://doi.org/10.3389/fpsyg.2019.00011.

Elias, F., 2015. Women's roles in integrated water resource management: A case study of the Mutale water user association, Limpopo, South Africa [WWW Document]. MSc dissertation, University of Monash. https://www.researchgate.net/publication/281026879_Women's_roles_in_integrated_water_resource_management_a_case_study_of_the_Mutale_water_user_association_Limpopo_South_Africa (accessed 7.8.22).

Fauconnier, I., Jenniskens, A., Perry, P., Fanaian, S., Sen, S., Sinha, V., Witmer, L., 2018. Women as change-makers in the governance of shared waters. Gland, Switzerland: IUCN, viii + 50pp.

Foster, T., 2013. Predictors of sustainability for community-managed handpumps in sub-Saharan Africa: Evidence from Liberia, Sierra Leone, and Uganda. *Environmental Science and Technology* 47, 12037–12046. https://doi.org/10.1021/ES402086N.

Garcia, S.L., 2019. Gender and Water [WWW Document]. Gender CC 2019. https://www.gendercc.net/fileadmin/inhalte/bilder/5_Gender_Climate/Gender-and-water-GenderCC-2019.pdf (accessed 7.12.22).

Gender, Water and Sanitation: A Policy Brief, 2006. UN Water.

Gender, Water and Sanitation Case Studies on Best Practices, 2006.

Kayser, G.L., Rao, N., Jose, R., Raj, A., 2019. Water, sanitation and hygiene: Measuring gender equality and empowerment. *Bulletin of the World Health Organization* 97(6), 438–440.

Khandker, V., Gandhi, V.P., Johnson, N., 2020. Gender perspective in water management: The involvement of women in participatory water institutions of Eastern India. *Water* 12. https://doi.org/10.3390/w12010196.

Kilsby, D., 2012. Putting a gender lens on WASH practice in Liquica, Timor-Leste. In Towards Inclusive WASH: Sharing evidence and experience from the field. WaterAid Australia.

Macarthur, J., Carrard, N., Willetts, J., 2020. WASH and gender: A critical review of the literature and implications for gender-transformative WASH research. *Journal of Water, Sanitation and Hygiene for Development* 10, 818–827. https://doi.org/10.2166/WASHDEV.2020.232.

Mitra, A., Rao, N., 2019. Gender, water, and nutrition in India: An intersectional perspective. *Water Alternatives* 12, 930–952.

Mthiyane, C., Mariga, I., Shimelis, H., Murugani, V., Morojele, P., Naidoo, K., Aphane, O., 2016. Empowerment of women through water use security, land use security and knowledge generation for improved household food security and sustainable rural livelihoos in selected areas in Limpopo. WRC Report No. 2082/1/15, Water Research Commission Report.

Najjingo Mangheni, M., Musiimenta, P., Boonabaana, B., Ann Tufan, H., 2021. Tracking the gender responsiveness of agricultural research across the research cycle: A monitoring and evaluation framework tested in Uganda and Rwanda. *Journal of Gender, Agriculture and Food Security* 6, 58–72. https://doi.org/10.19268/JGAFS.622021.4.

Narayan, D., 1995. The contribution of people's participation: Evidence from 121 rural water supply projects. ESD Occasional Paper No. 1. Washington, DC: World Bank.

Nigussie, L., Lefore, N., Schmitter, P., Nicol, A., 2017. Gender and water technologies: Water lifting for irrigation and multiple purposes in Ethiopia. International Livestock Research Institute, Nairobi, Kenya, p 36.

Njuki, J., Waithanji, E., Sakwa, B., Kariuki, J., Mukewa, E., Ngige, J., 2014. Can market-based approaches to technology development and dissemination benefit women smallholder farmers? A qualitative assessment of gender dynamics in the ownership, purchase, and use of irrigation pumps in Kenya and Tanzania. IFPRI Discussion Paper 01357, International Food Policy Research Institute.

Obuobie, E., Drechsel, P., Danso, G., Raschid-Sally, L., 2004. Gender in open-space Irrigated urban vegetable farming in Ghana. International Water Management Institute (IWMI), West Africa Office, Accra, Ghana, p 3.

Olango, T.M., Tesfaye, B., Catellani, M., Pè, M.E., 2014. Indigenous knowledge, use and on-farm management of enset (Ensete ventricosum (Welw.) Cheesman) diversity in Wolaita, Southern Ethiopia. *Journal of Ethnobiology and Ethnomedicine* 10. https://doi.org/10.1186/1746-4269-10-41.

O'Reilly, K., 2006. "Traditional" women, "modern" water: Linking gender and commodification in Rajasthan, India. *Geoforum* 37, 958–972. https://doi.org/10.1016/J.GEOFORUM.2006.05.008.

Pouramin, P., Nagabhatla, N., Miletto, M., 2020. A systematic review of water and gender interlinkages: Assessing the intersection with health. *Frontiers in Water* 2, 6. https://doi.org/10.3389/frwa.2020.00006.

Raney, T., Anríquez, G., Croppenstedt, A., Gerosa, S., Lowder, S., Matuscke, I., Skoet, J., Doss, C., 2011. The role of women in agriculture. ESA Working Paper No. 11-02. The Food and Agriculture Organization of the United Nations.

Richman Gambe, T., 2019. The gender dimensions of water poverty: Exploring water shortages in Chitungwiza. *Journal of Poverty* 23, 105–122. https://doi.org/10.1080/1087554 9.2018.1517399.

Saad, D., 2019. Why women's involvement is so vital to water projects' success -- or failure [WWW Document]. The Conversation. https://theconversation.com/why-womens-involvement-is-so-vital-to-water-projects-success-or-failure-115011 (accessed 7.8.22).

Saad, D., Byrne, D., Drechsel, P., 2017. Social perspectives on the effective management of wastewater. Physico-chemical wastewater treatment and resource recovery. https://doi.org/10.5772/67312.

Sayyed, J., 2018. 'Water wise women' program shatters stereotypes as female plumbers resolve Jordan's water scarcity - MVSLIM [WWW Document]. Mvslim. https://mvslim.com/water-wise-women-program-shatters-stereotypes-as-female-plumbers-resolve-jordans-water-scarcity/ (accessed 7.14.22).

Serwah, N.A., Sam, P., 2006. Gender mainstreaming and integration of women in decision-making: The case of water management in Samari-Nkwanta, Ghana. Wagadu.

Silva, B.B., Sales, B., Carolina Lanza, A., Heller, L., Rezende, S., 2020. Water and sanitation are not gender-neutral: Human rights in rural Brazilian communities. *Water Policy* 22, 102–120. https://doi.org/10.2166/wp.2020.126.

Stevenson, E.G.J., Ambelu, A., Caruso, B.A., Tesfaye, Y., Freeman, M.C., 2016. Community water improvement, household water insecurity, and women's psychological distress: An intervention and control study in Ethiopia. *PLoS One* 11. https://doi.org/10.1371/journal.pone.0153432.

Stewart, R., Wright, B., Smith, L., Roberts, S., Russell, N., 2021. Gendered stereotypes and norms: A systematic review of interventions designed to shift attitudes and behaviour. *Heliyon* 7. https://doi.org/10.1016/J.HELIYON.2021.E06660.

Sülün, E.E., 2018. Women, water resource management, and sustainable development: The Turkey-North cyprus water pipeline project. *Resources* 7. https://doi.org/10.3390/resources7030050.

Svetanoff, R.C., Ilobodo, I., 2022. Women and water: Lessons learned from a humanitarian intervention at the Igusi clinic, Matabeleland, Zimbabwe [WWW Document]. *The Journal of Gender and Water*. https://repository.upenn.edu/cgi/viewcontent.cgi?article=1091&context=wh2ojournal (accessed 7.12.22).

Taron, A., Drechsel, P., Gebrezgabher, S., 2021. Gender dimensions of solid and liquid waste management for reuse in agriculture in Asia and Africa. Resource Recovery & Reuse Series, International Water Management Institute (IWMI): Colombo, Sri Lanka, CGIAR Research Program on Water, Land and Ecosystems (WLE) 21.

Theis, S., Lefore, N., Meinzen-Dick, R., Bryan, E., 2018. What happens after technology adoption? Gendered aspects of small-scale irrigation technologies in Ethiopia, Ghana, and Tanzania. *Agriculture and Human Values* 35, 671–684. https://doi.org/10.1007/s10460-018-9862-8.

van Houweling, E., Hall, R.P., Diop, A.S., Davis, J., Seiss, M., 2012. The role of productive water use in women's livelihoods: Evidence from rural senegal. *Water Alternatives* 5, 658–677.

Yanıkkaya, B., Nairn, A.M. (Eds.), 2021. Multidisciplinary perspectives on women, voice, and agency. *Advances in Religious and Cultural Studies*. https://doi.org/10.4018/978-1-7998-4829-5.

10 Enhancing socio-ecological interactions to achieve sustainable decentralised sanitation systems
Why people are not using technical solutions

Betsie le Roux, Attie van Niekerk,
Erna Kruger, and Betty Maimela

10.1 INTRODUCTION

Universal access to water, sanitation, and hygiene (WASH) in many peri-urban and rural villages has not yet been achieved. Various technical solutions are available to solve the problem, but end-user acceptance of sanitation solutions is the main unsolved problem. Education and demonstrations to ensure the sustainable use of the given technology seldom succeed. Almost a third of the children in South Africa live under the "food poverty line" (that only allows enough for basic nutrition and no other essentials), and almost half live under the "lower bound poverty line" (that allows enough for essentials such as clothing but only if some nutritional costs are sacrificed. In the Limpopo Province, more than 30% of the population experienced hunger in 2013. The prevalence of stunting among boys and girls zero to three years of age was 26.9% and 25.9%, respectively (Said-Mohamed et al., 2015). According to the Mopani District Municipality (2019), 74% of households in the Mopani municipal area earn less than R1 100 per month. There is reason to think that poverty and hunger have worsened due to COVID-19. Modelled results from 44 African countries indicate that the expected rapid population growth will be the main cause of food insecurity and undernourishment throughout Africa. The effects of climate change were considered insignificant compared to the effect of population growth.

Universal access to WASH in the communal tenure villages in Limpopo, both peri-urban and rural, is still far from being achieved. The number of households with access to free basic sanitation is 1,360, and the number of backlogs is 63 times more,

DOI: 10.1201/9781003327615-10

at 86,388 (Mopani District Municipality, 2019). Rural households generally undertake their on-site arrangements, such as pit toilets. Several risks are associated with these pit toilets, including flooding, groundwater pollution, basic hygiene, the community's health and the safety of young children. The pit toilets require labour-intensive maintenance, and space becomes a problem each time one toilet is filled, and residents have to build another, especially in peri-urban areas where stands are small (Van Vuuren, 2014). The water supply is mostly insufficient (below 25 litres per person per day) because of the lack of pipeline reticulation and the widespread nature of the households in rural villages (Mopani District Municipality, 2019).

A circular economy, where waste products become resources that are used, can potentially solve many of these problems in rural and peri-urban communities and may result in significant improvements. Human excreta contains nutrients needed for food production and can be used to produce biogas for cooking or other energy needs. It is unclear whether these communities will be open to such possibilities. Still, as opposed to modern cultures' current linear take-make-dispose pathways, the African perspectives of cycles may be more open to adapting to circular economic practices. A new Water Research Commission project (van Niekerk et al., 2022) was initiated in April 2022 to investigate the potential of introducing DSPs of using human excreta as a resource for other important things household practices such as food and energy production. The development phase will be done in one rural and one peri-urban community in the Limpopo Province.

Many technologies are available for decentralised sanitation systems that can potentially be used in the service of rural and peri-urban communities in South Africa. Many of these technologies have been tested in previous projects, and some, like the pour-flush toilets, were relatively successful (Pillay and Bhagwan, 2021; Van Vuuren, 2014; Water Research Commission, 2014). However, in many cases, the reason for different solutions' successes or failures is unclear. Previous studies have engaged with the end users to determine their willingness to use different sanitation solutions. Still, as far as we know, there has not been a culture-driven approach to designing sanitation solutions. There have been thorough analyses of cultural and religious aspects and thought patterns of the end users that did not follow through to involve end users in the development of solutions. It seems as if the cultural information was not integrated into the solutions in, for example, studies by Taing et al. (2014) and Akpabio and Takara (2014). Technical concepts imported from another region without considering the complexity of the socio-cultural context have too often failed. Education and demonstrations to ensure the sustainable use of the given technology seldom succeed.

In African traditions, cultural, social and religious patterns are integrated. Modernisation typically delegates religious beliefs to the private life of individuals, separately from the larger social networks. All these patterns are in flux and can be expected to differ from one community to the other. Still, one can also expect that some trends are similar in certain communities, which makes it easier to transfer a solution that works in one community to a similar community with a few adjustments.

Decentralised sanitation and smallholder farming (SHF) have many characteristics of complex socio-ecological systems (SES), with many different actors and interconnected subsystems. This chapter explores the complex systems theory and how it can be applied to develop new ways of successfully integrating a DSP in a WEF framework. This WEF framework will define the requirements, opportunities and goals of the DSPs to be developed that will produce food and/or energy and healthy water resources.

10.2   METHODOLOGY

A literature review was conducted to collate existing knowledge on the African cultural and religious thought patterns on sanitation, food production, cooking and eating patterns and energy use and to better understand sanitation problems and practices such as open defecation. Reliable peer-reviewed publications, books, and other sources the authors were familiar with were perused. Further literature searches were undertaken by following citations and publications of known authors and using search words such as decentralised sanitation, sanitation in Africa and open defecation. Reports published by the WRC and the Pollution Research Group in KwaZulu-Natal were found on their respective websites. A literature review was also undertaken on the more recent advances in the complex systems theory. Interviews were conducted with people from the Mafarane village in Limpopo to determine their thoughts and feelings regarding sanitation in their communities.

10.2.1   THE COMPLEX SYSTEMS THEORY

Cilliers (2008) hesitated to define the complex systems theory because it is not simple enough to apply fixed characteristics. Nor is there a prescribed methodology to apply the complex systems theory. What is needed is for a scientist to develop a feeling and a certain attitude to engage with these systems (Cilliers, 2008; Preiser, 2019). Understanding complex systems theory in terms of how it differs from the traditional scientific methodology is more useful.

Complex versus complicated systems and holistic versus reductionist approaches. The reductionist approach is to comprehend a unified entity in terms of its components. Complicated systems, e.g., a computer, can be disassembled into their components and reassembled to operate in a specific way. To understand complicated systems, the reductionist approach can be used, where the components of a larger system can be studied to determine linear cause-and-effect relationships without the 'interference' of the rest of the system. Reductionism produces isolated technologies that operate in predictable ways (van Rooyen et al., 2020). The scientific method typically follows the reductionist approach because it is designed to study the components of a system to learn something about the system as a whole. However, the reductionist approach is inappropriate for complex systems (Cilliers, 2008). Suppose the isolated technologies that are produced through the reductionist approach are implemented in a context where they have to function within a larger, complex system. In that case, this larger system often interferes with the intended cause-and-effect relationships.

On the other hand, complex systems comprise several factors – things and thoughts – that interact and combine to produce unpredictable and even highly surprising outcomes. Emergence, defined as new system arrangements and behaviours, is a typical property of complex systems that occurs when the whole system has different and nonreducible properties to the properties of the system's components (Preiser, 2019; van Rooyen et al., 2020). Considering all these factors and their interactions requires a holistic approach in contrast to the reductionist approach.

Context and intervention. Complex systems are open systems, meaning the boundary between the system and its environment is unclear. Changing the context of a system will also change the system itself (Preiser, 2019). When explaining complexity theories, Jean Boulton stated that the emphasis should be on the context, not the intervention (Boulton, 2019). The complex systems theory perspective also requires that we bring social systems and ecological systems together in a way that they are "not just overlapping and interdependent, but inseparable. This perspective emphasizes that people, economies, societies, and cultures shape and are in turn shaped by ecosystems" (Reyers et al., 2018).

Scale: An SES is typically a dynamic cross-scale system where global decisions impact local conditions and emergence on local scales, impacting global conditions (Reyers et al., 2018). This creates difficulties when implementing sustainable solutions. Effective interventions depend on selecting the most appropriate scale to focus on without neglecting the interactions between the chosen scale and the scales above and below it.

Other relevant properties of complex systems: Characteristics of a SES that are relevant for our project include multiple perspectives, inter-relationships and communication, feedback loops; ideas emerging from the interactions, and continuous adaptation to change. Relationships are especially important in complex systems, and linear thinking, i.e., predictable cause-and-effect outcomes, should not be expected (Preiser, 2019). Development is done through an iterative process until a suitable solution emerges.

10.3 TRADITIONAL AFRICAN CULTURES AND THE WATER-ENERGY-FOOD NEXUS

In African traditional cultures, real life, well-being and moral values cannot be treated outside the spiritual, political, social and religious worldviews, and matters related to water and sanitation are better understood from the perspective of 'subjective' rather than 'objective' worldviews (Douglas, 2003). These worldviews, to a bigger or lesser degree, form the undertone that influences the life choices of the African people. The Nigerian Bishop and International Chairman of the Organization of African Instituted Churches, Daniel Okoh, said to Öhlmann et al. (2019):

> People from Sub-Saharan Africa … are highly religious… So, for Africa, because of the religious nature, you will always find a way of using it to get the … commitment of the people to the project, whatever it is. If it is water, it must be explained spiritually. If it is [an] agricultural project, it must be explained spiritually…. Honestly, if you do not do that, you will lose it.

Culture can be seen as a strategy to deal with the world in which people live and a way of associating with and interacting with each other and nature. Cultural aspects include practices consisting of behaviour patterns, the meaning people give to these patterns of behaviour and the associated feelings. We can discuss "patterns of culture" (Benedict, 1989). A practice is a subdivision of culture, an established way of doing certain things. Practices also have patterns and must fit into the larger cultural, social and ecological patterns. Our cars, for example, fit well into our cultural and social patterns but not the larger ecology. Practices, like cultures, are constantly changing, sometimes faster, sometimes slower. They emerge from the ongoing interaction and combination of multiple things and thoughts, such as traditions, modern technologies, ecological conditions, political and economic events, etc.

Water, energy and food are relevant topics in the African context, and it is necessary to understand these topics' cultural and religious context. In the following sections, we will give attention to African cultural patterns regarding energy, agriculture, food, cooking, cuisine, and sanitation.

10.3.1 ENERGY

In 2013, K. J. Wessels and a group of researchers reported that over 80% of households across sub-Saharan Africa (SSA) rely on biomass as their primary energy source. They calculated that at current levels of fuelwood consumption, biomass in many parts of the Limpopo region would be exhausted within 13 years. It further showed that it would require a 15% annual reduction in consumption for eight years to a level of 20% of households using fuelwood before the use of biomass would reach sustainable levels. They concluded that the severity of dwindling fuelwood reserves in African savannahs underscored the importance of providing affordable energy for rural communities (Wessels et al., 2013).

The rich significance of sitting and living around the fire for traditional family life shows how domestic energy use is integrated into household practices. However, there is a tendency to replace sitting around the fire with sitting in front of the TV. If that happens, the wood stove can potentially be replaced by biogas, depending on how the household as a system changes.

Decentralised sanitation is relevant for energy use in two ways: (i) saving energy to build and maintain centralised sanitation and treat wastewater, and (ii) the possibility to generate biogas from sewage through anaerobic digestion (AD). According to Msibi and Kornelius (2017), you need waste from approximately 205 chickens, 8 cows, 20 pigs, or 63 people to feed a 2,500 L/day biodigester, which is generally enough energy to satisfy the cooking requirements of one household. Therefore, biogas production is not feasible on a household scale. Msibi and Kornelius (2017) calculated that one non-sewered household generates enough greywater to feed on a 2,500 L/day biodigester, and the use of greywater is recommended. The use of biogas digestion is currently limited by a lack of supporting policies, unsuitable climates, limited support from the private sector, installation, operation and maintenance costs of the digesters, lack of technical knowledge and limited water availability (Msibi and Kornelius, 2017). In certain areas, the lack of feedstock for the digestor may also be a limiting factor. According to Meegoda et al. (2018), if a biodigester is overloaded, it could cause acidification and stop the microbial breakdown process. A biodigester needs fairly intensive management.

According to Lin et al. (2018), AD is economically more profitable than composting on a larger scale, while composting is more profitable on a smaller scale.

10.3.2 FOOD AND AGRICULTURE

10.3.2.1 Agriculture in the traditional African culture

In traditional African cultures, food production is regarded as being closely related to the fertility of people and the land. In this tradition, fertility is essentially a religious concept. Fertility is a manifestation of a mysterious life force. It depends on the relation with the ancestors and, ultimately, more remotely, on God. The modern approach requires harmony between forces and not control over nature in search of progress and a better future. In Chapter 3 of the recent book edited by Matholeni et al. (2020), Georgina Kwanima Boateng writes:

> Earth, therefore, is a woman and her fertility is revered because it is the source of sustenance and reproduction. The spiritual connection between Asaase Yaa (Mother Earth) and women in Africa cannot be overemphasised.

It is a long-standing tradition. Fifty years ago, Mother Earth was described as a dominant motif throughout modern African literature (Cartey, 1969). It is also a dominant motif in local cultures, specifically in agriculture. In 1938, Jomo Kenyatta, who later became the first president of Kenya, wrote:

> In Gikuyu life, the earth is so visibly the mother of all things animate, and the generations are so closely linked together by their common participation in the land, that agricultural ritual, and reverence for ancestral spirits, must naturally play the foremost part in religious ceremonial. Communion with the ancestral spirits is perpetuated through contact with the soil in which the ancestors of the tribe lie buried.... the earth is the most sacred thing above all that dwell in or on it... Ceremonies are performed to cause the rain to fall, to purify and bless the seeds, and again to purify the crops.
>
> *(Kenyatta, 1985)*

Rain is regarded in Zulu tradition as fertilisation of the earth by the sky, as a husband fertilises his wife. The earth cannot bear fruit if the rain does not work on it with water (Berglund, 1976). Berglund (1976) describes a ritual in Zulu culture to make the field fertile. The ritual contains many male and female symbols to ensure fertility. When Berglund asked a diviner about it, the diviner said that the field is "the mother from whom we eat." A male could not perform the ritual because "Men do not sow. They slaughter the animals when there is to be meat. But they do not sow."

10.3.2.2 Food in the traditional African context

When integrating decentralised sanitation with SHF practices, eating patterns are as important as sanitation patterns. The eating pattern, and where applicable, the market, will determine what food can be produced and what not.

In his study of the history of food and cuisine in Africa, James McCann (2010) emphasises two things: (i) African cooking and cuisines have formed over history and have expressed agility in keeping up with changing times, and (ii) food is deeply embedded in the culture. McCann (2010) frequently refers to the dynamism of African foods over the years. There is a rich variety of cuisines across different regions of the continent.

> Contact with world regions like the Indian Ocean rim (from at least the first century CE) and the Atlantic world (after 1500) brought many more challenges and opportunities that African cooks built into their stews, porridges, and breads.

This included the use of food that was borrowed from other continents, such as maize, bananas and spices. In different regions, different influences from elsewhere have combined with local cultures so that, in each place, some type of cooking has emerged that involves the layering of ideas, daily rituals of eating, ingredients, and methods of assembling foods for both public and private meals.

> Cuisine is a product of history, and a meal is a conjuncture of time, place, and particular ingredients. Globally, cooking and cuisine can be seen as a creative composition at the heart of all cultural expressions of ourselves. food, like dress, music, and art, carries deeper structures of cultural identity that form a marker of group coherence and solidarity—food helps define who we are.
>
> *(McCann, 2010)*

In African cultures, the most important rule/concern at any ceremony is the amount of food you prepare; one ought not to disappoint guests or starve them (Phasha et al., 2020). However, the emphasis on the agility of African cuisine indicates that the beliefs and taboos around food have not been as strict as it is, for example, in the Jewish religion.

Some traditions may have a negative impact if viewed from the outside, especially concerning the health and well-being of women. Traditional food taboos often have a bearing on the relationship between men and women. Lung'aho (2021) states that in SSA, pregnant women are forbidden from eating protein-rich foods such as eggs and snails for several reasons, including the fear that the child may develop bad habits. In some cultures, men eat before women and children. And boys may eat before girls. Africa's cooks were women, but they often served food to the men and children first and got to eat at the end when everyone else had had their fill. If food is scarce, a mother may sacrifice the food left after serving the men to the children. Community education and socio-behavioural change are needed to give equal priority to the nutrition of all family members.

Similarly, Chakona and Shackleton (2019) documented food taboos and beliefs among pregnant isiXhosa women. They found that cultural beliefs and food taboos followed by some pregnant women influence their food consumption, which impacts the health of mothers and children during pregnancy and immediately afterwards. Overall, 37% of the women reported one or more food practices shaped by local cultural taboos or beliefs. The most commonly avoided foods were meat products, fish, potatoes, fruits, beans, eggs, butternut and pumpkin, rich in essential micronutrients, protein and carbohydrates. Most foods were avoided for reasons associated with pregnancy outcome and labour and to avoid an undesirable body form for the baby.

10.3.3   SANITATION IN THE TRADITIONAL AFRICAN CONTEXT

10.3.3.1   Human excreta and the concept of impurity

Distinct cultures have different attitudes towards toilet systems and the treatment of human excreta. Our cultures and contexts structure the basic instinctive repulsion towards excreta into attitudes and treatment patterns (Warner et al., 2008). Each context is different, but certain trends are widely spread across SSA, such as a lack of proper sanitation for many and the impact of modernity, which conflicts with the spiritual forces operating in material things in African cultures.

The conflict between the secular worldview of modernity and the religious nature of African traditions is also relevant in the search for sustainable sanitation solutions. The meaning associated with such concepts as dirt, pollution, hygiene and disease evolves in relation to local cultural experiences, with special forms of values and risks. However, the ambivalence characterising such meanings underlies the sanitation and hygiene challenges. It leads to a disconnect between inner convictions and overt actions. In most reports across SSA, local knowledge and the equation of cleanliness with godliness and beauty sharply contrast with actual physical sanitation and hygiene practices and behaviours in many contexts and forms. Mphahlele (1962) blames the missionaries:

> What do those missionaries think? It all began when those desperate ladies taught us how to brush our teeth, wash with soap, and sleep with windows open to let in the fresh air - early to bed, early to rise, cleanliness is next to godliness.

Monnig (1978) emphasises the central role of "impurity," which is one of "a great variety of supernatural forces (that) may cause unfortunate events." However, these supernatural forces often act with and within natural causes. They do not necessarily preclude a person from treating an unfortunate event, such as sickness, naturally and supernaturally. A Sepedi word used to describe impurity is ditšhila, but the word is broadly used for excreta and also means sin. Literally, ditšhila means dirt, but it may be better translated as an impurity, and more particularly, ritual impurity. Conditions of impurity include:

> A woman giving birth, as well as the unborn child, the placenta and the hut where the birth has taken place… children who are born unnaturally, i.e., twins, malformed children, children who are born with teeth…The condition of ditšhila is closely connected with the critical changes of life, particularly with its beginning and its end… The impurity requires ritual cleansing.
>
> *(Monnig, 1978)*

During informal discussions with rural communities in Limpopo, people said that women and men are not meant to share the same toilet in the Tsonga traditions. It is a taboo, more specifically related to brides and grooms. It is seen as unclean and spiritually compromising to share as women need to sit, and men do not. However, sharing toilets between men and women in rural areas is no longer much of an issue, although separate toilets are still preferred. Mostly, people do not have enough money or resources to continue this practice. There is a belief that the bad odours in pit latrines are linked to bad spirits and that sharing toilets is spiritually compromising.

In his book Bantu Heritage, Junod (1938) does not refer to sanitation. On p. 93, it is said: "Bantu women have a strong sense of modesty, which is unfortunately deteriorating. They always take great care to choose a special place for bathing and the men who dare to approach this place are booed."

However, according to Berglund (1976), when healing is needed, faeces are often associated with the evil that has caused the illness. A prominent theme in the book *Zulu Thought-Patterns and Symbolism* (Berglund, 1976) is that of cleansing from evil by physically ejecting something from the body, like spittle, vomit and emetics (p. 292) and blowing out of medicines (p. 352). Ukuhlanza (to clean)

> … refers to vomiting and expulsion of faeces after an emetic or purge… All bodily excess, particularly faeces, which is vile, must be disposed of outside the homestead and, preferably, be buried. 'This thing is vile. A home is good. They do not agree. That is why it must be concealed somewhere at a distance from the homestead.' Zulu accepts this disposal of something vile as normal…. Evil which, on the other hand, is not expelled normally, must be cast out through acts such as enemas and vomiting. Today castor oil and a large number of other purges are obtainable in chemist shops and made use of extensively. There are Zulu who 'cleanse the stomach from poison' regularly every week, even more frequently, sometimes making use of both laxatives and enemas. In cases of sickness, disregarding the type, enemas, laxatives and vomiting are often automatically administered to the patient, especially if the sickness causes a rise in temperature. 'If the sick person is hot (i.e., runs a temperature) it is certain that there is a great medicine (i.e., sorcery) inside (him/her). Where does the medicine enter? Is it not through the mouth? So, it is in the stomach. That is why there must be vomiting and enemas. These things remove the poison which causes the sickness.'
>
> *(Berglund, 1976, pp. 328–329)*

> 'Cleaning the baby out' by intestinal washes is believed to cool the child down and protect it by purging harmful evil influences.
>
> *(van Andel et al., 2015)*

10.3.3.2 Open defecation

There are many reasons why open defecation is still practised. These include convenience, unavailable toilet facilities, poorly constructed and private toilets, water scarcity, a lack of toilet paper, and children not being trained to use toilets. Traditional beliefs and practices can promote open defecation. For example, some believe it is cleaner (Dittmer, 2009).

In SSA, there are cultural restrictions regarding human excreta and toilet use. Some of these beliefs concerning the location and type of a toilet facility have also increased open defecation practices in some places, for example:

- Excreta from different people should not be put on top of each other, which is what would happen in a communal pit latrine.
- Storing faeces underground is unacceptable in some cultures, as it is believed to pollute the soil where the ancestors are buried.
- In some cases, people fear using a communal pit latrine because witch doctors using excreta for harmful purposes will then know where to find it. This can be a more serious problem for women in their menstrual cycle

because blood is considered a powerful substance in witchcraft (Akpabio and Takara, 2014), which is further discussed in the next section.

10.3.3.3   Gender and generational considerations

Cases of treating children's faeces and other waste products with a tolerance have been widely reported; linked not only with the idea of inoffensiveness of child excreta, but such tolerance also carries spiritual implications for parents and potential parents. Every material element of child hygiene (sputum, faeces, urine and other child waste products) entails potential blessing, depending on how it is handled.

(Akpabio and Takara, 2014)

The literature reports on many cultural restrictions to women and their management of menstrual hygiene because menstruation is considered to be connected to evil spirits and curses (see discussion on the impurity in **Section 4.3.1**). During menstruation, women and girls are considered 'impure.' Women and girls need proper, hygienic facilities, privacy and water for washing their hands, bodies and clothes that were used. Where disposable pads are used, they need a disposal system. These requirements are seldom discussed in SSA because of the contempt with which the topic is considered. Women are often ashamed of their condition and hide their used cloths in unhygienic places, exposing them to diseases (Hickling and Hutton, 2014).

Similar thought patterns have been revealed during informal discussions with communities in the Limpopo Province. According to these people, women must dispose of all the sanitary essentials straight inside the toilet to avoid witchcraft being done to them, which correlates with what Akpabio and Takara (2014) said (refer to **Section 4.3.2**). Disposing of sanitary essentials in bins both in homesteads and public spaces is still seriously frowned upon. Women are expected to wash, burn or bury what they cannot dispose of in their toilets. For younger women, this is a cause of great embarrassment and also leads to unhygienic practices of resisting going to the toilet and disposing of sanitary essentials in inappropriate ways to try and hide these essentials from view. Few toilet systems are robust enough to handle objects like sanitary pads. Toilets from which the content is to be reused will not get rid of these items in the way these people expect. So, such beliefs will have important implications for the uptake of decentralised sanitation solutions.

Furthermore, current sanitation systems also make women more vulnerable because they may be attacked when they go out to use the toilet at night. Women often do not drink water to avoid having to go out at night, which can again cause health problems (Warner et al., 2008). Women must consider their safety when deciding where and when to defecate, and open defecation in between shrubs may feel like the safer option. Defecating before sunrise or after sunset may provide them with more privacy, but that may not be in favour of their safety.

However, in responding to this issue, just as with other issues, we must consider local cultures. In his book, *A Short History of African Philosophy*, Hallen (2009) argues that one reason for the unmeant negative impact of Western efforts to improve the quality of life in Africa is that the conceptual frameworks used to "understand"

African society have their origin in Western culture. This applies to concepts such as community, family and gender. For example, African female scholars such as Oyeronke Oyewumi, Ifi Amadiume and Nkiru Nzegwu.

> at various points and in the strongest terms reject 'feminism' as a Western-based and Western-oriented movement that has yet to demonstrate that it is prepared to reject the misrepresentations of African societies by Western scholarship and is prepared to learn from rather than dictate to the non-Western world.

Hallen continues:

> Western feminists strengthen the gendering of society in individualistic terms, while traditional African cultures put the community first, and give male and female equal and interdependent roles in the community, which makes it possible not to gender society. Amadiume, for example, blames Western feminists that their imposed systems erode all positive aspects of historical gains, "…leaving us impoverished, naked to abuse, and objects of pity to Western aid rescue missions".
>
> *(Quoted by Hallen (2009))*

These women protest against people from outside who impose their own ideas in their context. It highlights the importance of approaching people with a learning attitude rather than an authoritarian attitude when undertaking development projects.

10.3.3.4 Environmental, socio-economic and behavioural considerations

According to van Oel (2002), things to consider when selecting the best options for the improvement of sanitation (environmental, economic and social) can be summarised:

1. Geological subsurface considerations.
2. Access to water.
3. Affordability by the recipient community for capital as well as for maintenance costs.
4. Future upgrading must be considered.
5. The recipient community must be fully involved in the choice of a system.
6. To stimulate real involvement, the community must be trained to do the development work themselves wherever possible.
7. The local authority must have the institutional structure necessary for the operation and maintenance of the system.
8. A system must operate despite misuse by unsophisticated users and should require as little maintenance as possible (van Oel, 2002).

We generally agree with van Oel (2002) but would formulate two statements (5 and 7) differently. In our approach, we emphasise the design of a sanitation practice, which incorporates a technical tool or system, with the recipient community rather than just involving them in choosing a system (as mentioned under nr 5). In cases where the local authority does not have the institutional capacity to operate and maintain toilet systems, the statement under nr 7 is not feasible. Taing et al. (2014) have shown that poor maintenance is due to a lack of responsible people in situations where several households share one toilet. Therefore, the household should be able to construct and maintain its own toilet system. If the household owns the toilet and they can maintain it, there is a fair chance that they will take responsibility for it.

In general, in communal tenure villages in Limpopo, around 12% of the households have no sanitation structures and practice open defecation, either in their own yards or in the surrounding veld; a large proportion has pit latrines (around 56%), and some have ventilated improved pits. The use of toilet paper for anal cleansing is rare (~5%), with most people using newspaper (74%) and stones (21%). Water for hygiene and sanitation is unavailable, and people do not regularly wash their hands after using their toilets. Cleaning of the toilets and top structures is also not done regularly due to water scarcity. In addition, there is a tendency for the pit latrines to fill up as the sandy-loam structure and depth of the soil can lead to the liquid not draining away fast enough (van Oel, 2002).

In a study conducted in Mohlaletse village (Sekhukhune District Municipality), the following conclusions were reached through a community-based process. People needed individual household solutions. Dry sanitation options, such as ventilated improved pit latrines (VIP) and urine diversion dehydration toilets (UDDT), are the most suitable. In this respect, approximately 65% of participants are aware of the fertilisation potential of human excreta and are prepared to consider sanitation options where human excreta are used for fertilisation. For households that use toilet paper for anal cleansing and have a groundwater tap on their plot, two other systems are also adequate (the pour-flush toilet and the aqua-privy with soak-away) (van Oel, 2002).

A survey in a different and more recent study in the rural Makhwane village (also in the Sekhukhune District Municipality) showed a situation much unchanged from 2002. An additional hygiene challenge noted within the community was the fact that children under the age of 12 in most of the households (62%) were not allowed to use the toilet. This practice was attributed to the lack of improved sanitation facilities; parents feared their children were at risk of falling into and drowning in outside toilets, such as pit latrines (Budeli et al., 2020).

Besides the hygiene challenges, major sanitation issues were observed across the village. Approximately 41% of the households in the community did not have toilets installed in their yards, and 86% of the households use the open field as an alternative for a sanitation facility, while the remaining 13% share sanitation facilities with their neighbours. The findings show that 60% of householders safely dispose of their solid waste (e.g., diapers) in the pit latrines, while 6% dispose of soiled diapers in the streams, 9% in the field, 37% in a separate pit and 11% burn them. Only 25% of households use reusable diapers. Households using reusable diapers generally dispose of the faecal matter by soaking the soiled fabric diapers in water and discarding the wastewater in an open field (Budeli et al., 2020).

The literature often acknowledges that people living in rural and peri-urban areas desire full waterborne sanitation (Pillay and Bhagwan, 2021; Van Vuuren, 2014; Water Research Commission, 2014). Out of 275 households in dense settlements in eThekwini, Ekurhuleni and the City of Tshwane, approximately 70% indicated that they preferred waterborne sanitation in their homes (Martin and Pansegrouw, 2009). However, Martin and Pansegrouw (2009) did indicate that these people were willing to consider other options when they became aware of the costs associated with these flush toilets. It is not only in terms of sanitation that lower-income households aspire to have the things available to people in urban areas. Van Niekerk (2008) observed that poor households closer to the cities planted lawns and flower gardens instead of vegetables, in contrast to communities in more rural areas who often produce vegetables and maize in their gardens. Therefore, one must deal with these feelings of low-income households being inferior to those in urban areas.

10.4 DISCUSSION

10.4.1 POOR UPTAKE OF SOLUTIONS

Social acceptability is repeatedly mentioned as one of the most important problems in implementing decentralised sanitation systems (Martin and Pansegrouw, 2009; Odindo et al., 2016; Taing et al., 2014). One of the largest studies on user perception of UDDTs showed that UDDTs installed since 2001 are still mostly in use. Out of 15,983 households, 85% indicated that all members and 8% that some members in the household use the UDDTs. However, user satisfaction was low, with 70% unsatisfied with the UDDTs. The reasons for the unhappiness of the users were bad odours from the toilet (27%), lack of privacy from doors that do not close properly (22%), poor quality of materials used and construction workmanship (12%), the urine pipe not being connected correctly (12%), etc. (Roma et al., 2013). Key findings of another study looking at qualitative and quantitative indicators of the perceptions of UDDTs showed:

1. Although 97% of recipients use UDDTs, 95% of interviewees did not consider these toilets a permanent sanitation solution for their households and were waiting for waterborne sewage.
2. Most participants reported not identifying with the UDDT benefits, such as using urine as a fertiliser.
3. The participants felt that the UDDT was not sensitive to their comfort since one must always be mindful if your urine or faecal matter is going to the right place.
4. The fear of allowing children between two and five years to use the UDDT toilet was a highly discussed issue. Most participants reported that they discouraged their children from using the UDDT, and they practised open defecation instead.
5. Among users, 80% were not maintaining the UDDT properly. The findings reveal that females mainly clean the UDDTs in the household, including emptying the toilet. A small proportion of respondents reported that older females do the task of emptying the toilet because being in contact with faecal matter will bring bad luck to younger females.
6. The older generation preferred the VIP toilet because they are accustomed to it, it requires less responsibility from the user, and the user does not have to empty it, and

7. The doors, back cover, and seats were reported to be items that easily break, which made people feel that they were given cheap *toilets that were not customised to their reality* (Mkhize et al., 2017).

Personal conversations with community members indicated the following complaints regarding their current sanitation (VIP latrines):

1. People in the communities see VIP latrines as an inferior service because it compromises their dignity, the toilets have bad odour and lots of flies, which affects mostly women, and they fill up very quickly.
2. Community members also feel like the government is undermining them because they come from rural areas, and the government is more concerned about supporting urban populations. They are told that waterborne sewage is impossible and expensive, but they see flush toilets in the cities and towns. The government also refers to the VIPs that they provide as standard basic services, and thus, people believe there should be better, more permanent options available. Still, the officials choose not to assist them with these.
3. Women are expected to keep the toilets clean, even though this is an almost impossible task with pit latrines, as none of the surfaces can be properly disinfected. There is a strong belief that many women's infections are related to using these unsanitary toilets.
4. The fact that access to water for cleaning, rinsing, and hygiene is extremely limited and very often not unavailable at or close to the toilets is a further contributing factor to unsanitary conditions.
5. Most people prefer their toddlers not to use the toilet because they have accidentally fallen inside the pit latrines and are dying. Parents are then blamed for being irresponsible. In addition, the pedestals are not appropriate for small children. It means that toddlers and small children, more often than not, defecate in the open. Some people use small buckets as latrines for the smaller children.

Roma et al. (2013) concluded that successfully implementing the UDDTs depends on education and establishing the economic return from using urine excreta for agricultural purposes. We agree that reusing urine and excreta may be important for successfully implementing decentralised sanitation. However, instead of educating people, we believe in co-developing a practice with the end users, i.e., one must change the technology to fit into the socio-ecological context instead of changing the people to fit the requirements of the technology.

10.4.2 APPLYING THE PRINCIPLES OF THE COMPLEX SYSTEMS THEORY IN THE SHF CONTEXT

Application of the complex systems theory may seem arbitrary and uncertain, but some proven approaches can be used. The following general principles should be applied in the development of the Decentralised Sanitation Practice in the WEF nexus (DSP-WEF):

- Firstly, as Cilliers (2008) mentioned, it is important to engage with the SHFs with a learning attitude and not preconceived ideas of the solutions. The eventual integrated sanitation-with-small-holder-farming practice(s) must emerge from the interaction between the different role players. This aligns with the fact that relationships are important in complex systems. Understanding problems and potential solutions could not originate from a single person's perspective but should involve the perspectives of as many different parties as possible. The viewpoints of the SHFs themselves are especially critical in assessing problems and developing solutions. Together, all role players must develop a common vision. Limited interaction between the different role players during the development phase would prevent learning and the emergence of more beneficial outcomes (van Rooyen et al., 2017). This approach can be called transdisciplinary research, in which SHFs, relevant NGOs and government officials, markets and scientists, etc., put their heads together to solve an everyday problem. The co-development process should be done through an iterative process of conceptualising, testing, and adapting until the most suitable solution is found.
- As discussed in **Section 3**, it is important to develop solutions at the correct scale in a complex system. One disadvantage of a larger scale project, e.g., sewerage systems, is that the individual households depend on the larger system and good management. Small-scale solutions give the household more control to solve their own problems. Solutions at various scales (from single households to smaller and larger combinations of local households) should be tested during future development to better understand the needs and requirements of the target communities.
- We agree with Boulton (2019) that the focus should be on context rather than interventions. The context is shaped by its history and the existing economic, social, ecological, political and other issues. However, we would formulate it slightly differently: our emphasis is on how the eventual solution emerges from the interaction of all aspects, including the context and the intervention. The focus should be on how the relevant role players experience the intervention, contextual factors, and what meaning they give to it. Attending to how people experience something and what meaning they give to it is called a phenomenological approach (Aydin, 2007). Thus, the emphasis must be on how technical interventions are integrated into and become embedded in the SHF practices, which are embedded in the wider context.
- Complexity can either be (i) observed and analysed or (ii) one can "participate in and creatively co-construct the phenomenological experiences of everyday instances and encounters of a messy, complex reality" (Preiser, 2019). We follow this second approach, not disregarding the first.

10.4.3 ADDRESSING BEHAVIOURAL AND CULTURAL THOUGHT PATTERNS IN THIS STUDY

Problems such as safety risks and environmental pollution of sanitation in rural and peri-urban areas should be addressed. More invisible problems, such as religious and cultural thought patterns and how they affect sanitation and other topics related to the WEF nexus, are just as important to understand and must be central to the approach followed during future research. One problem continuously identified in the literature is the ambition of low-income communities to have lifestyles similar to higher-income urban communities. It indicates that the root of the problems extends to a much deeper socio-economic condition of inequality in the country. It also highlights that the more wasteful practices in urban communities must be addressed to improve the quality of life in rural and peri-urban areas.

It is uncertain to what extent traditional thought patterns and symbolism play a role in the target community of this project and to what extent modern ways of thinking have been acquired. Modern technology has been introduced, and the relationship with the land has been modernised. Still, it is uncertain what has remained of the traditional, cultural and religious thought patterns, what influence it will have in sanitation development projects and how it should be taken up in the practices that emerge out of the various interactions that should occur during a development project. The relevant stakeholders and role players must determine the answer to this question in each local context. The information on cultural thought patterns from the literature is, however, important for the following reasons:

1. It provides a theoretical framework that should be tested in each community.
2. It creates an awareness of the potential perspectives of the SHFs.
3. It creates respect for the culture and behaviour of the people.

From the literature, we can derive that traditionally, human excreta are seen as vile, something that has to be concealed somewhere at a distance from the homestead. Sickness is both a natural and/or a spiritual matter. It can be caused by evil through witchcraft or jealousy, but it often has a physical presence in the body that can be cleansed out by being expelled from the body. Laxatives and enemas can help to cleanse the body. Fieldwork is needed in any community where solutions are being developed to determine to what extent human excreta is experienced as an impurity in normal circumstances and if any rituals are involved in normal daily affairs. If human excreta is seen as impure and associated with evil, the question is how participants can be motivated to consider it a resource.

An interesting aspect of impurity (ditšhila) in African cultures is that it is closely connected with the critical changes of life, particularly with its beginning and its end. Life is often seen cyclically; when you are born, you come from the world of the ancestors, and when you die, you return to them. In modern African literature, birth and death are often taken up in the cycle of life so that life comes from death as a plant grows from a seed buried in the ground. The seed has to die for the plant to grow. That implies that impurity is also taken up in the circle of life. Could this thought pattern contribute to developing a circular agricultural pattern in which human excreta are taken up in the agricultural practice?

10.5 CONCLUSIONS

Nutrition and sanitation are two areas that can be linked in a circular economy. In both areas, there seem to be particular concerns regarding the health and safety of women and children. Ecological impacts are also concerned with current sanitation and energy consumption practices. Many projects have attempted to develop decentralised sanitation systems in rural and peri-urban communities to solve these problems. However, these technologies have often not been accepted by the end users. This could be due to the less obvious importance of cultural and religious thought patterns, how they influence the people's behaviours and their acceptance of the proposed solutions.

What is needed is a careful assessment of the cultures and the relevant feelings and thoughts of the low-income communities that need the solutions. Such an assessment will require the scientist to form a relationship with the end-user and involve them in developing the solution. Such an approach aligns with the complex systems theory, which guides how the development process should be undertaken.

REFERENCES

Akpabio, E.M., Takara, K. (2014) Understanding and confronting cultural complexities characterizing water, sanitation and hygiene in Sub-Saharan Africa. *Water International* 39, 921–932.

Aydin, C. (2007) De vele gezichten van de fenomenologie. *Tijdschrift Voor Filosofie* 69.

Benedict, R. (1989) *Patterns of culture*. Houghton Mifflin, Boston. First published in 1934.

Berglund, A.I. (1976) *Zulu thought-patterns and symbolism/Axel-Ivar Berglund*. Swedish Institute of Missionary Research, Uppsala.

Boulton, J. (2019) Systemic change, complexity and development. Interview with Belinda Reyers. Centre of complex systems in transition.

Budeli, P., Moropeng, R., Mpenyana-Monyatsi, L., Kamika, I., Momba, M. (2020) Status of water sources, hygiene and sanitation and its impact on the health of households of Makwane Village. Limpopo Province, South Africa. ResearchGate.net.

Cartey, W. (1969) *Whispers from a continent: The literature of contemporary black Africa*. Heinemann Educational Books, London.

Chakona, G., Shackleton, C. (2019) Food taboos and cultural beliefs influence food choice and dietary preferences among pregnant women in the Eastern Cape, South Africa. *Nutrients* 11, 2668.

Cilliers, P. (2008) Complexity theory as a general framework for sustainability science. In: M. Burns and A. Weaver (Eds), *Exploring sustainable science: A South African perspective*, Sun Press, Stellenbosch, pp. 39–57.

Dittmer, A. (2009) *Towards total sanitation: Socio-cultural barriers and triggers to total sanitation in West Africa*. WaterAid, Lusaka.

Douglas, M. (2003) *Purity and danger: An analysis of concepts of pollution and taboo*. Routledge, London.

Hallen, B. (2009) *A short history of African philosophy*. Indiana University Press, Bloomington.

Hickling, S., Hutton, G. (2014) Economics of inadequate sanitation in Africa, Sanitation and Hygiene in Africa: Where do We stand? Analysis from the AfricaSan Conference, Kigali, Rwanda. IWA Publishing, London, pp. 29–34.

Junod, H.P. (1938) *Bantu heritage*. Johannesburg Hortors Limited for the Transvaal Chamber of Mines, Pretoria.

Kenyatta, J. (1985) *Facing mount Kenya*. African Writers Series. Heinemann Educational Books, London (first published in 1983).

Lin, L., Xu, F., Ge, X., Li, Y. (2018) Improving the sustainability of organic waste management practices in the food-energy-water nexus: A comparative review of anaerobic digestion and composting. *Renewable and Sustainable Energy Reviews* 89, 151–167.

Lung'aho, M. (2021) Africa: Why culture is key in tackling malnutrition. SciDev.Net.

Mphahlele, E. (1962) *The African Image*. Faber & Faber, London, p 316.

Martin, R., Pansegrouw, P. (2009) *Development of a model for determining affordable and sustainable sanitation demand in dense settlements of South Africa*. Water Research Commission, South Africa.

Matholeni, N.P., Boateng, G.K., Manyonganise, M. (2020) *Mother earth, Mother Africa & African indigenous religions*. African Sun Media Stellenbosch.

McCann, J.C. (2010) *Stirring the pot: A history of African Cuisine (Africa in World History)*. Hurst and Co., London.

Meegoda, J.N., Li, B., Patel, K., Wang, L.B. (2018) A review of the processes, parameters, and optimization of anaerobic digestion. *International Journal of Environmental Research and Public Health* 15, 2224.

Mkhize, N., Taylor, M., Udert, K., Gounden, T., Buckley, C. (2017) Urine diversion dry toilets in eThekwini Municipality, South Africa: Acceptance, use and maintenance through users' eyes. *Journal of Water, Sanitation and Hygiene for Development* 7(1), 111–120.

Monnig, H.O. (1978) *The pedi* (Second edition). Van Schaik, Pretoria.

Mopani District Municipality (2019) Draft annual report 2018/19, Limpopo Province, South Africa.

Msibi, S.S., Kornelius, G. (2017) Potential for domestic biogas as household energy supply in South Africa. *Journal of Energy in Southern Africa* 28, 1–13.

Odindo, A.O., Bame, I.B., Musazura, W., Hughes, J.C., Buckley, C.A. (2016) Integrating agriculture in designing on-site, low cost sanitation technologies in social housing schemes. WRC Report No. TT 700/16. Water Research Commission, Pretoria.

Öhlmann, P., Gräb, W., Frost, M. (2019) African initiated Christianity and the decolonisation of development: Sustainable development in Pentecostal and independent churches. Routledge, London and New York, p 355.

Phasha, L., Molelekwa, G.F., Mokgobu, M.I., Morodi, T.J., Mokoena, M.M., Mudau, L.S. (2020) Influence of cultural practices on food waste in South Africa—A review. *Journal of Ethnic Foods* 7, 37.

Pillay, S., Bhagwan, J. (2021) SaNiTi–A WRC research strategy and response to transforming sanitation into the future. Working Paper. Water Research Commission, Pretoria.

Preiser, R. (2019) Identifying general trends and patterns in complex systems research: An overview of theoretical and practical implications. *Systems Research and Behavioral Science* 36, 706–714.

Reyers, B., Folke, C., Moore, M.-L., Biggs, R., Galaz, V. (2018) Social-ecological systems insights for navigating the dynamics of the Anthropocene. *Annual Review of Environment and Resources* 43, 267–289.

Roma, E., Philp, K., Buckley, C., Scott, D., Xulu, S. (2013) User perceptions of urine diversion dehydration toilets: Experiences from a cross-sectional study in eThekwini Municipality. *Water SA* 39, 305–312.

Said-Mohamed, R., Micklesfield, L.K., Pettifor, J.M., Norris, S.A. (2015) Has the prevalence of stunting in South African children changed in 40 years? A systematic review. *BMC Public Health* 15, 1–10.

Taing, L., Spiegel, A., Vice, K., Schroeder, M. (2014) *Free basic sanitation in informal settlements: An ethnography of so-called communal toilet use and maintenance*. Water Research Commission, Pretoria.

van Andel, T., van Onselen, S., Myren, B., Towns, A., Quiroz, D. (2015) "The medicine from behind": The frequent use of enemas in western African traditional medicine. *Journal of Ethnopharmacology* 174, 637–643.

van Niekerk, A., Le Roux, B., Kruger, E., Howard, M., Maimela, B. (2022) Scoping study towards developing a WEF framework for decentralised sanitation and peri-urban communities in Limpopo' Water Research Commission. Project in progress.

Van Niekerk, C.E. (2008) *The vegetation and land use of a South African township in Hammanskraal, Gauteng.* University of Pretoria.

van Oel, P. (2002) Improving Sanitation in Mohlaletse Village. Community based approach for the employment intensive construction of sanitation facilities as part of the Mohlaletse Youth Service Programme. M Civil Engineering. University of Twente, The Netherlands.

van Rooyen, A.F., Moyo, M., Bjornlund, H., Dube, T., Parry, K., Stirzaker, R. (2020) Identifying leverage points to transition dysfunctional irrigation schemes towards complex adaptive systems. *International Journal of Water Resources Development* 36, S171–S198.

van Rooyen, A.F., Ramshaw, P., Moyo, M., Stirzaker, R., Bjornlund, H. (2017) Theory and application of agricultural innovation platforms for improved irrigation scheme management in Southern Africa. *International Journal of Water Resources Development* 33, 804–823.

Van Vuuren, L. (2014) No flash in the pan: How pour flush toilets are driving away SA's sanitation backlog: Sanitation. *Water Wheel* 13, 16–20.

Warner, W., Heeb, J., Jenssen, P., Gnanakan, K., Conradin, K. (2008) Socio-cultural aspects of ecological sanitation. M4-2: PDF-Presentation. Aarau: Seecon.

Water Research Commission (2014) Sanitation: Two new sanitation technology innovations from the WRC. Technical Brief. WRC, Pretoria.

Wessels, K.J., Colgan, M., Erasmus, B.F.N., Asner, G., Twine, W., Mathieu, R., Van Aardt, J., Fisher, J., Smit, I. (2013) Unsustainable fuelwood extraction from South African savannas. *Environmental Research Letters* 8, 014007.

11 A WEF nexus–based planning framework to assess progress towards Sustainable Development Goals

Sylvester Mpandeli, Luxon Nhamo,
Stanley Liphadzi, Jennifer Molwantwa,
and Tafadzwanashe Mabhaudhi

11.1 INTRODUCTION

In various spheres of activity, nexus concepts have been introduced to increase the understanding of interconnected systems. Although the water-energy-food (WEF) nexus has been the dominant nexus type, other nexuses have emerged. In 2014, the United Nations' Sustainable Development Report addressed the aspects of sustainability through nexus concepts such as (i) the climate-land-energy-water nexus, (ii) the oceans-livelihoods nexus, (iii) the industrialisation-sustainable-consumption-and-production nexus, and (iv) the infrastructure-inequality-resilience nexus (UN, 2014). Some studies have developed the water-health nexus (Confalonieri and Schuster-Wallace, 2011), the water-milk nexus (Amarasinghe et al., 2012), the water-health-environment-nutrition nexus (Nhamo and Ndlela, 2021), urban nexus (Lehmann, 2018; Nhamo et al., 2021b), rural-urban nexus (Constant and Taylor, 2020), and the water-soil-waste nexus (Hülsmann and Ardakanian, 2014). Such an array of nexus types has led to new terms, such as nexus thinking and nexus planning, to move away from the dominance of the WEF nexus (Naidoo et al., 2021c; Nhamo and Ndlela, 2021).

The increasing focus on the nexus approach implies an interest in interactions, interrelationships, and interconnections, such as among sectors or activities, within practical and essential themes for society (Ghodsvali et al., 2019; Naidoo et al., 2021c). These interactions might form interdependencies, constraints, and synergies that arise when changes in one area affect others (Naidoo et al., 2021c; Nhamo et al., 2018a). Their impacts might be viewed as either positive or negative (Nhamo et al., 2018b). Holistic consideration of such interactions is often called 'nexus thinking or planning' or taking a nexus approach (Nhamo and Ndlela, 2021; Venghaus and

DOI: 10.1201/9781003327615-11

Hake, 2018). A study done in 2015 traced the use and development of nexus concepts related to natural resource use through the 1980s and 1990s (Scott et al., 2015). These included pairwise considerations of the bi-directional linkages between water and food, water and energy and energy and food (Hoff, 2011). By 2008, the WEF nexus was the clear focus of several authors, leading to Bonn's influential nexus conference in 2011 (Hoff, 2011). Since then, research interest and activity on nexus planning have accelerated exponentially (Liphadzi et al., 2021; Nhamo et al., 2021a), leading to such developments as the Water, Energy, and Food Security Nexus Resource Platform (https://www.water-energy-food.org/) and a Global Nexus Secretariat.

Several attempts have been made to develop WEF nexus analytical tools to examine the WEF resources' diverse interactions (Albrecht et al., 2018; FAO, 2014; McGrane et al., 2018; Waughray, 2011). Even so, and besides the various attempts to develop WEF nexus analytical tools, such tools have not been widely adopted as most of them continued pursuing sectoral or linear approaches, or there is no cross-sectoral integration and analysis that promote transformational change (Albrecht et al., 2018; Mabhaudhi et al., 2021; Nhamo et al., 2018a, 2019b). Existing WEF nexus models like the multiscale integrated analysis of societal and ecosystem metabolism (MuSIASEM) (Giampietro et al., 2013) tend to integrate sector models that include the Agricultural Production Systems Simulator, RENA's Preliminary Nexus Assessment Tool Soil and Water Assessment Tool, Water Energy Food Nexus Rapid Appraisal Tool, Water Evaluation and Planning System, and Physical, Economic, and Nutritional Water Productivity. The MuSIASEM model has been applied as a simulation tool for case studies in Mauritius, the Punjab, and South Africa (https://www.sciencedirect.com/science/article/abs/pii/S0360544208001965).

On the other hand, Daher and Mohtar developed a model (WEF Nexus Tool 2.0) that they applied to simulate and compare alternate scenarios for food self-sufficiency in Qatar (Daher and Mohtar, 2015). The other model is a linked set of economic and biophysical models that underlie the *Australian National Outlook 2015* (Hatfield-Dodds, 2015). However, recent studies have identified limitations with these models, particularly when linking populations' behaviour with their physical ecosystems (Albrecht et al., 2018; Bizikova et al., 2013; Nhamo et al., 2020a). These pioneer models lack the tools to evaluate synergies and trade-offs in an integrated way, prevent conflicts, reduce investment risks, and maximise economic returns. Additionally, nexus planning approaches are transformative models envisaged to address the interlinked contemporary challenges and replace the widely used linear models that have now reached their limits (Lehmann, 2018; Naidoo et al., 2021b; Nhamo and Ndlela, 2021). Therefore, pursuing sector-based and linear approaches within the context of the WEF nexus has the potential to continue producing optimum efficiencies in specified areas at the expense of the other linked sectors (Nhamo and Ndlela, 2021).

Thus, the main limiting factor with pioneer nexus planning models is the pursuit of sector-based models without a clear integration of the linked sectors (Nhamo et al., 2020a). These models fail to establish quantitative relationships between the WEF sectors and cannot explain and simplify the interlinkages among the sustainability indicators linking the sectors. For the WEF nexus to be a completely transformative approach, it needs a decision-support tool that assesses the interlinked sectors as a whole, eliminating a "silo" approach in resource planning, development, utilisation, and management (Mabhaudhi et al., 2021; Nhamo et al., 2020a). There is a need for an integrated WEF nexus tool capable of assessing resource development and utilisation in a holistic way (Mabrey and Vittorio, 2018; McGrane et al., 2018).

Therefore, although the WEF nexus is envisioned to harness the three interlinked global security concerns of access to water, sustainable energy, and food security, some gaps hinder the approach from fully operationalising (Naidoo et al., 2021c; Nhamo et al., 2019b). Critiques of the WEF nexus point to the volume of theoretical literature that has been published since 2015, but only focusing on the importance of the approach without the rigour of empirical evidence (Liu et al., 2017; Mpandeli et al., 2018; Nhamo et al., 2018b; Terrapon-Pfaff et al., 2018). In recent years, promising attempts have been made to develop practical and integrative models to establish the quantitative relationships between the three resources and indicate priority areas for intervention (Nhamo et al., 2020a).

Holistic management of WEF resources and eliminating the traditional sector-based linear approaches can improve existing efforts to enhance resource security and achieve sustainable development (Dargin et al., 2019; Díaz et al., 2015; Endo et al., 2017). Recent developments have seen the development of WEF nexus tools capable of establishing numerical relationships between the interlinked but different WEF sectors and providing decision-support tools to identify priority areas needing intervention (Mabhaudhi et al., 2021; Naidoo et al., 2021c; Nhamo et al., 2019b). The integrative WEF nexus analytical model developed by Nhamo et al. establishes the quantitative relationships between the WEF sectors, providing scientific evidence to operationalise the WEF nexus (Nhamo et al., 2019b). In a study done in South Africa, quantitative relationships among WEF sectors were established through a multi-criteria decision-making (MCDM) method to connect the sectors and establish an integrated WEF nexus index, classified either as sustainable or unsustainable (Nhamo et al., 2019b). The model identifies synergies and trade-offs to develop and manage resources from an informed point of view. This study adopts this analytical model by Nhamo et al. (2020) to develop a WEF nexus–based methodological framework to guide policy and decision-makers to assess progress towards achieving related SDGs.

11.2 MULTI-CRITERIA DECISION-MAKING AND THE WEF NEXUS

A complex scenario encountered when integrating and analysing the three WEF sectors simultaneously is their distinct science sources and the different units of measurement. This alone is complex enough to devise a unique tool to assess the three resources holistically. Amidst this complexity, a recent integrative model that analyses the WEF sectors in an integrated manner was developed by Nhamo et al. and piloted in South Africa (Nhamo et al., 2020a). The tool considers all three sectors equally and applies the MCDM to establish the interlinkages and interrelationships between the sectors. The model further informs policy and decision-makers on priority areas for immediate intervention, making it a decision-support tool (Naidoo et al., 2021c; Nhamo et al., 2020a).

The MCDM was chosen as it is an approach used mainly for structuring and making complex decisions and solving problems that involve multiple criteria or factors of different origins and with different units of measurement (Kumar et al., 2017). With the increasing complexity and diversity of managing resources and their continued depletion, degradation, and insecurity, the sectoral approach is no longer plausible (Liphadzi et al., 2021). The three interlinked WEF resources have been managed independently for a long time using linear approaches with some success; this is no longer plausible as challenges in any one of the three sectors also trigger a host of challenges in the other two sectors (Mabhaudhi et al., 2020). Linear models have now reached their limits, and therefore, there is an urgent need to adopt cross-sectoral transformative and systems models (Naidoo et al., 2021a, 2021c; Nhamo and Ndlela, 2021). Linear approaches promote sector-based resource management, creating optimum efficiencies in selected sectors at the expense of others, thus creating unnecessary tensions and duplicating activities (Mpandeli et al., 2018; Nhamo and Ndlela, 2021). The challenges have been compounded by the worsening climate change and other environmental and social drives of change (Mpandeli et al., 2018). Also, focusing on the developments in one sector only transfers challenges to other sectors (Nhamo et al., 2018a). The MCDM, therefore, becomes useful as it facilitates the integration of socio-economic, environmental, technical, and institutional issues related to resource management using transformative approaches (Liphadzi et al., 2021; Mabhaudhi et al., 2021). As a dynamic approach applicable in various fields, the MCDM is well-supported and widely used as a decision-support tool in various science fields (Kiker et al., 2005). These attributes of the MCDM facilitate linking the WEF nexus and SDGs as both use indicators (Mabhaudhi et al., 2021).

11.2.1 An overview of the Analytic Hierarchy Process (AHP)

Of the various MCDM methods available in the literature, the analytic hierarchy process (AHP) remains the most acceptable and widely used because of its robustness, as demonstrated by comparative studies on MCDM methods (de FSM Russo and Camanho, 2015; Tscheikner-Gratl et al., 2017; Velasquez and Hester, 2013). The completeness of the AHP has seen it being applied in various science fields that, include Environmental Sustainability, Economic Wellbeing, Sociology, Programming, Resource Allocation, Strategic Planning, and Project/Risk Management to integrate diverse and different indicators to monitor performance for benchmarking, policy analysis and decision-making (Cherchye and Kuosmanen, 2004; Dizdaroglu, 2017; Forman and Gass, 2001; Zanella et al., 2013). The AHP has provided useful results in these fields, and recently, the WEF nexus integrates distinct but interlinked indicators, which generally cannot be analysed through linear approaches.

Despite the subjective decisions intrinsic to the AHP, it remains pivotal in policy decisions and performance evaluation as it captures both subjective and objective assessments (Cherchye et al., 2007). The uncertainty embedded in the AHP analysis due to subjective considerations is substituted by incorporating reliable baseline data that indicates the real situation on the ground and the determination of the consistency ratio (Brunelli, 2014; Zhou et al., 2007).

11.2.2   A WEF NEXUS INTEGRATIVE MODEL

The WEF nexus indicators were integrated using the AHP, an MCDM method (Saaty, 1977; Triantaphyllou and Mann, 1995). The aim was to establish the numerical relationships among the WEF nexus components, simplify their intricate relationships, identify priority areas for intervention, minimise trade-offs and maximise synergies. The AHP, introduced by Saaty (1987), is a theory of measurement to derive ratio scales from discrete and continuous paired comparisons to help decision-makers set priorities and make the best decisions. The AHP comparison matrix is determined by comparing two indicators at a time using Saaty's scale, which ranges between 1/9 and 9 (Saaty, 1977).

11.2.3   WEF NEXUS SUSTAINABILITY INDICATORS

The essence of the WEF nexus is its integrated systems approach and cross-sectoral management of resources and its envisaged role in achieving sustainability in resource use and management (Mabhaudhi et al., 2021; Naidoo et al., 2021c). The emphasis is to ensure that the developments and transformations in one sector should only be executed after considering the impacts on the other sectors (Mpandeli et al., 2018; Nhamo et al., 2018b). As sustainability indicators are measurable parameters that evaluate progress towards sustainable development, they are a fitting yardstick to assess progress towards SDGs through the WEF nexus (Nhamo et al., 2020a). The link between SDGs and the WEF nexus is that both address indicators related to resource security, including availability, accessibility, self-sufficiency, and how they influence respective production (productivity) (Bizikova et al., 2013; Nhamo et al., 2020a). The resource security indicators (availability, accessibility, self-sufficiency, and productivity) constitute the key drivers in resource management (Flammini et al., 2017; Lee et al., 2012). The same indicators also form the basis of socio-ecological and environmental sustainability (Rasul and Sharma, 2016).

Within the WEF nexus sustainability indicators, some pillars support the indicators. These pillars are essential when determining the quantitative relationships among indicators. Each WEF nexus sustainability indicator and related pillars are considered when determining the numerical linkages of resource management.

11.3   LINKAGES BETWEEN WEF NEXUS AND SDGS 2, 6, AND 7

Since 2011, the WEF nexus has been promoted as an approach that enhances the cross-sectoral and integrated management of resources, ensuring that any planned developments in one sector should only be implemented after considering the impacts (synergies, trade-offs, and implications) in the other two sectors, as well as its ability to identify different interventional priorities (Mpandeli et al., 2018; Nhamo et al., 2018b, 2019c). Thus, the approach is concerned with resource sustainability and security, which are determined by factors such as availability, accessibility, self-sufficiency, and productivity, from which related indicators are defined and used to measure resource management and sustainability (Bizikova et al., 2014; Nhamo et al., 2019b). Each WEF sector has pillars that sustain respective indicators and contribute

TABLE 11.1

WEF nexus sustainability indicators and the related SDG indicators

Sector	WEF nexus indicator	Related SDG indicator
Water	Proportion of crops/energy produced per unit of water used (productivity)	6.4.1: Change in water-use efficiency over time
	Proportion of available freshwater resources per capita (availability)	6.4.2: Freshwater withdrawal as a proportion of available freshwater resources
Energy	Proportion of population with access to electricity (accessibility)	7.1.1: Proportion of population with access to electricity
	Energy intensity measured in terms of primary energy and GDP (productivity)	7.3.1: Energy intensity measured in terms of primary energy and GDP
Food	Prevalence of moderate or severe food insecurity in the population (self-sufficiency)	2.1.2: Prevalence of moderate or severe food insecurity in the population
	Proportion of sustainable agricultural production per unit area (cereal productivity)	2.4.1: Proportion of agricultural area under productive and sustainable agriculture

Source: Nhamo et al. (2019).

significantly when establishing numerical relationships between indicators during the pairwise comparison matrix of the AHP (Nhamo et al., 2020a). Thus, WEF nexus sustainability indicators and pillars (Table 11.1) are directly linked to related SDGs and are essential for evaluating SDGs implementation progress (Mabhaudhi et al., 2021; Nhamo et al., 2019c). Both the WEF nexus and SDGs serve the same purpose of ending poverty and achieving economically and environmentally sustainable outcomes (Naidoo et al., 2021c). The former serves as an approach to spearhead the implementation of WEF nexus–linked SDGs. Table 11.1 lists WEF nexus indicators and the related SDG indicators.

The emphasis is on indicators that directly fall under the WEF nexus framework and speak to the security of water, energy, and food resources and are framed to improve integrated efficiencies in resource use and management to attain sustainability. These WEF nexus attributes are reflected in SDGs 2, 6, and 7. The capability of the WEF nexus to establish an integrated numerical relationship between interlinked sectors and give an overall synopsis of resource management sustainability over time simplifies the monitoring and assessment of progress in implementing related SDGs indicators (Mabhaudhi et al., 2021).

11.3.1 Interlinkages between WEF resources and related SDGs

Access to WEF resources has long been recognised as the main driver of socio-economic development. Environmental changes and the three resources constitute the basic components for sustainable livelihoods, forming the basis for sustainable development (Allen and Prosperi, 2016; Nhamo and Ndlela, 2021). Therefore, any progress to meet SDGs is hinged on the extent to which WEF resources are planned, accessed,

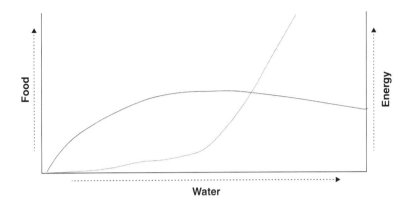

FIGURE 11.1   An imaginary example of the intricate interlinkages between water, energy and food resources over time and the impact on achieving related SDGs.

and managed (Mensah and Ricart Casadevall, 2019). As the three resources are interlinked, transitioning towards a more sustainable and resilient future requires integrated and transformative approaches that recognise that contemporary challenges and their solutions are also interrelated (Naidoo et al., 2021c). Thus, integrating the three basic resources is key to achieving sustainable development. Figure 11.1 demonstrates the interlinkages between the WEF resources and provides the rationale for systems thinking. It establishes the interconnectedness between the three resources, the need for holistic thinking, and the potential for "leverage points" for policy intervention.

The demand for water, energy, and food resources has increased over time due to the increasing population when the resource base has depleted due to unsustainable resource management practices. Food supply has not been at pace with population growth (Nhamo et al., 2019d), energy consumption will almost double by 2050 (Ferroukhi et al., 2015), and freshwater resources continue to deplete due to a combination of climate change and increased use particularly in irrigated agriculture (Turral et al., 2011). These changes, coupled with the continued use of linear or sectoral approaches, will only exacerbate the existing challenges and are a risk to achieving the SDGs. Given these challenges and the need for transformative approaches addressing today's interlinked and cross-sectoral challenges, the proposed framework is intended to provide policymakers with pathways to achieve integration across the policy cycle and enhance resilience and adaptation-building strategies at all levels and scales.

11.4   WEF NEXUS FRAMEWORK TO ASSESS SUSTAINABLE DEVELOPMENT

This study establishes a framework to guide policy and decision-makers to assess progress towards SDGs and enhance resource security. The framework is based on the WEF nexus analytical model developed by Nhamo et al. (2020a), which focuses on WEF nexus sustainability indicators and establishes the linkages between WEF nexus indicators and related SDGs indicators. This developed framework

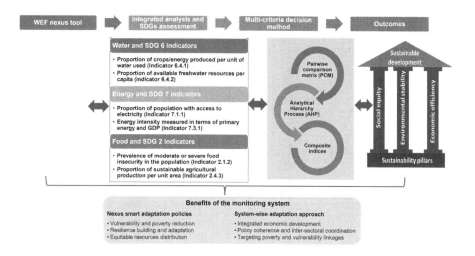

FIGURE 11.2   A WEF nexus–based framework to assess progress towards achieving related Sustainable Development Goals.

(Figure 11.2) provides the required pathways to assess the progress towards sustainable development over time.

The framework guides strategic policy formulation towards a sustainable future by establishing an integrated cross-sectoral numerical relationship among different but interlinked sectors. The framework guides coherent strategic decisions towards sustainability and resource security, indicating priority areas needing urgent interventions (Nhamo et al., 2020a).

The framework provides an overview of the interlinkages among different WEF nexus and SDGs sustainability indicators using the AHP, MCDM. The AHP determines composite indices for each indicator to relate the differing WEF sectors quantitatively. These attributes are critical in integrated and cross-sectoral resource use and management. The WEF nexus tool block (the block to the left of Figure 11.2) establishes the essence of the tool and what the WEF nexus approach is envisaged to achieve. The success of the WEF nexus as a framework for achieving sustainability relies on good governance that underpins the WEF nexus implementation towards socio-ecological sustainability (Bizikova et al., 2013; Hoff, 2011).

The block on integrated analysis and SDGs assessment (second block of Figure 11.2) establishes the linkages between WEF nexus indicators (Nhamo et al., 2019b) and related SDG indicators (https://unstats.un.org/sdgs/metadata/). An assessment of the progress towards achieving sustainability by 2030 (UNGA, 2015) through WEF nexus-related SDG indicators is achieved by evaluating resource use and management at a given time interval to achieve set SDG targets (Mabhaudhi et al., 2021; Naidoo et al., 2021c; Nhamo et al., 2021a). The progress towards a sustainable future is assessed by comparing changes over successive periodic intervals (Mabhaudhi et al., 2021). As the WEF nexus identifies priority areas for intervention, it becomes an integrated decision-support tool for implementing sustainability strategies and ensuring economic efficiency and social equity to achieve sustainability by 2030.

The third block (the MCDM block) represents the model adopted in the framework to quantitatively establish relationships and interlinkages among the WEF sectors at any given spatio-temporal scale (Mabhaudhi et al., 2021; Nhamo et al., 2020a). The fourth block represents the outcomes of the assessment, including social equity, environmental stability, and economic efficiency (Figure 11.2), the three pillars of sustainable development (UNGA, 2015).

The benefits derived from implementing the SDGs and assessing progress through the systematic and integrated approach of the WEF nexus are given at the bottom of the framework. The success in meeting SDG targets is assessed by the level achieved in promoting prosperity while protecting the planet by not exceeding the planetary boundaries in resource exploitation and use (Kimani-Murage et al., 2021; Nhamo et al., 2019c). Good governance is a prerequisite in resource management. It provides political will and reduces vulnerability and poverty, enhancing resilience and adaptation to climate change, equitable resource distribution, integrated economic development, policy coherence, and inter-sectoral coordination.

11.4.1 ASSESSING PROGRESS TOWARDS SDGS OVER TIME

Developing an integrated WEF nexus analytical framework to achieve sustainability is important for research and policy as it guides the formulations of coherent strategies that drive towards sustainability and resource security (Mabhaudhi et al., 2021). The WEF nexus analytical model (Nhamo et al., 2020a) facilitates understanding complex relationships amongst the interlinked sectors and informs a holistic and integrated approach to resource management in achieving SDGs. To achieve the SDGs and enhance resource security, and considering their complexity and multidisciplinary nature, the multi-criteria analysis framework provides the decision-support tools that facilitate quantifying the relationships among different but interlinked sectors (Nhamo et al., 2020a). The numerical representation of the interlinked WEF sectors informs strategic policy formulations that lead to sustainability. The level of progress in achieving SDG consists of an evaluation of the progress made towards set goals between different time frames (Figure 11.3).

The assessment includes measuring the gap between the previous years and the present status. In the example in Figure 11.3, the 2020 graph indicates progress towards achieving related SDG indicators from 2015. However, the progress could be positive or negative, and the model indicates areas needing immediate intervention in the case of negative progress.

The developed framework provides a holistic assessment and intervention by considering all the interlinked sectors equally, including stakeholders, footprints of water production, distribution, and allocation between the linked sectors, such as energy costs (Naidoo et al., 2021b). This is critical for long-term and sustainable management decisions. Environmental footprints provide insightful indicators that guide informed analysis of WEF resources by quantifying synergies and trade-offs along the whole supply chain (Vanham et al., 2019). Identifying these supply chain footprints (diet behaviour, reduction of food losses, and waste) links the WEF nexus with the Circular Economy approach. Both aim to reduce the effects of consumptive and degradative resource utilisation (Naidoo et al., 2021b). This also enhances

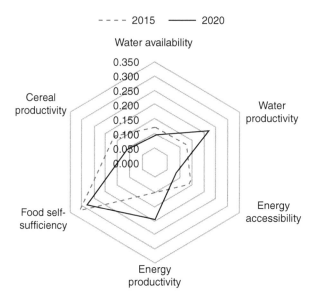

FIGURE 11.3 Management of resources over time and the progress towards related SDGs.

the notion that transformative approaches are interlinkages and inform each other (Naidoo et al., 2021c).

The WEF nexus–based analytical framework proposed in this study can identify and balance both synergies and trade-offs of an evidence-based nexus practice as it provides an overall overview of the current status of resource management. The framework is developed to be applicable at any spatio-temporal scale, depending on data availability (Nhamo et al., 2020b).

11.4.2 DATA SOURCES AND AVAILABILITY

Although the WEF nexus has been recognised as an important decision-support tool that assesses progress towards SDGs, and its application has been gathering momentum worldwide, its progress has been hampered by data unavailability and heterogeneity (Naidoo et al., 2021c). Data availability is a key input during the weighting of sustainability indicators in an AHP's pairwise comparison (Nhamo et al., 2019a). Data uniformity is critical for comparison purposes, particularly for different regions or countries. The challenges associated with data collection variations and storage include data disparity, mismatch, and plurality (Liu et al., 2017; Naidoo et al., 2021c). Data availability facilitates the assessment of trade-offs and synergies to minimise conflicts and is vital to sustainable development (Giampietro, 2018). Therefore, data availability forms the basis for determining indicator weights during the pairwise comparison in the AHP.

However, at the national and regional levels, WEF nexus-related data are often accessed from open-source databases like FAOSTAT, AQUASTAT, and World Bank Indicators. National statistical agents can also provide data at the national level.

Still, the challenge could emerge when comparing two or more countries as data collection methods and storage platforms differ from country to country. Remotely sensed data has, however, shown to be an important long-term data source for some sustainability indicators where data is not readily available (Giuliani et al., 2017; Makapela et al., 2015). The Landsat Mission, for example, has been providing uninterrupted land and atmospheric information since 1972.

11.5 THE SIGNIFICANCE OF THE FRAMEWORK

As a transformative and multicentric approach, the WEF nexus is envisaged to replace linear approaches or sector-based planning by providing more holistic and integrated solutions to resource management for sustainable development (Naidoo et al., 2021c; Nhamo and Ndlela, 2021). WEF nexus planning acknowledges that all sectors influence and drive each other and should be viewed, analysed, and implemented simultaneously without prioritising one above the others (Naidoo et al., 2021c; Nhamo et al., 2020a). These attributes promote integrated socio-economic development by informing sustainable development as resources are managed systematically and holistically (Naidoo et al., 2021c; Nhamo et al., 2020a). The developed WEF nexus framework can enhance resource security without compromising ecosystem services as nexus planning facilitates the understanding and systematic analysis of socio-ecological interactions in these three sectors and promotes human and environmental health (Naidoo et al., 2021c; Nhamo and Ndlela, 2021). The key is its capability to enhance a coordinated and sustainable cross-sectoral management of resources at various spatio-temporal scales (Naidoo et al., 2021c).

Thus, the WEF nexus is a consultative platform for stakeholder engagement where integrated solutions are deliberated and implemented (Ghodsvali et al., 2019). It is intended to break the traditional thematic silos as current developments indicate a need to transition towards transformative approaches from the current linear models that are shown to have reached their limits as evidenced by contemporary challenges that are interlinked and cross-sectoral (Naidoo et al., 2021c). Focusing on single sectors will only exacerbate existing challenges, as evidenced during the COVID-19 pandemic, where policy and decision-makers focused on health issues. Yet, the lockdowns resulted in company closures, unemployment, and worsening poverty, among other challenges (Nhamo and Ndlela, 2021). The stakeholder consultation through the WEF nexus has the potential to:

- Promote cross-sectoral engagements among interlinked sectors and formulate coherent strategies and policies that enhance resilience and adaptation.
- Identify cross-sectoral priority areas needing intervention and address synergies and trade-offs holistically.
- Promote cross-sectoral resource management and develop a shared understanding of the pertinent challenges, objectives, and scenarios.
- Develop shared frameworks, strategies, and policies that promote collaborative interventions through multi-sectoral and multiscale collaborations towards achieving the SDGs.

11.6   CONCLUSIONS

The WEF nexus has become a useful decision-support tool to guide integrated solutions in resource management, enhance resilience and adaptive capacities to climate change, and promote strategic resource governance. This study has developed a framework to assist policymakers in addressing how to achieve integration across the policy cycle and assessing levels of integration at any spatio-temporal scale. The framework guides strategic policy decisions and aids the quantification of WEF sectors' interactions. The framework provides the pathways to identify trade-offs and synergies and provides the lens through which priority areas for intervention are identified. This is critical for monitoring SDG progress and informing corrective measures needed for effective, sustainable development. Apart from being a decision-support tool for integrated resources management, the WEF nexus is an important pathway for addressing the challenges of poverty, unemployment, and inequality. The approach promotes integrated planning, decision-making, governance, and management of resources, the needed attributes to achieving simultaneous water, energy, food security, job and wealth creation, and sustainable natural resources management. The developed framework provides guidelines enhancing cross-sectoral cooperation and mitigating conflicts, increasing resource-use efficiencies. The common attributes between the WEF nexus and SDGS lead to the (a) promotion of sustainable and efficient resource use, (b) access to resources for vulnerable population groups, (c) maintenance and support of underlying ecosystem services, and (d) improving human wellbeing. Importantly, the WEF nexus approach addresses the five key elements of the SDGs, i.e. People, Planet, Prosperity, Peace, and Partnership, the main factors needed to achieve sustainability.

REFERENCES

Albrecht, T.R., Crootof, A., Scott, C.A. (2018) The Water-Energy-Food nexus: A systematic review of methods for nexus assessment. *Environmental Research Letters* 13, 043002.
Allen, T., Prosperi, P. (2016) Modeling sustainable food systems. *Environmental Management* 57, 956–975.
Amarasinghe, U.A., Shah, T., Smakhtin, V. (2012) Water–milk nexus in India: a path to a sustainable water future? *International Journal of Agricultural Sustainability* 10, 93–108.
Bizikova, L., Roy, D., Swanson, D., Venema, H.D., McCandless, M. (2013) The water-energy-food security nexus: Towards a practical planning and decision-support framework for landscape investment and risk management. International Development Research Centre (IDRC), Winnipeg, Manitoba, Canada.
Bizikova, L., Roy, D., Venema, H.D., McCandless, M., Swanson, D., Khachtryan, A., Borden, C., Zubrycki, K. (2014) Water-energy-food nexus and agricultural investment: A sustainable development guidebook. International Institute for Sustainable Development (IISD), Winnipeg, Canada.
Brunelli, M. (2014) *Introduction to the analytic hierarchy process.* Springer, Alto, Finland.
Cherchye, L., Kuosmanen, T. (2004) Benchmarking sustainable development: A synthetic meta-index approach. Research Paper, UNU-WIDER, United Nations University (UNU).
Cherchye, L., Moesen, W., Rogge, N., Van Puyenbroeck, T. (2007) An introduction to 'benefit of the doubt' composite indicators. *Social Indicators Research* 82, 111–145.
Confalonieri, U.E., Schuster-Wallace, C.J. (2011) Data integration at the water–health nexus. *Current Opinion in Environmental Sustainability* 6, 512–516.

Constant, N.L., Taylor, P.J. (2020) Restoring the forest revives our culture: Ecosystem services and values for ecological restoration across the rural-urban nexus in South Africa. *Forest Policy and Economics* 118, 102222.

Daher, B.T., Mohtar, R.H. (2015) Water–energy–food (WEF) Nexus Tool 2.0: Guiding integrative resource planning and decision-making. *Water International* 40, 748–771.

Dargin, J., Daher, B., Mohtar, R.H. (2019) Complexity versus simplicity in water energy food nexus (WEF) assessment tools. *Science of the Total Environment* 650, 1566–1575.

de FSM Russo, R., Camanho, R. (2015) Criteria in AHP: A systematic review of literature. *Procedia Computer Science* 55, 1123–1132.

Díaz, S., Demissew, S., Carabias, J., Joly, C., Lonsdale, M., Ash, N., Larigauderie, A., Adhikari, J.R., Arico, S., Báldi, A. (2015) The IPBES conceptual framework—Connecting nature and people. *Current Opinion in Environmental Sustainability* 14, 1–16.

Dizdaroglu, D. (2017) The role of indicator-based sustainability assessment in policy and the decision-making process: A review and outlook. *Sustainability* 9, 1018.

Endo, A., Tsurita, I., Burnett, K., Orencio, P.M. (2017) A review of the current state of research on the water, energy, and food nexus. *Journal of Hydrology: Regional Studies* 11, 20–30.

FAO (2014) *The Water-Energy-Food nexus: A new approach in support of food security and sustainable agriculture.* Food and Agriculture Organisation of the United Nations (FAO), Rome, Italy, p. 28.

Ferroukhi, R., Nagpal, D., Lopez-Peña, A., Hodges, T., Mohtar, R.H., Daher, B., Mohtar, S., Keulertz, M. (2015) *Renewable energy in the water, energy & food nexus.* The International Renewable Energy Agency (IRENA), Masdar City, pp. 1–125.

Flammini, A., Puri, M., Pluschke, L., Dubois, O. (2017) *Walking the nexus talk: Assessing the water-energy-food nexus in the context of the sustainable energy for all initiative.* FAO, Rome.

Forman, E.H., Gass, S.I. (2001) The analytic hierarchy process—An exposition. *Operations Research* 49, 469–486.

Ghodsvali, M., Krishnamurthy, S., de Vries, B. (2019) Review of transdisciplinary approaches to food-water-energy nexus: A guide towards sustainable development. *Environmental Science & Policy* 101, 266–278.

Giampietro, M. (2018) Perception and representation of the resource nexus at the interface between society and the natural environment. *Sustainability* 10, 2545.

Giampietro, M., Giampietro, M., Richard, J., Aspinall, R., Bukkens, S.G.F., Benalcazar, J.C., Diaz-Maurin, F., Flammini, A., Gomiero, T., Kovacic, Z., Madrid, C., Jesús, R.M., Tarik, S.T. (2013) An innovative accounting framework for the Food-energy-water nexus: Application of the MuSIASEM approach to three case studies, environment and natural resources. Food and Agriculture Aorganisation of the United Nations (FAO), Rome, Italy.

Giuliani, G., Chatenoux, B., De Bono, A., Rodila, D., Richard, J.-P., Allenbach, K., Dao, H., Peduzzi, P. (2017) Building an earth observations data cube: Lessons learned from the swiss data cube (sdc) on generating analysis ready data (ard). *Big Earth Data* 1, 100–117.

Hatfield-Dodds, S. (2015) Australian National Outlook 2015 Technical Report: Economic activity, resource use, environmental performance and living standards, 1970–2060, CSIRO Impact Science. CSIRO, Canberra, Australia, p. 56.

Hoff, H. (2011) Understanding the nexus: Background paper for the Bonn2011 Conference: The Water, Energy and Food Security Nexu. Stockholm Environment Institute (SEI), Stockholm, Sweden, p. 52.

Hülsmann, S., Ardakanian, R. (2014) White book on advancing a nexus approach to the sustainable management of water, soil and waste. UNU-FLORES, Dresden, Germany.

Kiker, G.A., Bridges, T.S., Varghese, A., Seager, T.P., Linkov, I. (2005) Application of multicriteria decision analysis in environmental decision making. *Integrated Environmental Assessment and Management* 1, 95–108.

Kimani-Murage, E., Gaupp, F., Lal, R., Hansson, H., Tang, T., Chaudhary, A., Nhamo, L., Mpandeli, S., Mabhaudhi, T., Headey, D.D. (2021) An optimal diet for planet and people. *One Earth* 4, 1189–1192.

Kumar, A., Sah, B., Singh, A.R., Deng, Y., He, X., Kumar, P., Bansal, R. (2017) A review of multi criteria decision making (MCDM) towards sustainable renewable energy development. *Renewable and Sustainable Energy Reviews* 69, 596–609.

Lee, B., Preston, F., Kooroshy, J., Bailey, R., Lahn, G. (2012) *Resources futures*. Chatham House, London.

Lehmann, S. (2018) Implementing the Urban Nexus approach for improved resource-efficiency of developing cities in Southeast-Asia. *City, Culture and Society* 13, 46–56.

Liphadzi, S., Mpandeli, S., Mabhaudhi, T., Naidoo, D., Nhamo, L. (2021) The evolution of the water–energy–food nexus as a transformative approach for sustainable development in South Africa, in: Muthu, S. (Ed.), *The water–energy–food nexus: Concept and assessments*. Springer, Kowloon, Hong Kong, pp. 35–67.

Liu, J., Yang, H., Cudennec, C., Gain, A., Hoff, H., Lawford, R., Qi, J., Strasser, L.d., Yillia, P., Zheng, C. (2017) Challenges in operationalizing the water–energy–food nexus. *Hydrological Sciences Journal* 62, 1714–1720.

Mabhaudhi, T., Mpandeli, S., Luxon Nhamo, V.G., Chimonyo, A.S., Naidoo, D., Liphadzi, S., Modi, A.T. (2020) Emerging water-energy-food nexus lessons, experiences, and opportunities in Southern Africa, in: Ahmad, V.-B.-H., David, S.T. (Eds.), *Environmental management of air, water, agriculture, and energy*. CRC Press, Boca Raton, Florida, p. 141.

Mabhaudhi, T., Nhamo, L., Chibarabada, T.P., Mabaya, G., Mpandeli, S., Liphadzi, S., Senzanje, A., Naidoo, D., Modi, A.T., Chivenge, P.P. (2021) Assessing progress towards sustainable development goals through nexus planning. *Water* 13, 1321.

Mabrey, D., Vittorio, M. (2018) Moving From theory to practice in the water-energy-food nexus: An evaluation of existing models and frameworks. *Water-Energy Nexus* 1(1), 17–25.

Makapela, L., Newby, T., Gibson, L., Majozi, N., Mathieu, R., Ramoelo, A., Mengistu, M., Jewitt, G., Bulcock, H., Chetty, K. (2015) Review of the use of earth observations remote sensing in water resource management in South Africa. Water Research Commission (WRC), Pretoria, South Africa, p. 153.

McGrane, S.J., Acuto, M., Artioli, F., Chen, P.Y., Comber, R., Cottee, J., Farr-Wharton, G., Green, N., Helfgott, A., Larcom, S. (2018) Scaling the nexus: Towards integrated frameworks for analysing water, energy and food. *The Geographical Journal,* 185(4), 419–431.

Mensah, J., Ricart Casadevall, S. (2019) Sustainable development: Meaning, history, principles, pillars, and implications for human action: Literature review. *Cogent Social Sciences* 5, 1653531.

Mpandeli, S., Naidoo, D., Mabhaudhi, T., Nhemachena, C., Nhamo, L., Liphadzi, S., Hlahla, S., Modi, A. (2018a) Climate change adaptation through the water-energy-food nexus in southern Africa. *International Journal of Environmental Research and Public Health* 15, 2306.

Naidoo, D., Liphadzi, S., Mpandeli, S., Nhamo, L., Modi, A.T., Mabhaudhi, T. (2021a) Post Covid-19: A water-energy-food nexus perspective for South Africa, in: Stagner, J., Ting, D. (Eds.), *Engineering for sustainable development and living: Preserving a future for the next generation to Cherish*. Brown Walker Press, Florida, p. 295.

Naidoo, D., Nhamo, L., Lottering, S., Mpandeli, S., Liphadzi, S., Modi, A.T., Trois, C., Mabhaudhi, T. (2021b) Transitional pathways towards achieving a circular economy in the water, energy, and food sectors. *Sustainability* 13, 9978.

Naidoo, D., Nhamo, L., Mpandeli, S., Sobratee, N., Senzanje, A., Liphadzi, S., Slotow, R., Jacobson, M., Modi, A., Mabhaudhi, T. (2021c) Operationalising the water-energy-food nexus through the theory of change. *Renewable and Sustainable Energy Reviews* 149, 10.

Nhamo, L., Mabhaudhi, T., Mpandeli, S. (2019a) A model to integrate and assess water-energy-food nexus performance: South Africa case study. World Irrigation Forum (WIF), Bali, Indonesia, p. 10.

Nhamo, L., Mabhaudhi, T., Mpandeli, S., Dickens, C., Nhemachena, C., Senzanje, A., Naidoo, D., Liphadzi, S., Modi, A.T. (2020a) An integrative analytical model for the water-energy-food nexus: South Africa case study. *Environmental Science and Policy* 109, 15–24.

Nhamo, L., Mabhaudhi, T., Mpandeli, S., Nhemachena, C., Senzanje, A., Naidoo, D., Liphadz, S., Modi, A.T. (2019b) Sustainability indicators and indices for the water-energy-food nexus for performance assessment: WEF nexus in practice - South Africa case study. Preprint 2019050359, 17.

Nhamo, L., Mabhaudhi, T., Mpandeli, S., Nhemachena, C., Sobratee, N., Naidoo, D., Liphadz, S., Modi, A. (2019c) Sustainability indicators and indices for the water-energy-food nexus for performance assessment: WEF nexus in practice. *MDPI Preprint*, 109, 15–24.

Nhamo, L., Matchaya, G., Mabhaudhi, T., Nhlengethwa, S., Nhemachena, C., Mpandeli, S. (2019d) Cereal production trends under climate change: Impacts and adaptation strategies in southern Africa. *Agriculture* 9, 30.

Nhamo, L., Mpandeli, S., Senzanje, A., Liphadzi, S., Naidoo, D., Modi, A.T., Mabhaudhi, T. (2021a) Transitioning toward sustainable development through the water–energy–food nexus, in: Ting, D., Carriveau, R. (Eds.), *Sustaining tomorrow via innovative engineering*. World Scientific, Singapore, pp. 311–332.

Nhamo, L., Ndlela, B. (2021) Nexus planning as a pathway towards sustainable environmental and human health post Covid-19. *Environment Research* 192, 110376.

Nhamo, L., Ndlela, B., Mpandeli, S., Mabhaudhi, T. (2020b) The water-energy-food nexus as an adaptation strategy for achieving sustainable livelihoods at a local level. *Sustainability* 12, 8582.

Nhamo, L., Ndlela, B., Nhemachena, C., Mabhaudhi, T., Mpandeli, S., Matchaya, G. (2018a) The water-energy-food nexus: Climate risks and opportunities in southern Africa. *Water* 10, 567.

Nhamo, L., Ndlela, B., Nhemachena, C., Mabhaudhi, T., Mpandeli, S., Matchaya, G. (2018b) The water-energy-food nexus: Climate risks and opportunities in Southern Africa. *Water* 10, 18.

Nhamo, L., Rwizi, L., Mpandeli, S., Botai, J., Magidi, J., Tazvinga, H., Sobratee, N., Liphadzi, S., Naidoo, D., Modi, A., Slotow, R., Mabhaudhi, T. (2021b) Urban nexus and transformative pathways towards a resilient Gauteng City-Region, South Africa. *Cities* 116, 103266.

Rasul, G., Sharma, B. (2016) The nexus approach to water–energy–food security: an option for adaptation to climate change. *Climate Policy* 16, 682–702.

Saaty, R.W. (1987) The analytic hierarchy process—What it is and how it is used. *Mathematical Modelling* 9, 161–176.

Saaty, T.L. (1977) A scaling method for priorities in hierarchical structures. *Journal of Mathematical Psychology* 15, 234–281.

Scott, C.A., Kurian, M., Wescoat, J.L. (2015) The water-energy-food nexus: enhancing adaptive capacity to complex global challenges. In: Kurian, M., Ardakanian, R. (Eds) *Governing the nexus*. Springer, Cham, pp. 15–38.

Terrapon-Pfaff, J., Ortiz, W., Dienst, C., Gröne, M.-C. (2018) Energising the WEF nexus to enhance sustainable development at local level. *Journal of Environmental Management* 223, 409–416.

Triantaphyllou, E., Mann, S.H. (1995) Using the analytic hierarchy process for decision making in engineering applications: some challenges. *International Journal of Industrial Engineering: Applications and Practice* 2, 35–44.

Tscheikner-Gratl, F., Egger, P., Rauch, W., Kleidorfer, M. (2017) Comparison of multi-criteria decision support methods for integrated rehabilitation prioritization. *Water* 9, 68.

Turral, H., Burke, J., Faurès, J.-M. (2011) *Climate change, water and food security*. Food and Agriculture Organization (FAO), Rome, Italy.

UN (2014) Prototype Global Sustainable Development Report. United Nations Department of Economic and Social Affairs (UNDESA), New York, p. 162.

UNGA (2015) Transforming our world: The 2030 Agenda for Sustainable Development, Resolution adopted by the General Assembly (UNGA). United Nations General Assembly, New York, p. 35.

Vanham, D., Leip, A., Galli, A., Kastner, T., Bruckner, M., Uwizeye, A., Van Dijk, K., Ercin, E., Dalin, C., Brandão, M. (2019) Environmental footprint family to address local to planetary sustainability and deliver on the SDGs. *Science of the Total Environment* 693, 133642.

Velasquez, M., Hester, P.T. (2013) An analysis of multi-criteria decision making methods. *International Journal of Operations Research* 10, 56–66.

Venghaus, S., Hake, J.-F. (2018) Nexus thinking in current EU policies–The interdependencies among food, energy and water resources. *Environmental Science & Policy* 90, 183–192.

Waughray, D. (2011) *Water securitythe water-food-energy-climate nexus: The World Economic Forum water initiative*. Island Press, Washington, DC.

Zanella, A., Camanho, A., Dias, T.G. (2013) Benchmarking countries' environmental performance. *Journal of the Operational Research Society* 64, 426–438.

Zhou, P., Ang, B., Poh, K. (2007) A mathematical programming approach to constructing composite indicators. *Ecological Economics* 62, 291–297.

12 Understanding the nexus between water, energy and food in the context of climate change adaptation
A river basin perspective

*Nosipho Dlamini, Aidan Senzanje,
Cuthbert Taguta, Tinashe Lindel Dirwai,
and Tafadzwanashe Mabhaudhi*

12.1 INTRODUCTION

In the past decade, increasing attention has been given to adaptation within the global climate change think-tank space (Huang et al., 2018). The recent International Panel on Climate Change Fifth and Sixth Assessment reports indicate that climate change will likely directly impact the water sector and agriculture, energy, domestic, and other sectors (Caretta et al., 2022). Past climate change trends for southern Africa have projected temperature increases and increased frequencies of extreme weather events in the form of droughts and floods (Olabanji et al., 2020). Therefore, climate change adaptation strategies are critically important, especially in developing regions like South Africa, because of their vulnerabilities emanating from limited resources and adaptive capacity and large dependency on climate-sensitive sectors such as agriculture, forestry, and fisheries (Filho et al., 2019; Rapholo and Makia, 2020). The formulation of adaptation strategies thus requires urgent attention anchored on transdisciplinary or cross-sectoral methods, owing to the complexity of climate change impacts and the fact that they cut across several sectors and levels (Nhamo et al., 2018).

In their 2030 vision, the World Health Organization acknowledged the significance of water supply resilience in climate change. The United Nations (UN) also acknowledged that effective water management is essential to maintaining sustainable development and advised strengthening institutional stability and boosting infrastructure investment to increase long-term climate change resilience (Butler et al., 2017). As much as there has been a progressive shift in the way water management is approached in research, with resilience and adaptation of water resources to climate

DOI: 10.1201/9781003327615-12

223

change gaining traction, delayed progress has been made in putting these concepts into practice (Butler et al., 2017; Caretta et al., 2022). Successful water-related climate change adaptation remains severely constrained by institutional barriers (Caretta et al., 2022). With water management practices partly shaped by legislation (Dube et al., 2021), the potential for integrated policy is limited by a lack of intersectoral coordination and communication within institutions and competing interests between water sectors (Caretta et al., 2022).

In South Africa, understanding a catchment's surface water hydrology has driven local water resource planning and management over the years (Dube et al., 2021). However, water resource management plans in certain South African catchments are vague and sectoral, negatively influencing other sectors. For example, the Olifants-Doorn Water Management Area (WMA), located on the west coast of South Africa, is a semi-arid river basin where the agricultural sector has been put at the forefront in terms of water governance; 95% of the WMAs' total water requirements are allocated to it, which, in turn, increased waterlogging, salination and the over-abstraction of aquifers (Knuppe and Meissner, 2016). Such outcomes have resulted in the degradation of land and water resources and the loss of associated ecosystem services (Knuppe and Meissner, 2016). A similar case is presented for the Breede River catchment in western South Africa, whereby the prioritization of the agricultural sector has caused a decline in water quality, noted by the concerningly high levels of phosphorus and chemical oxygen demand. As a result, poor water quality has reduced living standards and social well-being, as well as the availability of water for water-dependent sectors (Cullis et al., 2018).

Apart from being sectoral, studies and approaches to climate change adaptation are largely based on national, regional, and global scales, thus neglecting climate change-related dynamics and their implications at the basin level (Olabanji et al., 2020). With the multiple stresses imposed by climate change, adaptation requires comprehensive and multidisciplinary approaches, with coordination and a better understanding of the impacts between different sectors and at different scales (Lele et al., 2018). Although the likely impacts of climate change on water, energy, and food production have raised serious concerns and have been emphasized in the pursuit of appropriate adaptation measures, Mabhaudhi et al. (2018b) highlighted that the nexus among water, energy, and food in South Africa has not been well researched. This lack of coordination, communication, and collaboration within sectors may significantly impact the efficiency and effectiveness of policies, thus impeding suitable measures from being implemented and resulting in inefficient resource utilization (Altamirano et al., 2018; Nhamo et al., 2020b). Similarly, previous case studies on the WEF nexus approach in South Africa have focused on spatial scales other than the catchment scale, for example, regional (Mabhaudhi et al., 2019), national (Nhamo et al., 2020a), provincial (Adom et al., 2022; Simpson et al., 2019), local municipality (Hulley, 2015; Nhamo et al., 2020b), farm (Seeliger et al., 2018), and household (Ningi et al., 2021) scales. A few WEF nexus studies have delved into the catchments, including Berg (Western Cape), Keiskamma (Eastern Cape), and uMgeni (KwaZulu-Natal) (Methner et al., 2021; Midgley et al., 2014). This leaves a critical gap in understanding how water, energy and food interplay in catchments, a spatial scale wherein water resources management decisions are deliberated in South Africa (DWAF, 2007; DWAF and WRC, 1996; Materechera, 2012; Molobela, 2011).

The Buffalo River catchment, located in KwaZulu-Natal, South Africa, presents an excellent case study for assessing climate change and silo-based resource management's implications on WEF resources at the catchment scale. This catchment faces significant problems adapting to the multiple effects of climate change, notably regarding water management. According to uMgeni (2020), the water distribution system in the high-rainfall Buffalo River catchment (on average, 802 mm/annum) has not been able to meet demand in recent years, and the droughts of 2015/2016 aggravated the situation. uMgeni (2020) further declared that the yield of the Ntshingwayo Dam, the Buffalo River catchment's largest water source, will not be sufficient to supply the 2035 water demands. Conversely, Dlamini and Mostert (2019) highlighted that the Buffalo River catchment has surplus water which can be allocated; however, current allocation plans need to be revised. Water management plans must be revised to eliminate imbalances in water distribution among users to provide energy and water to the growing population of approximately 0.7 million and expand agricultural production beyond the current 53% of the potential irrigated areas being utilized; water management plans must be revised to eliminate imbalances in water distribution among users (Dlamini and Mostert, 2019). In this regard, emerging cross-sectoral approaches such as the WEF nexus could be useful.

The WEF nexus is generally characterized as a resource management strategy considering the interactions (synergies, trade-offs) among water, energy, and food for harmonisation and optimization (Mpandeli et al., 2018). In contrast to the water-centric Integrated Water Resources Management (IWRM) approach, the WEF nexus is poly-centric as it examines all resources equally and manages them more holistically (Simpson and Jewitt, 2019), thus enabling more integrated and cost-effective planning, decision-making, implementation, monitoring, and evaluation (Altamirano et al., 2018). The WEF nexus has been extensively researched, with numerous studies fixating their analysis on niches or sectors from a political, social, or scientific standpoint (Mpandeli et al., 2018).

Using the Buffalo River catchment in the KwaZulu-Natal province, South Africa, as a case study, we propose applying the WEF nexus as a viable tool for developing multi-sector climate change adaptation strategies for integrated and efficient resource management at the catchment level. This chapter discusses what adaptation means in today's climate-sensitive society and the possible impact on resource security. The chapter is organised as follows: the next section (Section 12.2) talks about adaptation to climate change from a WEF nexus lens. After that, Section 12.3 discusses the key challenges of water, food, and energy security. Section 12.4 presents the case study on the Buffalo River catchment. Lastly, we provide the conclusions and recommendations.

12.2   ADAPTATION TO CLIMATE CHANGE THROUGH THE WEF NEXUS

The principal objectives of adaptation are to increase the respective region's adaptive capacity and resilience and reduce vulnerability to climate and non-climatic changes

(Owen, 2020; Phan et al., 2020). Adaptation is inextricably linked to achieving sustainable water, energy, and food management, as it addresses water, energy, and food security challenges in unison (Rasul and Sharma, 2016).

Before the advent of the WEF nexus concept, the approach to climate change adaptation remained sectoral. Sectoral adaptation approaches potentiate maladaptation through reducing capacity or increasing risks in another area or sector, weakening resilience, and increasing vulnerability (Mpandeli et al., 2018), and the likelihood of compounded challenges in the future (Nhamo et al., 2018; Rasul and Sharma, 2016). For instance, establishing a large, clean and green renewable energy power plant (e.g., solar photovoltaic) on arable land in agro-based locations can lead to energy security at the expense of food security at a local scale (Brunet et al., 2020). Similarly, energy subsidies for groundwater extraction to cope with surface water shortages and uncertainty in water availability in agricultural production can lead to overexploitation and wastage of groundwater, increased demand for energy, and ultimately undermining WEF security (Rasul and Sharma, 2016). Thus, in resource management and development, there is a need to deliberately acknowledge that WEF sectors and resources do not exist in isolation but coexist in an inextricable web of connections termed the nexus.

Integrated approaches to resources management, such as IWRM and nexus thinking, seek to accelerate action that culminates in solutions, including efficient resource use and sustainable development (Benson et al., 2015; Smith and Clausen, 2015). The more established IWRM has been around for several decades and is defined as a process that promotes the coordinated development and management of water, land, and related resources to maximize the resultant economic and social welfare equitably without compromising the sustainability of vital ecosystems (GWP-TAC, 2000; Smith and Clausen, 2015; Stucki and Smith, 2011). While IWRM is a water-centric undertaking, the WEF nexus is a poly-centric approach that treats the energy and food sectors equally to the water sector in engagement while acknowledging their mutual interdependencies and coexistence in the ecosystem (Benson et al., 2015; de Loë and Patterson, 2017; Grigg, 2019; Smith and Clausen, 2015).

While the IWRM's default spatial scale is the hydrological basin (river, lake, groundwater), the WEF nexus seeks multiple scales, including the basin, depending on context and objectives (Benson et al., 2015; Sadeghi and Sharifi Moghadam, 2021; Stucki and Smith, 2011). Additionally, the major actors in IWRM emanate from the water sector, while the WEF nexus approach promotes co-ownership and engagement across all relevant sectors (Grigg, 2019; Sadeghi and Sharifi Moghadam, 2021; Smith and Clausen, 2015; Stucki and Smith, 2011). Thus, the WEF nexus is an 'out of the water box' and beyond-IWRM approach for sustainable development and management of WEF resources and the ecosystem that underpins their security (de Loë and Patterson, 2017).

The nexus approach aims to understand the strings linking WEF resources and ecosystems to maximize synergies and minimize trade-offs (Hoff, 2011). The WEF nexus approach also strives for the collective security of WEF resources, as visualized in Figure 12.1, through integrated and multisectoral resource management geared towards productive WEF systems while maintaining the integrity and sustainability

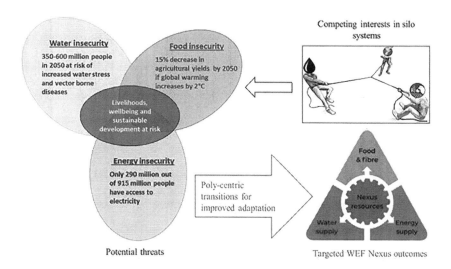

FIGURE 12.1   WEF nexus' approach to adaptation (Adapted from Nhamo et al. (2020b)).

of ecosystems (Nhamo et al., 2020a). Successful adaptation interventions are pinned on the ability to accommodate the issues of scaling across all the dimensions (out-scaling, up-scaling, and deep-scaling) (Holtermann and Nandalal, 2015). As the nexus approach is a system thinking approach that also encompasses socio-ecological systems, the WEF nexus tool can inform decision-makers for planning fit-for-purpose adaption strategies across scales (Holtermann and Nandalal, 2015).

A study by Nhamo et al. (2020b) applied the WEF nexus analytical tool to model the interrelatedness of individual WEF components in Sakhisizwe Local Municipality in the Eastern Cape Province, South Africa. The study revealed imbalances in resource utilisation and provided a holistic view of the usage and interconnectedness of WEF resources at a local level, as seen in Figure 12.2. The findings can be used to develop adaptation strategies centred on pragmatic trade-offs for effective resource utilisation. Wa'el A et al. (2017) investigated urban adaptation during Iraq's winter and summer seasons using the WEF nexus approach at the household scale. The study surveyed 419 households, and the results revealed the WEF nexus analytical tool successfully predicted the present and historical competing multiple water and energy uses. Wa'el A et al. (2017) echo the global mainstream discourse that adaptation strategies should function as a holistic and all-encompassing assessment of the interconnectedness and interlinkages.

On flexibility and adaptability, the WEF nexus can potentially address social inclusion through centreing adaptation measures around de-risking livelihoods and increasing the resilience of the vulnerable (Naidoo et al., 2021). For example, suppose WEF nexus analytical tools are applied at the field level (smallholder farming) to predict historical, current, and future water competing activities. In that case, careful planning can be employed to ensure water and energy adequacy. Improving water and energy availability reduces labour burdens and facilitates the scaling

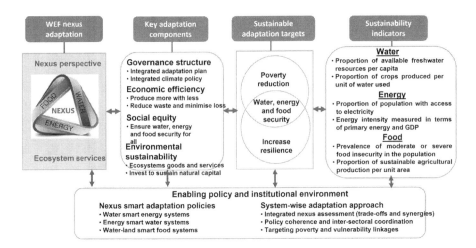

FIGURE 12.2   WEF nexus livelihoods adaptation and transformation framework (Nhamo et al., 2020b).

of food production systems for food security. Improved water availability leads to water adequacy that facilitates the horizontal and vertical diversification of food systems at a smallholder scale (Tavenner et al., 2019). Horizontal and vertical diversification incorporating crop-livestock systems as a part of climate change adaptation strategies can build resilience across scales and the social spectrum (Djurfeldt et al., 2018).

Several studies report on social inclusion, crop and income diversification to adapt to climate change and build resilience (Acevedo et al., 2020; Howden et al., 2007; Lin, 2011). WEF nexus methods provide opportunities to tackle multiple sustainable development goals (SDGs) 2 (zero hunger), 6 (clean water and sanitation), and 7 (affordable and clean energy), with synergies to SDGs 1 (no poverty), 5 (gender equality), 8 (decent work and economic growth), 12 (responsible consumption and production), 13 (climate action), 14 (life below water), and 15 (life on land) (Naidoo et al., 2021).

12.3   KEY CHALLENGES ON WEF RESOURCES SECURITY UNDER CLIMATE CHANGE

12.3.1   Water, Energy, and Food Security

Economies and society rely on water, energy and food for developmental growth and sustenance. These resources are required in adequate amounts to meet different uses. The level of adequacy of WEF is synonymous with the security of the three resources. Hence, water security, which is defined as the capacity of a population to secure sustainable access to water of adequate quantities and acceptable quality for sustaining livelihoods, human well-being, and socioeconomic development (UN-Water, 2013), ensures the protection against water-borne pollution and water-related disasters and preserves ecosystems in a climate of peace and political stability (Grey and Sadoff, 2007). Affordability, stability and safety are key pillars of water security (Mabhaudhi et al., 2019).

Energy security is the state in which all users always have access to a sufficient, stable, uninterrupted and reliable supply of energy from available sources at an affordable price (IEA, 2022a, 2022b). The supporting pillars for energy security include energy sufficiency, reliability, affordability and energy type (Ang et al., 2015; Mabhaudhi et al., 2019; Miller, 2017).

Food security refers to a state whereby all people have continuous physical, social and economic access to sufficient, safe and nutritious food to satisfy their dietary requirements and food preferences for an active, healthy life (FAO, 2014). Food security is driven by four dimensions or pillars of security, including food availability, accessibility, stability of supply, and utilisation (Nhamo et al., 2020b). Water, energy, and food underpin several SDGs, including 2, 6, 7 and those on poverty reduction, sustainable cities, human health and livelihoods, and the environment.

12.3.2 DRIVERS OF WEF INSECURITY

The global security of WEF resources is threatened by myriad drivers such as global population growth, rapid urbanization, changing diets and economic growth (FAO, 2022; Mabhaudhi et al., 2016; UN-Water, 2018; Yang et al., 2018). The global population has more than tripled since the middle of the 20th century; it hit 8 billion in 2022 and will reach 9.7 billion marks by 2050 (Smith and Clausen, 2015; UN-DESA, 2021, 2022). For instance, South Africa faces increasing demand for water, energy, and food due to population growth, urbanisation, and economic development (Mabhaudhi et al., 2018a; Nhamo et al., 2018). Therefore, this continuous population growth, accompanied by resource-intensive dietary changes, is anticipated to worsen the already-existing pressures on food production and security (UN-DESA, 2021; UN-Water, 2018).

The continued development and growth of economies in an urbanizing world increase resource demand (FAO, 2022; UN-DESA, 2019). For example, the business-as-usual future projections for 2030 and 2050 relative to 2012 predict demand increases of 40%–100% for energy and 50%–70% for food, which will require 20%–55% more water and 10%–30% more agricultural land (FAO, 2022; IRENA, 2015; UNESCO, 2022; WEC, 2013). Although the scenario space is wide and uncertain, it is agreed that food production and energy generation to meet these increased demands will further escalate environmental and sustainability challenges, degradation and depletion of natural resources, climate change and biodiversity loss (van Vuuren et al., 2019). South Africa is facing increased frequencies of climate change-induced extreme events such as droughts and floods, which undermine the reliable supply of WEF resources (Mabhaudhi et al., 2018a; Mpandeli et al., 2018; Nhamo et al., 2018). To withstand these current and future anticipated pressures, there is a need to ensure balanced and integrated people-nature-economy sustainable management of water, food and energy (UN-Water, 2018).

Potential challenges to climate change adaptation within a WEF nexus approach in South Africa include inequality (quality education and resources), sectoral approaches to resource management, weak governmental and institutional capacity, poverty, culture, and politics (Mabhaudhi et al., 2018b).

12.4   CASE STUDY: THE BUFFALO RIVER CATCHMENT

12.4.1   CLIMATE CHANGE CHALLENGES IN THE BUFFALO RIVER CATCHMENT

The Buffalo River catchment (Figure 12.3) forms part of the uThukela WMA in the warm, humid, high-elevation Drakensberg Mountain region. This catchment faces significant problems in adapting to the multiple effects of climate change, notably regarding water management. The catchment area receives the bulk of its yearly rainfall during the summer months, which averages 802 mm per annum (uMgeni, 2020). The flow direction is 339 km south-easterly from the eastern escarpment (Newcastle area) through the Amajuba and uMzinyathi District Municipalities, then confluences with uThukela River in the Msinga Local Municipality (Dlamini and Mostert, 2019; uMgeni, 2020). As such, the catchment provides water to the following local municipalities: (1) Newcastle Local Municipality, (2) Dannhauser Local Municipality, (3) Utrecht Local Municipality, and (4) Nquthu Local Municipality (uMgeni, 2020).

Water resources are the primary medium through which climate change impacts are felt (Nhamo et al., 2020a). According to Dlamini et al. (2022), the Buffalo River catchment is expected to experience increasing intensity and variability of precipitation throughout the 21st century, thus propelling increased evaporation, surface runoff, and fluctuations of reservoir storage volumes, consequently affecting water reliability for food, energy and domestic sectors provisions.

12.4.2   INCREASING POPULATION AND WATER DEMANDS

The population densities in the local municipalities within the Buffalo River catchment vary significantly. As per Table 12.1, Newcastle is the most densely populated area, growing at about 2% annually. The Utrecht local municipality is sparsely populated, with only 18 people per square metre. Therefore, due to population growth and density variations, climate change impacts water supply availability, allocations, and demand will vary (spatial and temporal) across the catchment.

TABLE 12.1
Population statistics of the local municipalities within the Buffalo River catchment

Local municipality	Area[1,2] (km²)	Population capacity[3]	Population growth rate[3] (%)	Population density per km[1]	Source
Newcastle	1 689	389 117	1.56	215	[1](Mahlaba, 2019), [3](StatsSA, 2016)
Utrecht	3 539	36 869	1.55	18.3	
Dannhauser	1 518	102 937	0.52	67.5	
Nquthu	1 962	171 325	0.81	84	[2](StatsSA, 2011), [3](StatsSA, 2016)
Total	8 708	700 248	1.22	79.83	

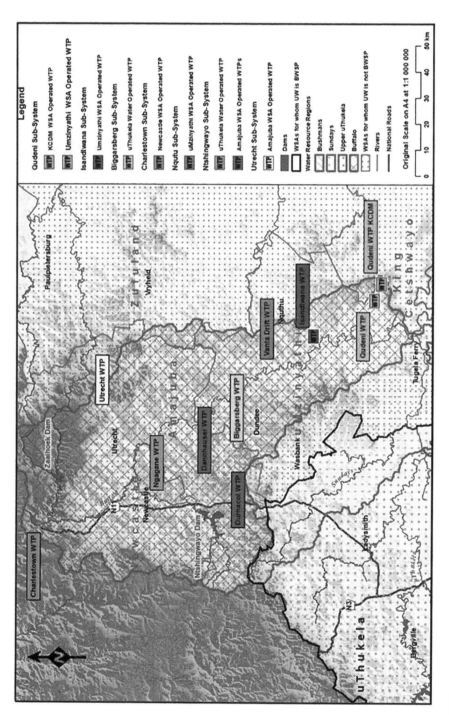

FIGURE 12.3   General layout of the Buffalo River catchment, South Africa (uMgeni, 2020).

Recent reports indicate that water supply schemes within the basin have been unable to sufficiently distribute water to their respective demand sites due to multiple drivers, including population growth (Dlamini and Mostert, 2019; Shabalala et al., 2020; uMgeni, 2020; uThukelaWater(Pty)Ltd, 2021) (uMgeni, 2020), poor resource management (e.g., pollution and degradation) (Wade, 2019) (Dlamini and Mostert, 2019) (uMgeni, 2020), economic growth (DAEARD, 2010), and climate change consequences have worsened such water deficits in this arid region (DAEARD, 2010; Lubega et al., 2019; Patrick, 2021; UNU-WIDER, 2016). For instance, the Ngagane water treatment plant (WTP), which is the largest WTP in the catchment, providing water to approximately 90% of the Newcastle Local Municipality's growing population, has been in deficit, and the droughts from 2015/2016 further exacerbated this deficiency in water supply (uMgeni, 2020). The required demand from the Ngagane WTP is 131.2 Ml/day, exceeding its design capacity of 130 Ml/day. By 2050, the Buffalo River catchment's total projected population water demand of 304 Ml/day is anticipated not to be catered for as it exceeds the total existing water supply infrastructure's capacity of approximately 192 Ml/day (uMgeni, 2020).

12.4.3 ENERGY AND AGRICULTURAL SECTORS' PRESSURES ON THE WATER SECTOR

To the best of our knowledge, no electricity production is currently taking place within the Buffalo River catchment. However, 24–27 million m^3 is allocated annually from the catchment's Zaaihoek Water Transfer Scheme to the Majuba Power Station for power plant cooling (uMgeni, 2020). The Majuba power station in the Upper Vaal WMA falls under the six South African power stations managed by Eskom, a state-owned public electricity utility. This diversion highly depends on Eskom's water demand projections and only occurs when required. Surplus water is transferred to the lower segments of the catchment, where it can be utilized for domestic and irrigation purposes (uMgeni, 2020).

The irrigation sector is the largest water consumer in the Buffalo River catchment, with requirements reaching 50 million m^3/annum, thus surpassing water use by domestic and industrial sectors (Dlamini and Mostert, 2019). Under climate change conditions, according to the global Agro-Ecological Zoning, assessment performed by the Food and Agricultural Organization (FAO) and a recent climate change study conducted by Dlamini et al. (2022), irrigation water requirements (IWR) in the Buffalo River basin are projected to increase due to reduced land productivity and crop suitability, particularly for soyabean, as its IWR is expected to double by the end of the 21st century. This poses a potential threat to the catchment since it will put more strain on the already-overburdened water supply system and the general economic development and population well-being.

12.4.4 THE NEED FOR INTEGRATED CLIMATE CHANGE ADAPTATION MEASURES

Implementing the WEF nexus thinking is essential for resource allocation and future resilience since the sectors are intimately dependent upon one another. Changes in one sector can profoundly impact an adjacent sector (Mabhaudhi et al., 2018b). In the case of the Buffalo River catchment case, increased intensities in precipitation and surface runoff projected under climate change are also set to impact water provisions and production outputs for agricultural and energy generation activities (Dlamini et al., 2022). Evidence-based decision-making is required to sustainably manage WEF resources under climate change.

Despite the challenges of the interconnected WEF resources, efforts to address them are not integrated. The challenges are highly complex in nature and constitute scenarios that can be addressed in an integrated manner (Rasul and Sharma, 2016). In the case of the Buffalo River catchment, each local and district municipality has developed integrated development plans, which address WEF issues in the respective area and potential adaptation strategies to climate change. Disjointed resource management can be unsustainable as these regions share water resources (Aklilu and Makalela, 2020).

In addition, the uThukela WMA and uMgeni Water institutes also proposed water management strategies under climate change. However, policy strategies addressing water security under climate change focus predominantly on increasing the water system's supply capacity to meet domestic water demands. Notwithstanding the importance of domestic water use, the agricultural and energy sectors are equally pivotal for poverty and vulnerability reduction. They must also be prioritized when developing integrated adaptation strategies to curb climate change impacts.

12.5   SYNERGIES AND TRADE-OFFS IN WEF NEXUS ADAPTATION STRATEGIES

Brunner et al. (2019) reported that the WEF nexus approach provides policymakers with the options for:

a. *"Synergies: whereby one intervention achieves multiple objectives,"*
b. *"Trade-offs: whereby a sector objective is rendered sub-optimal in favour of another that is optimized, and"*
c. *"Compromise: whereby all sectors accept a result that is less than perfect for one or more stakeholders for the sake of the common good."*

Identifying trade-offs, synergies, and compromises in the WEF nexus might bring new perspectives and prospects to minimize trade-offs and increase synergies for the development of effective adaptation strategies (McGrane et al., 2019; Rasul and Sharma, 2016), which is fundamental for developing regions which are prone to experiencing high vulnerabilities from climatic changes (Dlamini et al., 2022; Kurian et al., 2018).

The nexus approach's objectives for adapting to climate change are strongly linked and have many similarities. Hence, even though sector-specific, i.e., non-nexus, adaptation measures such as groundwater extraction, desalination plants, water-use-efficient irrigation technology, renewable energy, and growing biofuels on wasteland might have positive implications for water, energy, and food resources, they may also increase the nexus challenge (Rasul and Sharma, 2016). This can be both very challenging and costly (Bhaduri et al., 2015); therefore, in the intricate and uncertain climate change environment, non-nexus measures should be put into action if they have the potential to produce combinatorial win-win outcomes or serve as complementing actions that promote resource-use maximization and enhance WEF nexus solutions (Brunner et al., 2019; Diez-Borge et al., 2022).

Under climate change, well-established renewable energy sources allow a just transition toward a less carbon-intensive future while still attaining sustainable water and food sector development (Zhang et al., 2018). For example, in Siklesh Village in Nepal, a 100 kW micro-hydro plant was built in 1994 to provide electricity for domestic and agro-processing, aiming to reduce carbon emissions. The findings show that the micro-hydro plant boosted the establishment of agro-processing mills, hence increasing agricultural productivity (Guta et al., 2017). In Canada's Saskatchewan province, wind energy expansions produce synergies by offsetting thermal power reductions. That WEF nexus study highlighted that wind energy decreases greenhouse gas emissions and water consumption for cooling thermal power plants. It also improves groundwater conservation due to reduced groundwater demands (Wu et al., 2021).

Improving the efficiency of freshwater usage also provides the potential for cross-sector synergy under climate change (Rasul and Sharma, 2016). For instance, drip irrigation can create synergies between the water and food sectors in the Zhangye catchment in China. The intervention can significantly improve the fields' water-use efficiency amid climate change (Shen et al., 2022). Furthermore, trade-offs with the energy sector were addressed in this WEF nexus assessment, indicating that the high-cost input factors may surpass the benefits of increased crop yield and water-saving measures. Mulching with the plastic film was suggested as an additional measure to increase the water-saving benefits (Shen et al., 2022). It has been evidenced that using a piped irrigation system instead of a canal system can produce synergy between the water, energy, and food sectors in the Breede River catchment of South Africa by cutting electricity costs by 30% per 5%–10% increase in irrigated areas, while additionally improving water quality (Seeliger et al., 2018).

Adopting water management practices to produce more food and energy with fewer water resources is vital for climate change adaptation (Mpandeli et al., 2018). Ahmadaali et al. (2018) established that implementing water management strategies, which encompass crop pattern changes with increased irrigation efficiency, decreases water demands and improves agricultural and environmental sustainability in the Urmia Lake basin, located in the north-western regions of Iran. Additionally, adjusting water allocation strategies that encourage a balance between WEF sectors and domestic water usage is recommended for improved WEF sustainability in the catchment (Ahmadaali et al., 2018).

Similarly, surplus water in the Buffalo River catchment is available in the basin, which needs to be cautiously allocated (Dlamini and Mostert, 2019; Dlamini et al., 2022). From examining the effects of climate change and water resource policies on the water supply-demand relationship in the Buffalo River catchment, Dlamini (2022) found that existing water resources policy plans are centred around ensuring that more than 70% of domestic water demands are met. However, little to no improvements were modelled in closing the gap between agricultural water demands and supply. Less than 3% of irrigation and energy generation water demands are projected to be met throughout the 21st century.

Furthermore, Dlamini (2022) proposed long-term integrated water resources strategies to improve water allocations within the Buffalo River catchment to accommodate agricultural and energy water demands. The proposed strategies include:

(a) diverting excess water in densely populated municipalities like Newcastle and Dannhauser to more agriculture-intensive areas such as Nquthu and Utrecht local municipalities, (b) upgrading existing WTPs, such as the Ngagane WTP, so that they operate at optimum capacity, (c) constructing dams in the Ncandu and Ngxobongo rivers for increased water supply, and (d) increasing water abstractions of reservoirs during peak rainfall years.

As much as the proposed strategies' water allocation changes decrease water resources' reliability to provide domestic water demands, this trade-off was modelled to significantly improve the overall water provisions and equality in water distribution among the WEF sectors (Dlamini, 2022). To curb this anticipated decline in reliability in meeting domestic water demands, Dlamini (2022) strongly advocated for the use of multi-purpose dams to reduce the pressure on water supplies by increasing irrigation diversions and generating hydropower, as well as working with the communities in the catchment to further establish water demand management strategies in light of the catchment's limited land resources, climate change, and ecosystem degradation.

In light of the WEF nexus thinking, discussions around the synergies and trade-offs emerging from these strategies should involve researchers, policymakers and decision-makers in developing the Buffalo River catchment's WEF resources (Dlamini, 2022). Therefore, it is recommended that multiple stakeholder platforms be established in the Buffalo River basin to address better synchronization and integration of WEF development plans, policies, and procedures for improved service delivery.

Instead of the current fragmented management of water by local municipalities, the establishment of the Phongola-Umzimkulu Catchment Management Agency (CMA), which is part of the nine CMAs planned by South Africa's Department of Water and Sanitation to execute water resource management at the catchment level (Munnik, 2020), and that covers the Buffalo River catchment, is encouraged (Munnik, 2020). The reason is that CMAs are better equipped to cope with water allocations in light of droughts, current unpredictability, and climate change challenges. They provide a better opportunity to host multiple stakeholders when framing integrated adaptive management strategies (Munnik, 2020).

12.6 CONCLUSIONS

Climate change has significantly impacted developing regions due to their low adaptive capacity and, more importantly, the lack of integration of climate change adaptation in the respective regions' development plans. Due to climate change's complexity, unpredictability, and urgency, developing adaptation strategies sustainably should not only focus solely on mitigating its effects. Still, it should also consider the broader social frame of reference in which these changes are taking place and the consequential impact on the security of water, energy, and food. This chapter, therefore, elaborated on the WEF nexus approach, which encourages the integration of WEF resource sustainability under climate change.

From a basin perspective, better knowledge and understanding of the WEF nexus under climate change provides a practical possibility to coordinate nexus solutions to balance the sustainable use of resources and build resilience. In the case of the Buffalo River catchment case, different governmental bodies manage water resources to meet their domestic needs, which renders their resource management strategies unsustainable as the water supply system has proven unreliable for providing water for energy and food sectors. In such cases, coordinating policies for the sustainability of the WEF nexus through research, development, and practices are advocated for boosting sustainable livelihoods and climate resilience in catchment communities.

REFERENCES

Acevedo, M., Pixley, K., Zinyengere, N., Meng, S., Tufan, H., Cichy, K., Bizikova, L., Isaacs, K., Ghezzi-Kopel, K., Porciello, J. (2020) A Scoping Review of Adoption of Climate-Resilient Crops by Small-Scale Producers in Low- and Middle-Income Countries. *Nature Plants* 6, 1231–1241.

Adom, R.K., Simatele, M.D., Reid, M. (2022) Addressing the Challenges of Water-Energy-Food Nexus Programme in the Context of Sustainable Development and Climate Change in South Africa. *Journal of Water and Climate Change* 13, 2761–2779.

Ahmadaali, J., Barani, G., Qaderi, K., Hessari, B. (2018) Analysis of the Effects of Water Management Strategies and Climate Change on the Environmental and Agricultural Sustainability of Urmia Basin, Iran. *Water* 10, 160.

Aklilu, A., Makalela, K. (2020) Challenges in the Implementation of Integrated Development Plan and Service Delivery in Lepelle-Nkumphi Municipality, Limpopo Province. *Internationaal Journal of Economics and Finance Studies* 12, 1–15.

Altamirano, M., vanBodegom, A., vanderLinden, N., deRijke, H., Verhagen, A., Bucx, T., Boccalon, A., vanderZwaan, B. (2018) Operationalizing the WEF Nexus: Quantifying the Trade-Offs and Synergies between the Water, Energy and Food Sectors. Dutch Climate Solutions Research Programme, Netherlands, Europe.

Ang, B.W., Choong, W.L., Ng, T.S. (2015) Energy Security: Definitions, Dimensions and Indexes. *Renewable and Sustainable Energy Reviews* 42, 1077–1093.

Benson, D., Gain, A.K., Rouillard, J. (2015) Water Governance in a Comparative Perspective: From IWRM to a 'Nexus' Approach? *Water Alternatives* 8, 756–773.

Bhaduri, A., Ringler, C., Dombrowski, I., Mohtar, R., Scheumann, W. (2015) Sustainability in the Water-Energy-Food Nexus. *Water International* 40, 723–732.

Brunet, C., Savadogo, O., Baptiste, P., Bouchard, M.A., Rakotoary, J.C., Ravoninjatovo, A., Cholez, C., Gendron, C., Merveille, N. (2020) Impacts Generated by a Large-Scale Solar Photovoltaic Power Plant Can Lead to Conflicts between Sustainable Development Goals: A Review of Key Lessons Learned in Madagascar. *Sustainability* 12, 7471.

Brunner, J., Carew-Reid, J., Glemet, R., McCartney, M., Riddell, P. (2019) Measuring, Understanding and Adapting to Nexus Trade-Offs in the Sekong, Sesan and Srepok Transboundary River Basins. International Union for Conservation of Nature and Natural Resources, Ha Noi, Viet Nam.

Butler, D., Ward, S., Sweetapple, C., Astraraie-Imani, M., Diao, K., Farmani, R., Fu, G. (2017) Reliable, Resilient and Sustainable Water Management: The Safe & SuRe Approach. *Global Challenges* 1, 63–77.

Caretta, M., Mukherji, A., Arfanuzzaman, M., Betts, R., Gelfan, A., Hirabayashi, Y., Lissner, T., Liu, J., Lopez-Gun, E., Morgan, R., Mwanga, S., Supratid, S. (2022) Water, in: Portner, H., Roberts, D., Tignor, M., Poloczanska, E., Mintenbeck, K., Alegria, A., Craig, M., Langsdorf, S., Loschke, S., Moller, V., Okem, A., Rama, B. (Eds.), *Climate Change 2022: Impacts, Adaptation and Vulnerability*. Cambridge University Press, Cambridge and New York, pp. 551–712.

Cullis, J., Rossouw, N., DuToit, G., Petrie, D., DeClercq, G., Horn, A. (2018) Economic Risk Due to Declining Water Quality in the Breede River Catchment. *Water SA* 44, 464–473.

DAEARD (2010) KwaZulu-Natal State of the Environment 2004: Inland Aquatic Environment Specialist Report. Department of Agriculture, Environmental Affairs and Rural Development, KwaZulu-Natal Provincial Government, Pietermaritzburg, South Africa.

de Loë, R.C., Patterson, J.J. (2017) Rethinking Water Governance: Moving beyond Water-Centric Perspectives in a Connected and Changing World. *Natural Resources Journal* 57, 75–99.

Diez-Borge, D., Garcia-Moya, F., Rosales-Asensio, E. (2022) Water Energy Food Analysis and Management Tools: A Review. *Energies* 15, 1146.

Djurfeldt, A.A., Dzanku, F.M., Isinika, A.C. (2018) *Perspectives on Agriculture, Diversification, and Gender in Rural Africa: Theoretical and Methodological Issues.* Oxford University Press, Oxford.

Dlamini, N. (2022) Developing Integrated Climate Change Strategies using the Water-Energy-Food Nexus Approach: A Case Study of the Buffalo River Catchment, South Africa University of KwaZulu-Natal, Pietermaritzburg, South Africa.

Dlamini, N., Mostert, R. (2019) Closure Plan in Support of the Environmental Authorisation for the Proposed Prospecting Right Application for Coal, Pseudocoal, Anthracite, Sand and Clay on the Remainder of the Farm 15454, Remainder of the Farm Highvake 9311, Remainder and Portion 1 of the Farm Lowvale 15596, Remainder of the Farm Ormiston 8195, Remainder and Portion 3 of the Farm Langklip 10711 and Remainder of the Farm 16763, Under Dannhauser Local Municipality, Kwa-Zulu Natal Province. Thikho Resources, Johannesburg, South Africa.

Dlamini, N., Senzanje, A., Mabhaudhi, T. (2022) Climate Change Impacts on Water Reliability: A Case Study of the Buffalo River Catchment, Thukela Water Management Area, South Africa. Manuscript under review.

Dube, R., Dube, B., Managa, R., Malan, A., Ramulondi, D., Ramathuba, T. (2021) Integrated Catchment Management: From Source to Receptor. Water Research Commission, Pretoria, South Africa.

DWAF (2007) Guidelines for the Development of Catchment Management Strategies: Towards Equity, Efficiency and Sustainability. Department of Water Affairs and Forestry (DWAF), Pretoria, South Africa.

DWAF, WRC (1996) The Philosophy and Practice of Integrated Catchment Management: Implications for Water Resources Management in South Africa (WRC Report No TT 81/96). Department of Water Affairs and Forestry (DWAF) and Water Reserach Commission (WRC), Pretoria, South Africa.

FAO (2014) The Water-Energy-Food Nexus: A New Approach in Support of Food Security and Sustainable Agriculture. Food and Agriculture Organization (FAO), Rome, Italy.

FAO (2022) The State of the World's Land and Water Resources for Food and Agriculture – Systems at breaking point. Main report. Food and Agriculture Organization of the United Nations (FAO), Rome, Italy.

Filho, W., Balogun, A., Olayide, O., Azeiteiro, U., Ayal, D., Munoz, P. (2019) Assessing the Impacts of Climate Change in Cities and their Adaptive Capacity: Towards Transformative Approaches to Climate Change Adaptation and Poverty Reduction in Urban Areas in a Set of Developing Countries. *Science of the Total Environment* 692, 1175–1190.

Grey, D., Sadoff, C.W. (2007) Sink or Swim? Water Security for Growth and Development. *Water Policy* 9, 545–571.

Grigg, N.S. (2019) IWRM and the Nexus Approach: Versatile Concepts for Water Resources Education. *Journal of Contemporary Water Research & Education* 166, 24–34.

Guta, D., Jara, J., Adhikari, N., Chen, Q., Gaur, V., Mirzabaev, A. (2017) Assessment of the Successes and Failures of Decentralized Energy Solutions and Implications for the Water-Energy-Food Security Nexus: Case Studies from Developing Countries. *Resources* 6, 24.

GWP-TAC (2000) Integrated Water Resources Management. Global Water Partnership, Stockholm, Sweden.

Hoff, H. (2011) Understanding the Nexus. Background Paper for the Bonn2011 Conference: The Water, Energy and Food Security Nexus, Stockholm.

Holtermann, T., Nandalal, K. (2015) The Water–Energy–Food Nexus and Climate Change Adaptation. *Change and Adaptation in Socio-Ecological Systems* 2, 118–120.

Howden, S., Soussana, J.-F., Tubiello, F., Chhetri, N., Dunlop, M., Meinke, H. (2007) Adapting Agriculture to Climate Change. *Proceedings of the National Academy of Sciences of the United States of America* 104, 19691–19696.

Huang, S., Wortmann, M., Duethmann, D., Menz, C., Shi, F., Zhao, C. (2018) Adaptation Strategies of Agriculture Water Management to Climate Change in the Upper Tarim River Basin, NW China. *Agricultural Water Management* 203, 207–224.

Hulley, S.M., (2015) The Food-Energy-Water-Land-Biodiversity (FEWLB) Nexus through the Lens of the Local Level: An Agricultural Case Study. Department of Environmental and Geographical Science University of Cape Town, Cape Town, South Africa.

IEA (2022a) Energy Security: Ensuring the Uninterrupted Availability of Energy Sources at an Affordable Price. International Energy Agency (IEA) Paris, France.

IEA (2022b) Energy security: Reliable, affordable access to all fuels and energy sources. International Energy Agency (IEA), Paris, France.

IRENA (2015) Renewable Energy in the Water, Energy and Food Nexus. International Renewable Energy Agency (IRENA), Abu Dhabi.

Knuppe, K., Meissner, R. (2016) Drivers and Barriers Towards Sustainable Water and Land Management in the Olifants-Doorn Water Management Area, South Africa. *Environmental Development* 20, 3–14.

Kurian, M., Portney, K., Rappold, G., Hannibal, B., Gebrechorkos, S. (2018) Governance of Water-Energy-Food Nexus: A Social Network Analysis Approach to Understanding Agency Behaviour, in: Hulsmann, S. (Ed.), *Managing Water, Soil and Waste Resources to Achieve Sustainable Development Goals*. Springer, Berlin, Germany, pp. 125–147.

Lele, S., Srinivasan, V., Thomas, B., Jamwal, P. (2018) Adapting to Climate Change in Rapidly Urbanizing River Basins: Insights from a Multiple-Concerns, Multiple-Stressors, and Multi-Level Approach. *Water International* 43, 281–304.

Lin, B.B. (2011) Resilience in Agriculture through Crop Diversification: Adaptive Management for Environmental Change. *BioScience* 61, 183–193.

Lubega, M.J., de Lavenne, A., Thuresson, J., Byström, M., Strömqvist, J., Bartosova, A., Arheimer, B. (2019) The Impact of Climate Change on the Water Quality in the KwaZulu-Natal Province of South Africa, 1st EGU General Assembly, EGU2019, Proceedings from the conference held 7-12 April, 2019 in Vienna, Austria, id.17049.

Mabhaudhi, T., Mpandeli, S., Madhlopa, A., Modi, A.T., Backeberg, G., Nhamo, L. (2016) Southern Africa's Water–Energy Nexus: Towards Regional Integration and Development. *Water* 8, 235.

Mabhaudhi, T., Mpandeli, S., Nhamo, L., Chimonyo, V., Nhemachena, C., Senzanje, A., Naidoo, D., Modi, A. (2018a) Prospects for Improving Irrigated Agriculture in Southern Africa: Linking Water, Energy and Food. *Water* 10, 1881.

Mabhaudhi, T., Nhamo, L., Mpandeli, S., Nhemachena, C., Senzanje, A., Sobratee, N., Chivenge, P.P., Slotow, R., Naidoo, D., Liphadzi, S., Modi, A.T. (2019) The Water–Energy–Food Nexus as a Tool to Transform Rural Livelihoods and Well-Being in Southern Africa. *International Journal of Environmental Research and Public Health* 16, 2970.

Mabhaudhi, T., Simpson, G., Badenhorst, J., Mohammed, M., Motongera, T., Senzanje, A., Jewitt, G. (2018b) Assessing the State of the Water-Energy-Food (WEF) Nexus in South Africa. Water Research Commission, Pretoria, South Africa.

Mahlaba, N. (2019) Newcastle Local Municipality Annual Report 2018/19, Newcastle, South Africa.

Materechera, F.M. (2012) Towards Integrated Catchment Management: Challenges surrounding implementation in the Gamtoos River Catchment, Environmental Geography. Nelson Mandela Metropolitan University, Gqeberha, South Africa.

McGrane, S., Acuto, M., Artioli, F., Chen, P., Comber, R., Cottee, J., Farr-Wharton, G., Green, N., Helfgott, A., Larcom, S., McCann, J. (2019) Scaling the Nexus: Towards Integrated Frameworks for Analysing Water, Energy and Food. *The Geographical Journal* 185, 419–431.

Methner, N., Midgley, S.J.E., Price, P., Ningi, T., Nxumalo, N.P., Rebelo, A.J., Stuart-Hill, S.I., Zhou, L., Mjanyelwa, V., Taruvinga, A. (2021) Exploring the Evidence of Water-Energy-Food Nexus Linkages to Sustainable Local Livelihoods and Wellbeing in South Africa. Water Research Commission (WRC), Pretoria, South Africa.

Midgley, S., New, M., Spelman, S.-S., Parker, K., (2014) The Food-Energy-Water-Land-Biodiversity (FEWLB) Nexus and Local Economic Development in the Berg River Catchment: Framework and Description. African Climate and Development Initiative, University of Cape Town, Cape Town.

Miller, B.G. (2017) 16- The Future Role of Coal, in: Miller, B.G. (Ed.), *Clean Coal Engineering Technology* (Second Edition). Butterworth-Heinemann, Massachusetts, USA, pp. 757–774.

Molobela, I. (2011) Management of Water Resources in South Africa: A Review. *African Journal of Environmental Science and Technology* 5, 993–1002.

Mpandeli, S., Naidoo, D., Mabhaudhi, T., Nhemachena, C., Nhamo, L., Liphadzi, S., Hlahla, S., Modi, A. (2018) Climate Change Adaptation through the Water-Energy-Food Nexus in Southern Africa. *International Journal of Environmental Research and Public Health* 15, 2306.

Munnik, V. (2020) The Reluctant Roll-out of Catchment Management Agencies. Water Research Commission Research Report, Pretoria, South Africa.

Naidoo, D., Nhamo, L., Mpandeli, S., Sobratee, N., Senzanje, A., Liphadzi, S., Slotow, R., Jacobson, M., Modi, A., Mabhaudhi, T. (2021) Operationalising the Water-Energy-Food Nexus through the Theory of Change. *Renewable and Sustainable Energy Reviews* 149, 111416.

Nhamo, L., Mabhaudhi, T., Mpandeli, S., Dickens, C., Nhemachena, C., Senzanje, A., Naidoo, D., Liphadzi, S., Modi, A. (2020a) An Integrative Analytical Model for the Water-Energy-Food Nexus: South Africa Case Study. *Environmental Science and Policy* 109, 15–24.

Nhamo, L., Ndlela, B., Mpandeli, S., Mabhaudhi, T. (2020b) The Water-Energy-Food Nexus as an Adaptation Strategy for Achieving Sustainable Livelihoods at a Local Level. *Sustainability* 12, 8582.

Nhamo, L., Ndlela, B., Nhemachena, C., Mabhaudhi, T., Mpandeli, S., Matchaya, G. (2018) The Water-Energy-Food Nexus: Climate Risks and Opportunities in Southern Africa. *Water* 10, 567.

Ningi, T., Taruvinga, A., Zhou, L., Ngarava, S. (2021) Household water-energy-food security nexus: Empirical evidence from Hamburg and Melani communities in South Africa. *International Journal of Development and Sustainability* 10, 315–339.

Olabanji, M., Ndarana, T., Davis, N. (2020) Impacts of Climate Change on Crop Production and Potential Adaptive Measures in the Olifants Catchment, South Africa. *Climate* 9, 6.

Owen, G. (2020) What Makes Climate Change Adaptation Effective? A Systematic Review of the Literature. *Global Environmental Change* 62, 102071.

Patrick, H.O. (2021) Climate Change and Water Insecurity in Rural uMkhanyakude District Municipality: An Assessment of Coping Strategies for Rural South Africa. *H2Open Journal* 4, 29–46.

Phan, T., Smart, J., Sahin, O., Stewart-Koster, B., Hadwen, W., Capon, S. (2020) Identifying and Prioritising Adaptation Options for a Coastal Freshwater Supply and Demand System Under Climatic and Non-Climatic Changes. *Regional Environmental Change* 20, 1–14.

Rapholo, M., Makia, L. (2020) Are Smallholder Farmers' Perceptions of Climate Variability Supported by Climatological Evidence? A Case Study of a Semi-Arid Region in South Africa. *International Journal of Climate Change Strategies and Management* 12, 571–585.

Rasul, G., Sharma, B. (2016) The Nexus Approach to Water-Energy-Food Security: An Option for Adaptation to Climate Change. *Climate Policy* 16, 682–702.

Sadeghi, S.H., Sharifi Moghadam, E. (2021) Integrated Watershed Management Vis-a-Vis Water–Energy–Food Nexus, in: Muthu, S.S. (Ed.), *The Water–Energy–Food Nexus: Concept and Assessments*. Springer, Singapore, pp. 69–96.

Seeliger, L., DeClercq, W., Hoffmann, W., Cullis, J., Horn, A., DeWitt, M. (2018) Applying the Water-Energy-Food Nexus to Farm Profitability in the Middle Breede Catchment, South Africa. *South African Journal of Science* 114, 1–10.

Shabalala, I., Zungu, N., Ntombela, P., Ngobesse, M., Moloi, L., Hoffman, L. (2020) Integrated Development Plan 2020/21. Nquthu Local Municipality Executive Committee, KwaZulu-Natal, South Africa.

Shen, Q., Niu, J., Liao, D., Du, T. (2022) A Resilience-Based Approach for Water Resources for Water Resources Management over a Typical Agricultural Region in Northwest China under Water-Energy-Food Nexus. *Ecological Indicators* 144, 109562.

Simpson, G., Jewitt, G. (2019) The Development of the Water-Energy-Food Nexus as a Framework for Achieving Resource Security: A Review. *Frontiers in Environmental Science* 7, 1–9.

Simpson, G.B., Badenhorst, J., Jewitt, G.P.W., Berchner, M., Davies, E. (2019) Competition for Land: The Water-Energy-Food Nexus and Coal Mining in Mpumalanga Province, South Africa. *Frontiers in Environmental Science* 7, 86.

Smith, M., Clausen, T.J. (2015) Integrated Water Resource Management: A New Way Forward (A Discussion Paper of the World Water Council Task Force on IWRM). World Water Council (WWC).

StatsSA (2011) Stastics by Place: Nquthu Local Municipality. Statistics South Africa, Pretoria, South Africa.

StatsSA (2016) Provincial Profile: KwaZulu-Natal Community Survey 2016. Statistics South Africa, Pretoria, South Africa.

Stucki, V., Smith, M. (2011) Integrated Approaches to Natural Resources Management in Practice: The Catalyzing Role of National Adaptation Programmes for Action. *Ambio* 40, 351–360.

Tavenner, K., Van Wijk, M., Fraval, S., Hammond, J., Baltenweck, I., Teufel, N., Kihoro, E., De Haan, N., Van Etten, J., Steinke, J. (2019) Intensifying Inequality? Gendered Trends in Commercializing and Diversifying Smallholder Farming Systems in East Africa. *Frontiers in Sustainable Food Systems* 3, 10.

uMgen (2020) Infrastructure Master Plan 2020 (2020/2021–2050/2051): Buffalo System. UMgeni Water, Pietermaritzburg, South Africa.

UN-DESA (2019) World Urbanization Prospects: The 2018 Revision (ST/ESA/SER.A/420). United Nations, Department of Economic and Social Affairs, Population Division, New York.

UN-DESA (2021) Global Population Growth and Sustainable Development (UN DESA/POP/2021/TR/NO. 2), New York.

UN-DESA (2022) World Population Prospects 2022: Summary of Results (UN DESA/POP/2022/TR/NO. 3). United Nations Department of Economic and Social Affairs, Population Division, New York.

UNESCO (2022) Groundwater: Making the Invisible Visible. The United Nations World Water Development Report 2022, Paris, France.

UNU-WIDER (2016) Potential Impacts of Climate Change on National Water Supply in South Africa. United Nations University World Institute for Development Economics Research (UNU-WIDER), Helsinki, Finland.

UN-Water (2013) Water Security and the Global Water Agenda: A UN-Water Analytical Brief. United Nations University Institute for Water, Environment & Health (UNU-INWEH), Hamilton, Canada, https://hdl.handle.net/20.500.12870/4206.

UN-Water (2018) Water, Food and Energy: A UN-Water Analytical Brief. UN-Water, Geneva, Switzerland, https://www.unwater.org/sites/default/files/app/uploads/2018/10/WaterFacts_water_food_and_energy_sep2018.pdf

uThukelaWater(Pty)Ltd (2021) uThukela Water (Pty) Ltd Draft Annual Report 2021, in: Khambule, N. (Ed.). uThukela Water (Pty) Ltd, Newcastle, South Africa.

van Vuuren, D.P., Bijl, D.L., Bogaart, P., Stehfest, E., Biemans, H., Dekker, S.C., Doelman, J.C., Gernaat, D.E.H.J., Harmsen, M. (2019) Integrated Scenarios to Support Analysis of the Food–Energy–Water Nexus. *Nature Sustainability* 2, 1132–1141.

Wade, M. (2019) Management of Multiple Stressors to the Lower Reach of the Thukela River Ecosystem, Ecological Sciences, School of Life Sciences, College of Agriculture, Engineering and Science. University of KwaZulu-Natal, Pietermaritzburg, South Africa.

Wa'el A, H., Memon, F.A., Savic, D.A. (2017) An Integrated Model to Evaluate Water-Energy-Food Nexus at a Household Scale. *Environmental Modelling & Software* 93, 366–380.

WEC (2013) World Energy Scenarios: Composing Energy Futures to 2050. World Energy Counci, London.

Wu, L., Elshorbagy, A., Pande, S., Zhuo, L. (2021) Trade-offs and Synergies in the Water-Energy-Food Nexus: The Case of Saskatchewan, Canada. *Resources, Conservation and Recycling* 164, 105192.

Yang, J., Yang, Y.C.E., Khan, H.F., Xie, H., Ringler, C., Ogilvie, A., Seidou, O., Djibo, A.G., van Weert, F., Tharme, R. (2018) Quantifying the Sustainability of Water Availability for the Water-Food-Energy-Ecosystem Nexus in the Niger River Basin. *Earth's Future* 6, 1292–1310.

Zhang, X., Li, H., Deng, Z., Ringler, C., Gao, Y., Hejazi, M., Leung, L. (2018) Impacts of Climate Change, Policy and Water-Energy-Food Nexus on Hydropower Development. *Renewable Energy* 116, 827–834.

13 Catalysing cleaner production systems

Benchmarking with the COVID-19 lockdowns in South Africa

James Magidi, Luxon Nhamo, Edward Kurwakumire, Webster Gumindoga, Sylvester Mpandeli, Stanley Liphadzi, and Tafadzwanashe Mabhaudhi

13.1 INTRODUCTION

Most global challenges, including air pollution, global warming, and health risks, are inextricably linked and can be traced back to anthropogenic activities (Patella et al., 2018). As widely established, GHG are the major drivers of climate change responsible for more frequent and intense extreme weather events (heat waves, droughts, floods, and cyclones), which in turn pose serious human and environmental health risks (Barrett et al., 2015). The sectors that emit most of the GHGs include energy, transport, industrial production, waste management, agriculture, forestry and other land-use activities (Ritchie and Roser, 2017). Emissions from processes within these sectors produce, among others, fine particulate matter ($PM_{2.5}$), a mixture of solid particles and liquid droplets found in the air that we breathe (Hill et al., 2009). $PM_{2.5}$ includes short-lived climate pollutants (SLCPs) such as black carbon and ground-level ozone. Other industrial pollutants in the atmosphere include tropospheric ozone gases (O_3), sulphur oxides (SO_2 and SO_3), nitrogen dioxide (NO_2), and benzo(a)pyrene (BaP) (Chen et al., 2007). Globally, atmospheric pollution causes over 6 million deaths yearly. A quarter of lung cancer cases, heart attacks and strokes cost 0.3% of the world's gross domestic product in healthcare expenditures and reduce workplace productivity (Anenberg et al., 2019). Additionally, more than ten times as many people die due to atmospheric pollution than are killed in road accidents (Ritchie and Roser, 2017).

The high concentration of gasses in the atmosphere mainly affects the most vulnerable in densely populated areas where ground-level ozone events are intensifying and causing respiratory diseases (Ghorani-Azam et al., 2016). This is particularly of great concern with the emergence of the Severe Acute Respiratory Syndrome

DOI: 10.1201/9781003327615-13

Coronavirus-2 (SARS-CoV-2), which causes the COVID-19 disease. People who already have respiratory health challenges and associated underlying conditions are at higher risk of COVID-19 (Schultze et al., 2020). Timely intervention to reduce air pollution through regulatory frameworks and cleaner production mechanisms are mitigatory measures of climate change and a means of improving human health and reducing healthcare costs (Wu et al., 2016). Another important benefit of reducing atmospheric pollution is the lessening of the rate of climate change and, consequently, its impacts (Sierra-Vergas and Teran, 2012). As highlighted in a study by Sofia et al., targeted sectors for cost-effective interventions to reduce air pollution include transportation, household combustion, waste management, agriculture, and industry, as they emit the highest pollutants into the atmosphere (Sofia et al., 2020). Adopting smart and clean technologies that promote the circular economy and contribute to reducing air pollution has the benefit of reducing global warming by as much as 0.6°C by 2050 while annually preventing 2.4 million premature deaths from ambient air pollution by 2030 as a result of the reduction in anthropogenic emissions (Vandyck et al., 2018; Xu and Ramanathan, 2017).

As interlinked and cross-cutting drivers generally cause health and environmental challenges, sector-based interventions risk exacerbating the challenges in other sectors (Meacham et al., 2016; Nhamo et al., 2020). For example, air pollution and climate change cannot be separated as air pollution is the main driver of climate change, and a warming globe increases the presence of pollutants in the atmosphere (Jacob and Winner, 2009). As air pollution and climate change are intricately interlinked, efforts to address them must be driven by cross-sectoral and unifying policy frameworks and circular and transformative approaches. This requires broad involvement and closer coordination between major stakeholders and sectors, including energy, mining, transport, environment, health, industry, agriculture, water and finance, which currently operate in silos (Nhamo et al., 2018).

On the other hand, air pollution does not only affect source areas but travels widely, crossing international boundaries and continents, yet it is generally regulated at the national level (Chen et al., 2017). Air pollution affects the state of climate worldwide, resulting in global warming. Addressing the challenges associated with air pollution at a local level presents huge gaps in monitoring, data collection and enforcement of emission controls. Furthermore, fragmented and locally based policies present an extra challenge as they are ineffective in reducing air pollution (Kjellstrom et al., 2006). Best practices to reduce air pollution require regional or global regulatory and policy frameworks, like the Paris Climate Agreement, to monitor, measure and report emissions in an integrated manner (Rogelj et al., 2016). While countries should operate within their ceiling of anthropogenic emissions or planetary boundaries, they should have targets for reducing pollutants through Nationally Determined Emission Contribution, which contributes to lessening global warming. Furthermore, this brings a need to align local policies with international frameworks. In response to this, many countries and members of the United Nations have developed climate change policies that relate to the United Nations Framework Convention on Climate Change and the Paris Agreement (Pauw et al., 2019).

As the challenges associated with atmospheric pollution are multifaceted, polycentric and multidisciplinary, they must be addressed through circular and transformative approaches (Manisalidis et al., 2020). Adopting circular models and embracing smart technologies such as the Internet of Things (IoT), Big Data and Data Analytics can curb pollution and inform coherent policy formulations that lead to resilience to various climate change vulnerabilities. Circular models are steadily replacing linear models in addressing today's cross-cutting challenges and implementing cross-sectoral interventions (Nhamo and Ndlela, 2021). This is particularly relevant in air pollution reduction initiatives as there are always trade-offs between pollution mitigation, sustainable development, and health objectives (Liu et al., 2019). Addressing these trade-offs requires smart policy frameworks informed through transformative processes (nexus planning, circular economy, one health, sustainable food systems, strategic foresight, horizon scanning, and scenario planning) to reduce risk. These transformative approaches are strategic processes that contribute towards slowing down near-term air pollution and mitigating climate change impacts (Nhamo et al., 2020).

There is, therefore, an urgent need for policy and decision-makers to operationalise smart strategies capable of reducing environmental pollution, including regulating the open burning of agricultural waste and accelerating the transition away from fossil fuel-driven vehicles towards electric ones and other low-emissions transportation systems. As the use of biomass for domestic purposes contributes about 25% of global carbon emissions, increasing access to clean fuels and devices for cooking, heating and lighting will lead to significant reductions in both carbon emissions and the household air pollution that cause about 3.8 million deaths worldwide each year (Ritchie and Roser, 2017). Thus, the concept of the circular economy becomes an essential pathway towards enhancing cleaner production strategies through waste reduction and allowing resources to remain in circulation for longer periods (Heshmati, 2017).

The COVID-19 economic lockdowns were meant to reduce the spread of the SARS-CoV-2 pandemic (Dong et al., 2020; Hamzelou, 2020), but the reduced industrial production resulted in reduced industrial emissions and air pollution for long periods outside the festive periods (WHO, 2020). However, the fundamental challenge is to continue with industrial production and avail resources necessary for human well-being within permissible pollution levels. Operating within the planetary boundaries requires smart technologies and the circular economy model to reduce waste and enhance sustainability (Naidoo et al., 2021). Pursuing traditional production, consumption and disposal models only increases the emission of toxic substances into the atmosphere, risking environmental and human health and the frequency and intensity of extreme weather events. This chapter used multi-spectral remote sensing to assess changes in NO_2 and $PM_{2.5}$ over South Africa between April 2021 and April 2020, before and during the COVID-19 economic lockdown. The aim was to develop pathways informed by circular models, promote sustainable interventions, reduce risk from atmospheric and enhance resilience-building initiatives.

13.2 MAJOR SOURCES OF POLLUTION IN SOUTH AFRICA

Air pollution hotspots in South Africa are identified in urban and coal mining areas due to the high concentration of manufacturing industries and thermal power generation, respectively (IEA, 2016). The high pollution levels in urban areas are

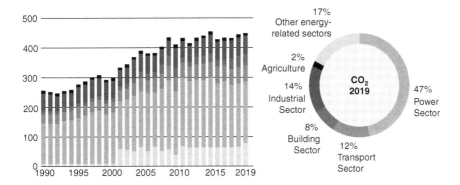

FIGURE 13.1 Incremental CO_2 emissions by sector from 1990 to 2019 in South Africa from fuel combustion ($MtCO_2$/year). Power generation remains the main CO_2 contributing sector.

Source: Climate Transparency, 2020.

compounded by the high traffic volumes and the use of fossil energy for domestic purposes (Figure 13.1) (ClimateTransparency, 2020). The industrial sector, energy generation, and automobile industry contribute the most to atmospheric pollution in South Africa (IEA, 2016). More than 90% of South Africa's electricity is generated from the combustion of coal, contributing to approximately 47% of total carbon emissions (Figure 13.1) that contain approximately 1.2% sulphur and up to 45% ash (IEA, 2016), accounting for high pollutant emissions in Mpumalanga Province.

Coal combustion is the major contributor of fine particulate matter ($PM_{2.5}$) in the air and is the main cause of acid rain in South Africa (Adesina et al., 2020). For instance, NO_2 levels in Cape Town are significantly higher than those measured in Kolkata in India and surpass the World Health Organization's annual mean guideline for air quality standard of 50 micrograms per cubic metre (IEA, 2016). Nearly 80% of NO_2 emissions in urban areas emanate from motor vehicles, and the remaining 20% from petrol and metal refining, coal power stations, and food processing. In addition to pollution from coal combustion, the use of coal and biomass for domestic purposes and coal-heated boilers in hospitals and factories contribute to air pollution (Perera, 2018).

13.2.1 DATA SOURCES FOR ATMOSPHERIC POLLUTION

At the global level, atmospheric pollution is being monitored at regular intervals using multi-spectral remote sensors such as the land remote-sensing satellite system, moderate resolution imaging spectroradiometre, and Sentinel (Seltenrich, 2014; Zheng et al., 2019). Satellite observations of tropospheric NO_2 have been conducted since 1995 by the Global Ozone Monitoring Experiment satellite instrument, designed to observe the various gases in the Earth's stratosphere and troposphere (Burrows et al., 1999). The Copernicus Program, managed by the European Space Agency, provides several Earth observation satellites for mapping and monitoring the Earth's physics and chemistry (McCabe et al., 2017). The launch of the Copernicus Sentinel-5 Precursor's (5P) spectrometre in 2017, which is called TROPOspheric

Monitoring Instrument (TROPOMI), allowed the monitoring of the density of a variety of atmospheric gases, aerosols, and clouds (Theys et al., 2017). The TROPOMI is a multi-spectral scanner with many spectral bands that utilises the ultraviolet, visible, near-infrared, and shortwave infrared to monitor ozone (O_3), methane (CH_4), formaldehyde (HCHO), carbon monoxide (CO), nitrogen dioxide (NO_2), and sulphur dioxide (SO_2) (Omrani et al., 2020; Theys et al., 2017).

These data are freely available via the Copernicus Open Access Data Hub and were accessed through the Google Earth Engine (GEE), a cloud-based computing platform where most spatial datasets are stored (Gorelick et al., 2017). Data from the Sentinel 5P was used to assess NO_2 and other aerosols in South Africa from April 2019 to April 2020. The maximum monthly NO_2 derived from the TROPOMI sensor was computed using GEE. The maximum value was used in the analysis to eliminate the off-nadir recordings, which might have a low value of NO_2 due to the distances between the surface and the sensor (Theys et al., 2017).

Monthly maximum emissions of NO_2 from April 2019 to April 2020 were computed for South Africa using data extracted from GEE. The maximum NO_2 value for each province was analysed using the raster cell statistics in R programming. The difference between April 2019 and March 2020 for all provinces was computed to determine the impact of the COVID-19 lockdown. The difference between April 2020 and April 2019 was calculated to assess whether COVID-19 affected the natural trends.

13.3 POLLUTION BEFORE, DURING, AND AFTER THE COVID-19 LOCKDOWNS

Air pollution, mainly NO_2 and $PM_{2.5}$, was assessed at the national level in South Africa between April 2019 and May 2020 (Figures 13.1 and 13.2) to cover the period during the economic shutdowns introduced due to the COVID-19 pandemic. Restrictions and lockdown regulatory measures rendered the closure of many industries except those that provided essential services such as thermal power stations for electricity generation, water, and sanitation, among others; however, they operated at limited capacity.

13.3.1 CHANGES IN NO_2 POLLUTION LEVELS

As depicted in Figure 13.2, levels of NO_2 are noticeably high in Gauteng and the surrounding provinces of Mpumalanga, Limpopo, North-West, and the Free State pre-lockdown events, with considerable reductions in NO_2 emissions observed during the selected lockdown months (April, May, and June 2020). The observed high levels of NO_2 emissions in Mpumalanga Province in the northeastern part of South Africa are particularly related to coal mining and thermal power generation, as earlier studies have shown (Yoro and Sekoai, 2016). Gauteng Province, the industrial and economic hub of South Africa, has the largest population; the mean NO_2 for the period under review was 47.84 µmol/m² where the maximum value was recorded in June 2019 (64.71 µmol/m²) and the minimum value in April 2020 at 41.07 µmol/m².

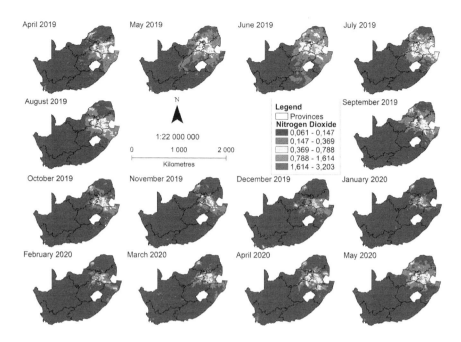

FIGURE 13.2 Variation in NO_2 concentration in South Africa before, during and after the COVID-19 lockdown. Gauteng, Mpumalanga, Limpopo, North-West, and Free State provinces contribute most of NO_2 due to high industrial activity, coal mining, and thermal power generation.

Notable decreases occurred during the hard lockdown (alert level 5) between April and June 2020. Based on the COVID-19 five-level alert system implemented by the South African government to manage the ease of lockdown, alert level 5 is the most stringent and alert level 1 the most relaxed. However, it is important to note that domestic fossil fuel use for heating and cooking peaked during the winter period, contributing to the high NO_2 content in Gauteng Province (Figure 13.2). The peaks and lows in the concentration of pollutants indicate the variations in NO_2 concentration over time as pollutant presence in the atmosphere is subject to vehicular volume and density of manufacturing industries. Low pollutant concentration levels occurred mainly during off-peak periods, such as the festive seasons when industrial activity was low. However, the longest recorded period of low NO_2 and other pollutants concentration was during the COVID-19 pandemic-induced lockdown (between March and May 2020).

However, after the easing of the lockdowns and subsequent opening of industries and an increase in vehicular volume, there was a sudden spike in the levels of NO_2 from 130.65 µmol/m^2 in March 2020 to 603.95 µmol/m^2 in June 2020 in the Gauteng Province. This confirms the contribution of manufacturing industries to atmospheric pollution, acid rain, water degradation and climate change. Exposure to NO_2 and aerosols poses health risks that manifest through several biological responses, including decreased lung function, indirect effects on oxidative stress and inflammatory responses, and inflammatory cytokine stimulation leading to inflammatory injury, among other risks (Comunian et al., 2020; Miller, 2020). In South Africa, these health challenges are prevalent in urban areas (Matooane et al., 2004).

13.3.2 CHANGES IN AEROSOLS POLLUTION LEVELS

The presence of aerosols over South Africa has become a severe human health and climate change concern, as evidenced by the degradation of air quality, coupled with challenges related to visibility impairment on the roads as well as the resultant increase in the intensity and frequency of extreme weather events (Kwon et al., 2020; Manisalidis et al., 2020).

Atmospheric aerosols (sulphate, nitrate, ammonium, organic carbon, elemental carbon, and mineral elements) absorb and diffuse solar and longwave radiations emitted from the Earth's surface. This process alters the surface's atmospheric radiation budget (Thandlam and Rahaman, 2019). Due to the critical function of aerosols in cloud condensation, changes in their composition can alter clouds' macro and micro characteristics, causing negative radiative impacts that result in the greenhouse effect (Christensen et al., 2020; Ren-Jian et al., 2012). Aerosol particle concentrations reduced drastically during the COVID-19-induced lockdown (Figure 13.3). However, KwaZulu-Natal Province remained the most aerosol-contributing province in South Africa.

The spider graph (Figure 13.4) demonstrates sulphur dioxide (SO2) changes per province during the 2020 COVID-19 economic lockdown. The trends indicate reduced SO_2 during March and April as there were reduced vehicular volumes,

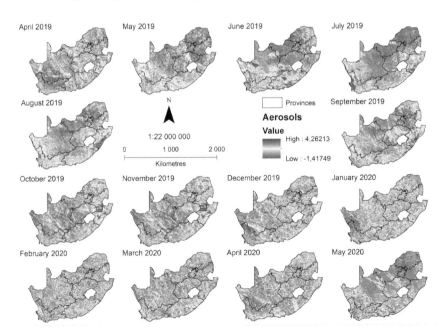

FIGURE 13.3 Variation in aerosols (mol/m^2) presence in the atmosphere in South Africa before, during and after the COVID-19 lockdown. Aerosol presence drastically dropped during the COVID-19 lockdown, particularly between March and May 2020.

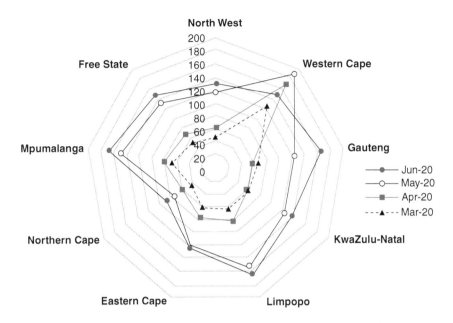

FIGURE 13.4 Variations in SO$_2$ over South Africa by province during the 2020 lockdown. SO$_2$ reduced significantly during March and April 2020 due to reduced vehicular presence.

confirming the contribution of vehicles to atmospheric pollution. However, the Gauteng and Western Cape remained the central SO$_2$ contributing provinces in South Africa. The analysis showed that the province emitting the most pollutants is Mpumalanga, with an average of 199.03 μmol/m^2, followed by Gauteng with 187.70 μmol/m^2, then Limpopo with an average of 92.91 μmol/m^2. The remaining provinces have an average below 75, except for Free State, KwaZulu-Natal, and Western Cape, with 78.42, 87.58, and 81.75 μmol/m^2, respectively.

13.4 IMPACT OF AIR POLLUTION ON HUMAN AND ENVIRONMENTAL HEALTH

Air pollution has become a major health risk as it is the cause of serious toxicological impacts affecting both human and environmental health (Ghorani-Azam et al., 2016). It has been the major source of acid rain, degrading terrestrial and aquatic ecosystems since the advent of the industrial revolution (Manisalidis et al., 2020). The risk posed by air pollution on human health is so huge that it is attributed to over 5 million deaths each year worldwide, representing 9% of deaths globally (IHME, 2018). In South Africa alone, air pollution and respiratory-related diseases (heart disease, stroke, lower respiratory infections, lung cancer, diabetes, and chronic obstructive pulmonary disease) killed over 23,000 people in 2017 alone (IHME, 2018). Figure 13.5 is a map indicating the death rates from air pollution-related illnesses across the globe measured as the number of deaths per 100,000 people per country. The highest air pollution death rates are experienced in Sub-Saharan Africa and South Asia.

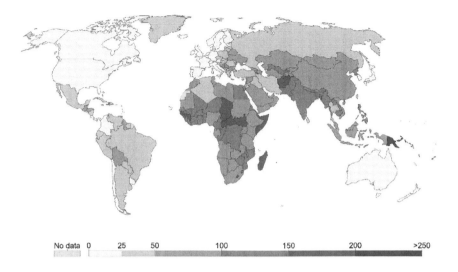

FIGURE 13.5 Global death rates from air pollution in 2017, measured as the number of deaths per 100,000 population.

Source: Institute for Health Metrics and Evaluation (IHME).

TABLE 13.1

Total number of affected people by pollution-related disasters in South Africa (2000–2020)

Disaster type	No. of affected people
Flash flood	4,100
Bacterial and other viral diseases	111,960
Drought	18,450,000
Total	18,566,060

Source : EM-DAT (Guha-Sapir et al., 2021).

It is now well-acknowledged that air pollution and climate change are closely linked, as pollution is just the other side of the same coin degrading Earth's quality of life (Manisalidis et al., 2020). Black carbon, methane, tropospheric ozone, aerosols, and other pollutants heavily affect incoming sunlight, causing some atmospheric changes, temperature increases and heatwaves (Fiore et al., 2015). The atmospheric changes caused by air pollution result in the incidence and prevalence of infectious diseases (Manisalidis et al., 2020). Previous studies have indicated that disease outbreaks' duration, timing, and intensity strongly correlate with climatic and environmental changes (Caminade et al., 2019; Manisalidis et al., 2020; Wu et al., 2016).

In South Africa, there has been an increase in air pollution-related disasters and infectious diseases (Nhamo and Ndlela, 2021), as shown in Table 13.1 (Guha-Sapir et al., 2020). South Africa has seen an increase in climate-related disasters since 2000, affecting over 18.5 million since 2000 (Table 13.1) (Guha-Sapir et al., 2020). Malaria and other viral diseases have resurgent recently as these pathogens are climate-sensitive (Abiodun et al., 2020). Increasing temperatures reduce the pathogen incubation period of parasites and viruses and contribute to the vector's geographic distribution changes (Caminade et al., 2019). The increasing spread, intensity, and frequency of epidemics are associated with air pollution and climate change.

13.5 STRATEGIC PATHWAYS TOWARDS REDUCED AIR POLLUTION

After identifying and assessing different pollutants in South Africa and the associated health risks, a conceptual framework was developed to provide pathways and smart solutions that lead to acceptable pollution levels without impacting industrial production. The conceptual framework (Figure 13.6) is developed around five thematic areas that include (i) drivers of change, (ii) pollution risks, (iii) responses and recovery, (iv) financing, and (v) adoption of smart solutions. The thematic areas were identified from the literature, particularly from research projects that address pollution-related challenges (Li et al., 2019; Park et al., 2020; Rao et al., 2017; Reis et al., 2022). Each theme is composed of actionable pathways to drive the transformational change towards Sustainable Development Goals (SDGs), particularly Goals 3 (good health and well-being), 6 (clean water and sanitation), 7 (affordable and clean energy), 11 (sustainable cities and communities), and 13 (climate action) with synergies with the other remaining goals.

Air pollution and climate change are intricately interlinked and are associated with the sectors that contribute the most GHGs. These sectors include transport, agriculture, energy, industry, and waste management. These are the same sectors that emit fine particulate matter ($PM_{2.5}$) and the other main pollutants found in the atmosphere (Barwise and Kumar, 2020). Other major atmospheric pollutants include SLCPs such as black carbon and ground-level ozone that pose risks to human health (Yamineva and Liu, 2019). The same anthropogenic activities disturbing the Earth's

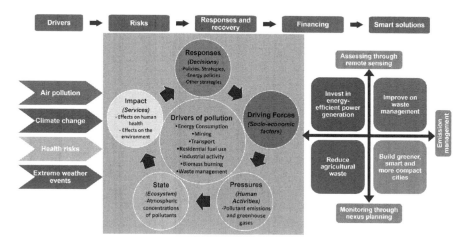

FIGURE 13.6 Transformational pathways towards cleaner environments, reduced air pollution, attainment of a circular economy and sustainable development.

climate also produce the most pollutants into the atmosphere, becoming the greatest threat to environmental and human health (Rhind, 2009).

Climate change and air pollution are the greatest threats to human health and livelihoods through heatwaves, environmental degradation (the major cause of the emergence of novel infectious diseases), and droughts and increasing temperatures that favour the emergence of pests and diseases (Manisalidis et al., 2020). We, therefore, establish the interlinkages between air pollution, climate change, human health and extreme weather events and provide the pathways for smart solutions that lead to sustainable development (Figure 13.6). Closely linking climate change and air pollution in policy and decision-making allows the formulation of cross-sectoral strategies that simultaneously mitigate interlinked challenges in a holistic manner (Nhamo et al., 2020). At the same time, inefficient, polluting, and energy-intensive practices drive both crises; hence, strategies aimed at reducing air pollution that create co-benefits for both climate and health are required. Likewise, measures to reduce emissions typically lead to reductions in co-emitted pollutants such as particulate matter.

The conceptual framework (Figure 13.6) is based on intricately interconnected drivers of change and processes. Any intervention to mitigate any drivers, like air pollution, should consider all the contributing elements driving the increasing pollutants and climate change. Mitigatory interventions require integrated and transformative approaches that include the circular economy and one health, other than pursuing linear models that have now reached their threshold and are no longer appropriate for addressing today's intricately connected problems (Alcayaga et al., 2019; Naidoo et al., 2021). In this digital world dominated by globalisation, the circular economy approach is powered by digital technologies (DTs), including the IoT, Big Data, and Data Analytics. These circular economy enablers facilitate tracking the flow of products, components, and materials, availing the resultant data for improved resource management and decision-making across different stages of the production cycle (Kristoffersen et al., 2020). Powered by DTs, the circular economy provides pathways towards sustainable growth, efficient resource use, good health and employment opportunities while saving the environment and natural resources (Kalmykova et al., 2018).

13.6 CATALYSING CLEANER PRODUCTION SYSTEMS

Pollution levels reduced drastically during the COVID-19 lockdowns, but the levels picked up again as soon as the lockdown regulations were relaxed. The COVID-19 pandemic revealed how pollution levels could go down when there is limited industrial production, which is presently linear. However, while there is an urgent need to reduce pollution levels, producing resources and products should proceed to meet human needs. There is, therefore, an urgent need to implement and operationalise circular production systems that embrace smart technologies and enhance cleaner environments. These initiatives are catalysed by adopting environmentally friendly energy sources and embracing circular strategies that include reducing, reusing, repairing, recycling, restoring and industrial symbiosis (Kristoffersen et al., 2020) (Velenturf and Purnell, 2021). Smart systems and technologies that include product service systems (PSS) and performance models could provide the pathways to fast-track the linkages between the circular economy and the IoT, speed the needed transformational change and attain the green economy (Ingemarsdotter et al., 2019; Naidoo et al., 2021). PSS offer various environmental benefits by capitalising on DTs and related connectivity to enhance resource use efficiency, extend product lifespan and close material loops (Alcayaga et al., 2019; Chauhan et al., 2022). By optimising product use and value preservation over time, industrialists can maximise profits from all value creation opportunities provided by PSS and spur a long-lasting revenue stream (Chauhan et al., 2022).

Thus, research on reducing air pollution should consider cross-sectoral analysis that includes socio-economic factors (the driving forces), human activities (the pressures), environmental and human health, the impact, and the responses (Figure 13.6). Such a transformational approach results in smart solutions and circular modelling through investment in renewable energy, improved waste management and recycling, transformation to green cities, and the reduction of agricultural waste towards sustainable agricultural systems. The conceptual framework, based on four thematic areas (drivers of change, the risks, responses and recovery, financing, and smart solutions), provides integrated solutions to mitigate existing challenges and provide the pathways towards achieving SDGs.

13.7   STRATEGIES TO REDUCE AIR POLLUTION

As already alluded, interventional strategies to mitigate air pollution require integrated and transformative approaches to address challenges across sectors and provide smart solutions towards a circular economy and cleaner production systems. Cross-sectoral cooperation at different levels (city, regional, national, and international) is critical to addressing air pollution and its challenges in an effective manner. There is a need for societal changes to shift towards coherent and smart policies and investments that support cleaner production, transport modes, and power generation, as well as energy-efficient housing and municipal waste management to reduce air pollution. Besides the potential of these transformational interventions in improving human health, they also reduce GHGs. Smart interventions translate into sustainable development, catalyse economic development, and turn urban areas into centres for human development and climate change adaptation. Table 13.2 provides proposed strategies that can be adopted to mitigate and reduce air pollution.

13.8   CONCLUSIONS

The COVID-19-induced lockdowns have highlighted the contribution of anthropogenic activities to atmospheric pollution, GHG emissions and, ultimately, global warming. Pollution levels were drastically reduced during the lockdown as industrial production and vehicular emissions were reduced. Transformational change is urgently needed to expedite the circular economy concept to achieve sustainable development by 2030. This is particularly urgent as air pollution has become a major health risk affecting millions worldwide. This chapter has produced a guiding framework for strategic and coherent policy formulations to enhance the maximum co-benefits for human health and climate. The challenge requires cross-sectoral

TABLE 13.2
Sector-specific strategies that can be adopted to reduce and mitigate air pollution

Sector	Mitigatory strategies
Agriculture	• Promote the use of renewable sources of energy for irrigation. • Installation of thermal screens for mushroom production to increase energy use efficiency. • Application of manures to reduce the need for inorganic fertiliser (utilising agricultural waste to produce manure for fertilising soils). • Use of excess agricultural waste to produce biogas for energy purposes within the sector (reduces demand for electricity, where coal is being used). • Avoid burning agricultural fields. • Promote healthy diets that are low in red and processed meat but rich in plant-based foods. • Capture of methane gas emitted from waste sites as an alternative to incineration.
Transport	• Introduce solar-powered cars. • Subsidy on purchases of electric vehicles. • Implement tougher vehicle emissions and efficiency standards. • Promote the use of public transport or promote carpooling. • Improve the efficiency of urban public transport. • Promote walking and cycling networks in cities. • Introduction of public transport powered using green energy. • Shifting to more efficient and cleaner heavy-duty diesel vehicles and low-emissions vehicles and fuels. • Promote the use of fuels with reduced sulphur content. • Using electric rail systems to transport raw materials from extraction plants to manufacturing plants and final products to wholesalers.
Cities	• Promote walking and cycling to and from work. • Promote the concept of compact cities, which are energy efficient. • Promote the use of public transport and invest in bus rapid transit and light rail. • Create green spaces that remove particulate matter and reduce the heat island effect. • Improve waste management and the capture of methane gas emitted from waste sites as an option for incineration. • Improving social housing delivery to minimise informal settlements. • Use of renewable energy for powering public infrastructure. • Planting more trees to lower average temperatures. • Planting carbon sequestrating vegetation. • Promoting smart street lighting initiatives.
Housing	• Promote the use of renewable sources of energy for lighting, heating, and cooking. • Replace the use of kerosene for household use with alternative, cleaner sources of energy. • Use of energy-efficient light bulbs and appliances. • Improve household and commercial energy efficiency through insulation and passive design principles such as natural ventilation and lighting. • Promote green rooftops. • Promote the application of artificial intelligence for efficient energy use.

(Continued)

**TABLE 13.2
(Continued)**

Sector	Mitigatory strategies
Waste management	• Drive urban areas towards the urban circular economy. • Reduce waste through waste separation, recycling, and reuse. • Promote biological waste management such as anaerobic waste digestion to produce biogas. • Adopt combustion technologies and formulate strict emission controls where incineration is unavoidable. • Use waste as energy sources in some industries such as cement production (landfills can be greatly reduced, also improving air quality).
Industry	• Adopting cleaner production technologies to reduce industrial emissions. • Switch to affordable and cleaner energy sources. • Encourage the recovery and use of gas released during fossil fuel production. • Promote the use of regenerative thermal oxidisers to destroy pollutants before they are released into the environment. • Provision of incentives by the government to motivate the implementation of cleaner production technology in different economic sectors. • Promote sustainable disposal of industrial waste. • Reuse of industrial waste that can be used in the combustion process, such as waste plastic and oil. • Aligning environmental management plans to the national environmental objectives. • Requesting sectors to report on their current emissions and target reductions as prescribed by the Climate Change Bill of South Africa.
Energy generation	• Discourage the use of environmentally unfriendly sources of energy like oil and coal for large-scale energy production. • Promote the use of low-emission fuels and renewable combustion-free power sources such as solar and wind. • Promoting investment in the green energy sector. • On-grid integration of electricity from renewable energy forms to supplement the load generation from power plants.
Policy	• To design the environmental legislation and ensure enforcement. • To design controls on emissions and ensure adherence by sectors. • To enforce emission ceilings per sector. • To ensure polluting sectors contribute to environmental rehabilitation.

coordination of different but interlinked sectors and stakeholders to develop and operationalise effective and integrated actions that improve public health through waste and pollution reduction. The proposed framework provides mitigatory pathways towards climate change resilience and adaptation. The most critical pathway is a unified governance and policy framework that promotes the reduction of air pollution and the promotion of the right to clean air. This is possible through transformational change, adopting smart technologies, and embracing circular models instead of linear approaches that focus on single sectors and forget the interconnectedness of systems. The recommended integrated and transformative approaches require public-private engagements and individual actions to reduce the effects of pollutants on human health and accelerate climate change. Interventions require communities to shift from the 'norm' towards a circular economy and use renewable energy sources. The changes include engineering solutions drastically reducing emissions from cooking stoves and vehicles. The increased frequency and intensity of the emergence of novel infectious diseases such as COVID-19 requires urgent and proactive interventions to reduce the risk of air pollution to human and environmental health.

REFERENCES

Abiodun, G.J., Adebiyi, B., Abiodun, R.O., Oladimeji, O., Oladimeji, K.E., Adeola, A.M., Makinde, O.S., Okosun, K.O., Djidjou-Demasse, R., Semegni, Y.J. (2020) Investigating the Resurgence of Malaria Prevalence in South Africa between 2015 and 2018: A Scoping Review. *The Open Public Health Journal* 13, 1.

Adesina, J., Piketh, S., Qhekwana, M., Burger, R., Language, B., Mkhatshwa, G. (2020) Contrasting Indoor and Ambient Particulate Matter Concentrations and Thermal Comfort in Coal and Non-coal Burning Households at South Africa Highveld. *Science of the Total Environment* 699, 134403.

Alcayaga, A., Wiener, M., Hansen, E.G. (2019) Towards a Framework of Smart-circular Systems: An Integrative Literature Review. *Journal of Cleaner Production* 221, 622–634.

Anenberg, S., Miller, J., Henze, D., Minjares, R. (2019) A Global Snapshot of the Air Pollution-Related Health Impacts of Transportation Sector Emissions in 2010 and 2015. International Council on Clean Transportation (ICCT), Washington, DC, p. 55.

Barrett, B., Charles, J.W., Temte, J.L. (2015) Climate Change, Human Health, and Epidemiological Transition. *Preventive Medicine* 70, 69–75.

Barwise, Y., Kumar, P. (2020) Designing Vegetation Barriers for Urban Air Pollution Abatement: A Practical Review for Appropriate Plant Species Selection. *Climate and Atmospheric Science* 3, 1–19.

Burrows, J.P., Weber, M., Buchwitz, M., Rozanov, V., Ladstätter-Weißenmayer, A., Richter, A., DeBeek, R., Hoogen, R., Bramstedt, K., Eichmann, K.-U. (1999) The Global Ozone Monitoring Experiment (GOME): Mission Concept and First Scientific Results. *Journal of the Atmospheric Sciences* 56, 151–175.

Caminade, C., McIntyre, K.M., Jones, A.E. (2019) Impact of Recent and Future Climate Change on Vector-Borne Diseases. *Annals of the New York Academy of Sciences* 1436, 157.

Chauhan, C., Parida, V., Dhir, A. (2022) Linking Circular Economy and Digitalisation Technologies: A Systematic Literature Review of Past Achievements and Future Promises. *Technological Forecasting and Social Change* 177, 121508.

Chen, J., Li, C., Ristovski, Z., Milic, A., Gu, Y., Islam, M.S., Wang, S., Hao, J., Zhang, H., He, C. (2017) A Review of Biomass Burning: Emissions and Impacts on Air Quality, Health and Climate in China. *Science of the Total Environment* 579, 1000–1034.

Chen, T.-M., Kuschner, W.G., Gokhale, J., Shofer, S. (2007) Outdoor Air Pollution: Nitrogen Dioxide, Sulfur Dioxide, and Carbon Monoxide Health Effects. *The American Journal of the Medical Sciences* 333, 249–256.

Christensen, M.W., Jones, W.K., Stier, P. (2020) Aerosols Enhance Cloud Lifetime and Brightness along the Stratus-to-Cumulus Transition. *Proceedings of the National Academy of Sciences* 117, 17591–17598.

ClimateTransparency (2020) Climate Transparency Report: Comparing G20 Climate Action and Responses to the Covid-19 Crisis. Climate Transparency, Berlin, Germany, p. 72.

Comunian, S., Dongo, D., Milani, C., Palestini, P. (2020) Air Pollution and Covid-19: The Role of Particulate Matter in the Spread and Increase of Covid-19's Morbidity and Mortality. *International Journal of Environmental Research and Public Health* 17, 4487.

Dong, E., Du, H., Gardner, L. (2020) An Interactive Web-based Dashboard to Track COVID-19 in Real Time. *The Lancet Infectious Diseases,* 20(5), 533–534.

Fiore, A.M., Naik, V., Leibensperger, E.M. (2015) Air Quality and Climate Connections. *Journal of the Air & Waste Management Association* 65, 645–685.

Ghorani-Azam, A., Riahi-Zanjani, B., Balali-Mood, M. (2016) Effects of Air Pollution on Human Health and Practical Measures for Prevention in Iran. *Journal of Research in Medical Sciences* 21, 65.

Gorelick, N., Hancher, M., Dixon, M., Ilyushchenko, S., Thau, D., Moore, R. (2017) Google Earth Engine: Planetary-Scale Geospatial Analysis for Everyone. *Remote Sensing of Environment* 202, 18–27.

Guha-Sapir, D., Below, R., Hoyois, P. (2020) The International Disaster Database. Centre for Research on the Epidemiology of Disasters (CRED): The CRED/OFDA International Disaster Database, Louvain, Brussels, Belgium. https://www.emdat.be/.

Hamzelou, J. (2020) World in Lockdown. *New Scientist* 245, 7.

Heshmati, A. (2017) A Review of the Circular Economy and Its Implementation. *International Journal of Green Economics* 11, 251–288.

Hill, J., Polasky, S., Nelson, E., Tilman, D., Huo, H., Ludwig, L., Neumann, J., Zheng, H., Bonta, D. (2009) Climate Change and Health Costs of Air Emissions from Biofuels and Gasoline. *Proceedings of the National Academy of Sciences* 106, 2077–2082.

IEA (2016) Energy and Air Pollution: World Energy Outlook Special Report 2016. International Energy Agency (IEA), Paris, France, p. 266.

IHME (2018) Global Burden of Disease Study Results 2017. Global Burden of Disease Collaborative Network, Institute for Health Metrics and Evaluation (IHME), Seattle.

Ingemarsdotter, E., Jamsin, E., Kortuem, G., Balkenende, R. (2019) Circular Strategies Enabled by the Internet of Things—A Framework and Analysis of Current Practice. *Sustainability* 11, 5689.

Jacob, D.J., Winner, D.A. (2009) Effect of Climate Change on Air Quality. *Atmospheric Environment* 43, 51–63.

Kalmykova, Y., Sadagopan, M., Rosado, L. (2018) Circular Economy–From Review of Theories and Practices to Development of Implementation Tools. *Resources, Conservation and Recycling* 135, 190–201.

Kjellstrom, T., Lodh, M., McMichael, T., Ranmuthugala, G., Shrestha, R., Kingsland, S. (2006) Air and Water Pollution: Burden and Strategies for Control, in: Jamison, D.T., Breman, J.G., Measham, A.R. (Eds.), *Disease Control Priorities in Developing Countries.* 2nd edition. The International Bank for Reconstruction and Development/The World Bank, Washington, DC, pp. 817–832.

Kristoffersen, E., Blomsma, F., Mikalef, P., Li, J. (2020) The Smart Circular Economy: A Digital-enabled Circular Strategies Framework for Manufacturing Companies. *Journal of Business Research* 120, 241–261.

Kwon, H.-S., Ryu, M.H., Carlsten, C. (2020) Ultrafine Particles: Unique Physicochemical Properties Relevant to Health and Disease. *Experimental & Molecular Medicine* 52, 318–328.

Li, N., Chen, W., Rafaj, P., Kiesewetter, G., Schöpp, W., Wang, H., Zhang, H., Krey, V., Riahi, K. (2019) Air Quality Improvement Co-benefits of Low-Carbon Pathways toward Well below the 2° C Climate Target in China. *Environmental Science & Technology* 53, 5576–5584.

Liu, J.-Y., Fujimori, S., Takahashi, K., Hasegawa, T., Wu, W., Takakura, J.Y., Masui, T. (2019) Identifying Trade-offs and Co-benefits of Climate Policies in China to Align Policies with SDGs and Achieve the 2°C Goal. *Environmental Research Letters* 14, 124070.

Manisalidis, I., Stavropoulou, E., Stavropoulos, A., Bezirtzoglou, E. (2020) Environmental and Health Impacts of Air Pollution: A Review. *Frontiers in Public Health* 8, 14.

Matooane, M., John, J., Oosthuizen, R., Binedell, M. (2004) Vulnerability of South African Communities to Air Pollution, 8th World Congress on Environmental Health, Durban, South Africa. Document Transformation Technologies, Durban, South Africa.

McCabe, M.F., Rodell, M., Alsdorf, D.E., Miralles, D.G., Uijlenhoet, R., Wagner, W., Lucieer, A., Houborg, R., Verhoest, N.E., Franz, T.E. (2017) The Future of Earth Observation in Hydrology. *Hydrology and Earth System Sciences* 21, 3879.

Meacham, M., Queiroz, C., Norström, A.V., Peterson, G.D. (2016) Social-Ecological Drivers of Multiple Ecosystem Services: What Variables Explain Patterns of Ecosystem Services across the Norrström Drainage Basin? *Ecology and Society* 21(1), 14.

Miller, M.R. (2020) Oxidative Stress and the Cardiovascular Effects of Air Pollution. *Free Radical Biology and Medicine,* 151, 69–87.

Naidoo, D., Nhamo, L., Lottering, S., Mpandeli, S., Liphadzi, S., Modi, A.T., Trois, C., Mabhaudhi, T. (2021) Transitional Pathways towards Achieving a Circular Economy in the Water, Energy, and Food Sectors. *Sustainability* 13, 9978.

Nhamo, L., Mabhaudhi, T., Mpandeli, S., Dickens, C., Nhemachena, C., Senzanje, A., Naidoo, D., Liphadzi, S., Modi, A.T. (2020) An Integrative Analytical Model for the Water-Energy-Food Nexus: South Africa Case Study. *Environmental Science and Policy* 109, 15–24.

Nhamo, L., Ndlela, B. (2021) Nexus Planning as a Pathway towards Sustainable Environmental and Human Health Post Covid-19. *Environment Research* 192, 110376.

Nhamo, L., Ndlela, B., Nhemachena, C., Mabhaudhi, T., Mpandeli, S., Matchaya, G. (2018) The Water-Energy-Food Nexus: Climate Risks and Opportunities in Southern Africa. *Water* 10, 567.

Omrani, H., Omrani, B., Parmentier, B., Helbich, M. (2020) Spatio-temporal Data on the Air Pollutant Nitrogen Dioxide Derived from Sentinel Satellite for France. *Data in Brief* 28, 105089.

Park, S., Kim, S.J., Yu, H., Lim, C.-H., Park, E., Kim, J., Lee, W.-K. (2020) Developing an Adaptive Pathway to Mitigate Air Pollution Risk for Vulnerable Groups in South Korea. *Sustainability* 12, 1790.

Patella, V., Florio, G., Magliacane, D., Giuliano, A., Crivellaro, M.A., Di Bartolomeo, D., Genovese, A., Palmieri, M., Postiglione, A., Ridolo, E. (2018) Urban Air Pollution and Climate Change: "The Decalogue: Allergy Safe Tree" for Allergic and Respiratory Diseases Care. *Clinical and Molecular Allergy* 16, 1–11.

Pauw, P., Mbeva, K., Van Asselt, H. (2019) Subtle Differentiation of Countries' Responsibilities under the Paris Agreement. *Palgrave Communications* 5, 1–7.

Perera, F. (2018) Pollution from Fossil-Fuel Combustion Is the Leading Environmental Threat to Global Pediatric Health and Equity: Solutions Exist. *International Journal of Environmental Research and Public Health* 15, 16.

Rao, S., Klimont, Z., Smith, S.J., Van Dingenen, R., Dentener, F., Bouwman, L., Riahi, K., Amann, M., Bodirsky, B.L., van Vuuren, D.P. (2017) Future Air Pollution in the Shared Socio-economic Pathways. *Global Environmental Change* 42, 346–358.

Reis, L.A., Drouet, L., Tavoni, M. (2022) Internalising Health-economic Impacts of Air Pollution into Climate Policy: A Global Modelling Study. *The Lancet Planetary Health* 6, e40–e48.

Ren-Jian, Z., Kin-Fai, H., Zhen-Xing, S. (2012) The Role of Aerosol in Climate Change, the Environment, and Human Health. *Atmospheric and Oceanic Science Letters* 5, 156–161.

Rhind, S. (2009) Anthropogenic Pollutants: A Threat to Ecosystem Sustainability? *Philosophical Transactions of the Royal Society B: Biological Sciences* 364, 3391–3401.

Ritchie, H., Roser, M. (2017) Cause of Death. *Our World in Data*, Oxford. https://ourworldin-data.org/causes-of-death.

Rogelj, J., Den Elzen, M., Höhne, N., Fransen, T., Fekete, H., Winkler, H., Schaeffer, R., Sha, F., Riahi, K., Meinshausen, M. (2016) Paris Agreement Climate Proposals Need a Boost to Keep Warming Well below 2 C. *Nature* 534, 631–639.

Schultze, A., Walker, A.J., MacKenna, B., Morton, C.E., Bhaskaran, K., Brown, J.P., Rentsch, C.T., Williamson, E., Drysdale, H., Croker, R. (2020) Risk of COVID-19-related Death among Patients with Chronic Obstructive Pulmonary Disease or Asthma Prescribed Inhaled Corticosteroids: An Observational Cohort Study Using the OpenSAFELY Platform. *The Lancet Respiratory Medicine* 8, 1106–1120.

Seltenrich, N. (2014) Remote-sensing Applications for Environmental Health Research. NLM-Export.

Sierra-Vergas, M.P., Teran, L.M. (2012) Air Pollution: Impact and Prevention. *Respirology* 17, 1031–1038.

Sofia, D., Gioiella, F., Lotrecchiano, N., Giuliano, A. (2020) Mitigation Strategies for Reducing Air Pollution. *Environmental Science and Pollution Research* 27, 19226–19235.

Thandlam, V., Rahaman, H. (2019) Evaluation of Surface Shortwave and Longwave Downwelling Radiations over the Global Tropical Oceans. *SN Applied Sciences* 1, 1–25.

Theys, N., Smedt, I.D., Yu, H., Danckaert, T., Gent, J.V., Hörmann, C., Wagner, T., Hedelt, P., Bauer, H., Romahn, F. (2017) Sulfur Dioxide Retrievals from TROPOMI Onboard Sentinel-5 Precursor: Algorithm Theoretical Basis. *Atmospheric Measurement Techniques* 10(1), 119–153.

Vandyck, T., Keramidas, K., Kitous, A., Spadaro, J.V., Van Dingenen, R., Holland, M., Saveyn, B. (2018) Air Quality Co-benefits for Human Health and Agriculture Counterbalance Costs to Meet Paris Agreement Pledges. *Nature Communications* 9, 1–11.

Velenturf, A.P., Purnell, P. (2021) Principles for a Sustainable Circular Economy. *Sustainable Production and Consumption* 27, 1437–1457.

WHO (2020) Coronavirus Disease 2019 (COVID-19): Situation Report–67. https://www.who.int/docs/default-source/coronaviruse/situation-reports/20200327-sitrep-67-covid-19.pdf

Wu, X., Lu, Y., Zhou, S., Chen, L., Xu, B. (2016) Impact of Climate Change on Human Infectious Diseases: Empirical Evidence and Human Adaptation. *Environment International* 86, 14–23.

Xu, Y., Ramanathan, V. (2017) Well below 2 C: Mitigation Strategies for Avoiding Dangerous to Catastrophic Climate Changes. *Proceedings of the National Academy of Sciences* 114, 10315–10323.

Yamineva, Y., Liu, Z. (2019) Cleaning the Air, Protecting the Climate: Policy, Legal and Institutional Nexus to Reduce Black Carbon Emissions in China. *Environmental Science & Policy* 95, 1–10.

Yoro, K.O., Sekoai, P.T. (2016) The Potential of CO_2 Capture and Storage Technology in South Africa's Coal-fired Thermal Power Plants. *Environments* 3, 24.

Zheng, Z., Yang, Z., Wu, Z., Marinello, F. (2019) Spatial Variation of NO_2 and Its Impact Factors in China: An Application of Sentinel-5P Products. *Remote Sensing* 11, 1939.

14 Enhancing the resilience and adaptation of the education sector through intention awareness

Lindiwe Carol Mthethwa

14.1 INTRODUCTION

The right to education for in-service teachers as university students is crucial in cracking major essential human rights while warranting full and equal participation. Students' rights and ways are critical in their emergence, development, and transformation. Unfortunately, the literature confirmed the institutional dimension of the transformative approach with the dragging of lecturers and in-service teachers' dimensions in the uplifting sustainable socio-economic transformation. Grooming students who are in-service teachers towards their studies after a long history of marginalisation calls for intensive strategies that will be infused into their modules. The COVID-19 pandemic has caused global actions, which became the utmost peculiar interventions for higher education research actions and findings. It cannot be overlooked that the challenges posed by the pandemic called for drastic actions from the national systems, government, and public spaces with accurate continuous academic engagement for the reparative future. This drove Higher Education institutions to reimagine how knowledge, learning, and teaching can drive future education, intervention, and humanity.

This chapter proposed the enhancement of resilience and adaptation through IA, wherein a collective matrix is designed with columns that echo its usage. The setting in question demonstrates the existing authenticity and the gap between the possible imminent scenario and the existing reality, which informs the ladders and stages to be taken. These steps are acknowledged to boost the future scenario and the construction of solid pieces of ladders toward the future scenario. Engagement is informed by the transformative Afrocentric mentoring models using scenario planning. The collective matrix in this regard is designed to be completed by in-service teachers and lecturers while the transformative Afrocentric models and the institution's vision guide all. It is argued that the intentional awareness of in-service teachers and lecturers could enhance and reduce the informational burden on the stakeholders. This will be done while promoting effectiveness in reparative strategies for enhancing climate change education.

DOI: 10.1201/9781003327615-14

This chapter conceptualises the notion of in-service teachers' and lecturers' epistemological access in crafting their curriculum. In crafting this, scenario planning is used with systems thinking for analysis. The core components of systems thinking underpin the scenario planning discussions, which involve a shared understanding of the problem, forethought, and coordinated actions (Sinnot, 2020). The epistemological understanding has confirmed this as it is narrowed in the study conducted by Ballim (2015). He pointed out the important finding: the university as an institution abdicating its responsibility to "teach properly".

Further, Ballim's reviews were based on two valuable opinions. First, he proposed tacitly that providing epistemological access is the task of the university and not of academic development/extended programmes. This could mean that everybody who enters the university has to be initiated into the construction of academic knowledge within specific disciplinary fields (Lange, 2017). Interestingly, this practice suggested that starting at an undergraduate level, there should be an exercise to make transparent the 'black box' of knowledge construction among in-service teachers. To be precise, in-service teachers would not only be taught to be able to make knowledge in their modules but would also acquire the behaviours, practices, and identities expected from them as engineers, doctors, historians, etc.

Second, Ballim identified the inversion of the notion of epistemological access as something that lecturers need to be helped with to understand the variety of ways of knowing and making sense of the world that their students have, and which can constitute the point of entry for epistemological access to university knowledge instead of its opposite (Ballim, 2015).

The commitment to ensuring equitable quality education and promoting lifelong learning opportunities for all is underscored by Sustainable Developmental Goal (SDG) number 4 (UNDP, 2019). Nhamo (2021) attested to the significance of SDG 4, especially in the localisation of SDG in Higher Education Institutions (HEI) and sustaining new initiatives in the core mandates of any university in research, teaching, engaged scholarship, and institutional operations. This is what Filho et al. (2019) described as the 'third mission' where there is mainstreaming into external stakeholders and society engagement platforms, which universities have long been lagging at. The connection between what the curriculum addressed as sustainability competencies needs to be reflected in the teaching and the learning process (Nhamo, 2021).

In this articulation of scenario planning, it is important to bear in mind that there will be the coining of the past, the present, and the future, initially crafted in the virtual space. This becomes the pillar of engagement of all relevant stakeholders in teaching and learning while alerting them using IA. About effective teaching and learning, this chapter claims scenario planning, which spans how consultation, communication, and adaptation will be done. Scenario planning serves as an essential functional reparative meaning towards a resilient future as it involves a collective matrix of the stakeholders using systems thinking. This calls for new educational development that enhances the competitiveness of organisations, students, and lecturers. A hopeful possibility is aggravated by communicating transition now and again while establishing key aspirations of fruitful teaching and learning that is more relevant and worthwhile. Reflections in education predicted this as a good predictor of the type of nation raised due to its reciprocal educational nature.

14.2   THE ROLE OF INTENTION AWARENESS AND SCENARIO PLANNING IN EDUCATION

In-service training students need to understand clearly the social context of their soon-coming clients who are faced with low levels of education and poor living standards, which hinders human development in Southern Africa (Facer & Selwyn, 2021; UNESCO, 2017, 2021). The rapidly growing population, political instability (security crises), and the COVID-19 pandemic exacerbated the situation.

IA is a drastic step toward a vision declaration and illustrating the scenarios (Kymäläinen et al. 2014). It requires some signalling to transform from desire, where every action will be demonstrated (Howard & Cambria, 2013). This could be an approach that extricates human intention as a premise for developing intention-aware systems and practically engaged interactions (Kymäläinen et al., 2014); four basic components of IA form the beliefs, desires, and intentions (BDI) model, which is concluded by the events that form part of scenario planning in education (Howard & Cambria, 2013). Intention-aware-based systems can offer an advantage over situation-aware-based systems. It builds on research in personalisation, activity, and behaviour analysis, as well as situation and context awareness, and combines it with agent communication and artificial intelligence (AI) techniques. IA builds on research in personalisation, activity and behaviour analysis, and situation and context awareness, and lastly, combines it with agent communication and AI techniques (Kymäläinen et al., 2014). This has proved to be the birth of the new interaction paradigm.

IA can reduce the informational burden on humans without limiting the effectiveness of the theory. This was one strategy for improving situation awareness in human-centric environments. IA was used as "the process of integrating actors' intentions into a unified view of the surrounding environment" (Howard & Cambria, 2013, p. 1). Education systems are for particular cultures, in particular times and environments; thus, appropriate pedagogic purposes are met.

In this regard, should an intentional stance exist in a system under human presence, this takes us to the next phase in the process, where the system state and its components are identified. The myriad factors like beliefs, goals, wants, previous commitments, fears, and hopes enable one to craft the appropriate attitude. Hence, Howard and Cambria (2013) grouped attitudes into three which are:

a. Information attitudes: an actor's preconceived notions towards information about the surrounding environment.
b. Proactive attitudes: attitudes that direct a mind state to favour action.
c. Normative attitudes: obligations and permissions [ibid].

These attitudes have their roots in the relationships among BDI. These relationships are anticipated to project an inherently desired teaching and learning context.

Scenario planning is a contemporary method that has recently gained popularity, intersecting academia, public and private sectors, and policy constructors (O'Neill et al., 2020; Serrano et al., 2018). Interestingly, it is suitable for coping with uncertainties in today's rapidly changing world, where learning happens through strategic conversations (Sardesai et al., 2021). Enabler of the consensus in considering the probability of a certain future. Organisations and the public are ensured with the possibility of ensuring innovation and flexibility. According to Boerjeson et al. (2006), they stipulated three main scenario categories, which are predictive, explorative, and normative. Predicative scenarios address what will happen; explorative scenarios "what can happen" and normative scenarios "how can a specific goal be achieved"?

14.3 AIMING AT PROMOTING TRANSFORMATIVE AGENDA IN HIGHER EDUCATION

South Africa's past has been rooted in violence, which has become the midwife of history (Ndlovu-Gatsheni & Ndlovu, 2021). Within the field of transformative sustainability education, there could be an explicit emphasis on fostering learning conditions that can shift traditions, expectations, and frames of reference while allowing new openings for change (Azoulay, 2021; Mezirow, 2003). Planning for the future is what, in many cases, the institutions believe that they do (Rieley, 1997). The literature presented higher institutions being embraced by a lack of integrity, caring and emotional support (Clark & Wallace, 2018). Nevertheless, the power of decision-making has to be shifted from lecturers as academics to administrators and managers, emphasising handling universities as business entities aiming at profitability rather than sustainability (Habib, 2016). The chapter's most important aspect is dragging students into the "construction of a new social contract for education" (Azoulay, 2021). Recently, "the everyday reality of inequality filters into macrocosms of university life, classrooms, and power dynamics, where deficit approach to individual struggle and failure has been one response" (Calitz, 2018, p. 36). In South Africa (SA), everyday tasks, beginning with handling fragile objects, require a high degree of situation awareness and a spatial attitude (Howard & Cambria, 2013). For many centuries, Africa had an identity of being a wild goose with a lack of intellectual output (Mazrui, 2001, 2004; Okeke, 2012). Heath and O'Donoghue (2021) claim that contemporary environmental knowledge involves systems thinking as an accurate approach to understanding social-ecological systems since it unpacks cause-and-effect processes and circularity in a system.

Surprisingly, the need to mainstream climate change knowledge into education and training curricula has long been discussed (Kutywayo et al., 2022). Hence, it was stated that climate change education should be part of the broader framework of education for sustainable development and should equip South African citizens to re-orient society towards social, economic and ecological sustainability (Kutywayo et al., 2022).

Thus, in 2019, mainstreaming climate change into secondary and tertiary education curricula has been identified as a key action in the National Climate Change Adaptation Strategy (RSA DEA, 2019). Ulmer and Wydra's (2020) study on 16 African countries became one of few works investigating the state of sustainability activities in African HEIs. Findings proposed a stance on sustainability in Africa rather than converging on negative circumstances. Meanwhile, in February 2023, there will be an interactive workshop at the Researcher to Reader conference, which will explore How we can better communicate SDG-related research to professionals and policymakers with a mixed audience of publishers, librarians, and researchers.

Contemporary discussions are emerging across the field of literacy, gender equality, and citizenship education, which shift a discourse towards sustainable development. This has been witnessed to accelerate involvement in climate change education, understanding political unrest, and solving societal conflicts (Facer, 2021).

14.4   UNDERSTANDING THE REPARATIVE FUTURE

Sustainability in a higher education context comprises four fields, namely, teaching, research, institutional service, and community engagement. Sustainability as the main drive toward a reparative future in developing countries is sparsely researched (Filho et al., 2019; Ulmer & Wydra, 2020).

The concept of 'future' is intimately and ubiquitously associated with education, yet this relationship remains poorly conceptualised in mainstream educational thought. Constructing a resilient, inclusive, and democratic culture by forecasting, imagining, planning, and building together is the way to go. This could be tailor-made to construct futures that distinguish and hunt for mending historical prejudices. Immediate interventions have vast opportunities for risks, but preparedness and thorough focus widen the scope of intervention with intelligence.

Scenario planning, combined with systems thinking, enables one to plan for every possibility while remaining agile to respond to every need and its outcomes (Sinnot, 2020).

Understanding contemporary challenges, policies, and changes directs to a reparative future (Gungordu et al., 2017; Lotz-sisitka et al., 2021). It has been highlighted that rural areas, particularly in the Global South, remain the centres of land grabbing, increasing "de-agrarianisation", food insecurity, cultural loss, extreme poverty, and climate change insensitivity (Chigbu, 2015, p. 1068). Attesting to Mazrui (2001), in-service teachers will be what they have been raised to become because of what lecturers think they are. Actions like self-flattering and self-promotion have eroded, wherein Africans were known for their self-praises which narrated their achievements and downfalls. Nevertheless, the new tone of ruralisation needs to be emphasised to understand the reparative future. As a means of transformation, the vision needs to be drafted and more over being communicated. Hence, thought is not thought unless it is also written (Mazrui, 2001, p. 99). However, the chapter claims that going back to the roots of communicating the vision among the community members in an Afrocentric mode will sustain the vision. Therefore, thoughts are worthless when not communicated and shared. The era for shifting the focus from researching the gaps to existing activities is now (Ulmer & Wydra, 2020). Moreover, African Higher Education Institutions have been identified as dragging in improving their sustainability (ibid). HEI curriculum should inform praxis of unlike urbanisation; ruralisation is portrayed as the solution provider where among lecturers and in-service teachers, there is "heritageisation" of rural ways of living (Chigbu, 2015).

In higher education, sustainability comprises four fields: teaching, research, institutional operations, and community outreach (Ulmer & Wydra, 2020). Literature confirmed the value of transformation; hence, it offers an opportunity for researching and rethinking how suitable and effective educational practices may be.

Transformation in learning within the education context requires for sustainability the commitment of faculty, lecturers, and in-service teachers. With their efforts, motivation, and innovative ideas, change in content and methods can materialise. All stakeholders must share real-world challenges where research is needed with lecturers and in-service teachers.

Further, the co-creation of research agendas and collaboration between lecturers and in-service teachers on research to inform practical action is the foundation of all. This will provide evidence-based practical information on actions to take and avoid in practice and policy-making. HEI's responsibility is to solve the myriad of problems (Okeke, 2012).

14.5   THE MAIN STAKEHOLDERS IN THE COLLECTIVE MATRIX

True Africanisation involves the identification of neglected impediments like the involvement of main stakeholders (GNI, 2022), identity building (Okeke, 2012), and revisiting humanity (Kamwendo, 2016).

It has been argued that "in trying to create a more sustainable world, we need to better look at how ecosystems work and become competent system thinkers" (Lotz-Sisitka & Lupele, 2017, p. 13). Hoover and Harder (2014) pointed out that there is limited and often no adequate institutional support and incentives for those lecturers willing to integrate SDG into their activities in HEI. This hinders the process of voluntarily enhancing research activities to mitigate contextual challenges. As the young generation, the service teachers are expected to be pushed into the front. The nation would have sat up when its young people lost their ability to stand up straight or speak.

In-service teachers in this chapter are the main stakeholders; hence, their participation, equality, and capabilities are what Calitz claims for in the South African context where she stipulated that:

> In an increasingly competitive system, how do students use their agency to navigate university life with limited personal resources and academic preparation? Despite significant personal and structural obstacles, what is the individual able to achieve using her agency?
>
> *(2018, p. 1)*

Scenario planning enables stakeholders to focus on future planning while all members are involved in the thinking process and using collective thinking to influence change (Sinnot, 2020).

14.6   THE DIFFERENCE BETWEEN CONSULTATION AND COMMUNICATION IN HIGHER EDUCATION

Gone are the times when African students in HEI were experiencing the zone of non-being included. In a collection that explores 'how it feels not to belong, Intruders' unpacked the feelings of Africans who are not free in Africa (Mashigo, 2018). Her collection particularly used current terms and advocated the future in an African style.

In the reconstruction struggle, "Africanisation demands a re-narration of the African existence" (Okeke, 2012). There must be a clear distinction between consultation and communication in constructing the new meaning of being in HEI. In this exercise, the meaning of being an African will be crafted, which involves training the mind and reflecting more on the contemporary structures to pave the way to a clear destiny. In maintaining a dialogue as a conversation, the maximum requirement is to expand the willingness to listen to others without bringing or dismissing their subjective views or lessening their value (Waghid, 2018). Such engagements paved the platform for "an ongoing re-thinking of the Environmental and Sustainability Education (ESE) research-policy interface as co-engaged processes of research-and-policy engagement with the potential to reduce the negative consequences" (Lotz-Sisitka et al., 2021).

It cannot be underestimated that Africa's intellectual authenticity has been challenged by Marxism since Mazrui (2001) believes it has more roots in Western ideology. Literature proved that students globally face ESD misconceptions about the understanding function of the ozone layer, the logic behind ozone layer thinning, and the environmental strains due to this (Gungordu et al., 2017). In SA, the National Climate Change Adaptation Strategy (RSA DEA, 2019) outlines objectives for building resilience and adaptation capacity and promoting climate change awareness. This involved the adaptation responses into development objectives, policy, planning, and implementation. However, it was imperative to ensure that resources and systems were in place to enable the implementation of climate change responses. Vulnerability reduction, climate change governance, and health and well-being promotion.

The Africana philosophy drives this communication as a species of thought. It involves theoretical questions raised by critical engagements with ideas in Africana cultures and their hybrid, mixed, or creolised forms worldwide. Worldwide, the concept of 'education' is agreed to refer to 'lead out' or 'to raise'. Students rely upon old or skillful people for guidance to reach a certain level of understanding. According to Adeyemi and Adeyinka (2003), African education included preparations (guiding youth according to their future roles); functionalism (initiation, imitation, and oral communication); communalism (communal spirit of life, work, and raising children); and wholisticism (promotion of multi-skills to prepare children to become productive citizens).

The approach used in this study is verified to have origins of conversations, communications, and consultations as not embraced by fun. Interestingly, the creative approach of scenario planning and systems thinking to the imagined future would be full of fun, relaxed space with the possibility of losing atmosphere to enhance the informal flow of discussions (Sinnot, 2020).

14.7 AFROCENTRIC MENTORING MODELS AS MEANS OF LEADING OR MENTORING IN-SERVICE TEACHERS

The COVID-19 pandemic has propelled the verge of a new era in education where curriculum demands a drastic shift towards reconstruction, transformation, and reinvolvement. Natural resource depletion and climate change could be understood well with greater involvement of higher education research (Filho et al., 2019). The reason behind promoting Afrocentric mentoring is that the aim of the philosophy of education is not to "go practical"; it is, rather, to facilitate understanding. To accomplish this, traditional education elements need to be revisited (Adeyemi & Adeyinka, 2003). Afrocentrism, in this regard, is not a philosophy for nationwide action but for critical activities (Horsthemke, 2018).

Literature confirms that a discussion from the collective matrix enhances the formulation of scenario narratives while enabling active engagement from students and lecturers (Sinnot, 2020). Education should be ranged to the world of work which fits well with humans in any society. Consultations and discussions ensure active engagement between teachers and students since "mentoring is not a straightforward extension of being a school teacher" (Arnold, 2006). Lecturers as mentors can develop modes of mentoring and diverse means of ensuring various resources. It is likely that modes of mentoring and the attendant intellectual histories also differ across generations, cultures, and international boundaries (Dyer & McKean, 2016).

Education is a reliable means to groom students toward reaching adulthood. They are taught to become the principals of their classrooms. Students must be aware of their intellectual heritage (Dyer & McKean, 2016). This applied methodology, called scenario planning, differs from a pure prognosis or forecast. Instead, it provides several possible future scenarios on how the macro surroundings for supply chains. It might look like a time horizon until 2030 (Sardesai et al., 2021). Lack of student involvement could result in resistance, a powerful tool to impede change. Students as pre-service teachers have the potential to integrate important transformational tools for teachers, namely idealised influence (ability to provide a sense of mission using good moral standards and ethical behaviour to drag others), inspirational motivation (communication by paving commitment to others, and inspire others for the buy into the institutional vision), intellectual stimulation (prompts creative and innovative in problem-solving skills), and personal decisive skills (ability to pursue own goals while motivating others) (Shava & Heystek, 2021).

14.8   LECTURER'S ROLE IN SCENARIO PLANNING IN THEIR TEACHING THAT IS TRANSFORMATIVE

There should be a model that lecturers could use to incorporate transformation using scenario planning. This approach will have no transcultural standards to judge the main culture that would be best promoted via education (Horsthemke, 2018). In giving birth to this model, O'Brien and Sygna's (2013) three Ps, which are Personal, Political, and Practical, lands in its way; see Figure 14.1.

The three spheres of transformation provide a space for students to reconnoitre how beliefs, values, and worldviews drive how they relate to challenges brought by political and practical spheres (Leichenko et al., 2022). The fourth 'P', standing for Possibilities, is suggested as an extension to this model. The lecturers are responsible for teaching students how to shape the future by connecting the dots of the possibilities of creativity, innovative skills, and resiliency when well-connected have an impact on sustainable living and energised humanity.

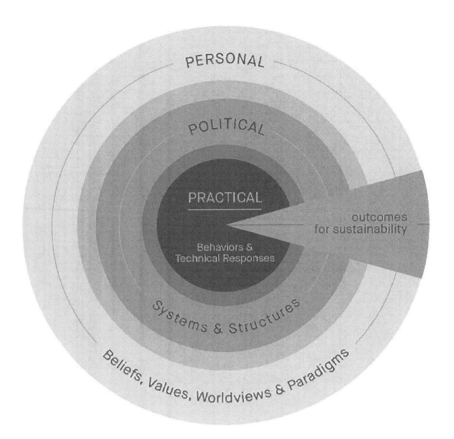

FIGURE 14.1   The three spheres of transformation (Adopted from Leichenko et al., 2022; O'Brien & Sygna, 2013).

14.9   CONCLUSION

The issue discussed in this chapter is a valuable contemporary; hence, a journey towards being truly Africanised is at stake. The development of good practice habits targets the betterment of Africa's HEIs. Scenario planning has been identified as a genuine method for generating thoughts while using system thinking as the lens to inform how moving forward is done. Education cannot exist in its perils without considering the environment and social needs. This chapter claims the deliberative action of a collective matrix will enhance the reparative future since lecturers and students will target stipulated scenarios to create possibilities in life.

REFERENCES

Adeyemi, M. B., & Adeyinka, A. A. (2003). The Principles and Content of African Traditional Education. *Education Philosophy and Theory*, 35(4), 425–440.

Arnold, E. (2006). Assessing the Quality of Mentoring: Sinking or Learning to Swim. *ELT Journal*, 60(2), 117–124.

Azoulay, A. (2021). *Reimagining a New Social Our Futures Contract for Together Education.* UNESCO.

Ballim, Y. (2015). Reflections on "Epistemological Access" and the Analytical Frameworks that May Guide Institutional Responses to Student Learning in South African Higher Education. Unpublished.

Boerjeson, L., Hoejer, M., Dreborg, K. H., Ekvall, T., & Finnveden, G. (2006). Scenario Types and Techniques: towards a User's Guide. *Future*, 38, 723–739.

Calitz, T. (2018). *Enhancing the Freedom to Flourish in Higher Education Participation, Equality, and Capabilities.* Routledge.

Chigbu, U. E. (2015). A Tool for Rural Transformation. *Development in Practice Ruralisation*, 25(7), 1067–1073. https://doi.org/10.1080/09614524.2015.1071783.

Clark, S. G., & Wallace, R. L. (2018). The Integrity Problem in Higher Education: Description, Consequences, and Recommendations. *Higher Education Review*, 50(2), 126–151.

Dyer, W. G., & McKean, A. E. (2016). Learning to "Know Oneself" through an Intellectual Genogram: A New Approach to Analyzing Academic Careers. *Academy of Management Learning and Education*, 15(3), 569–587. https://doi.org/10.5465/amle.2014.0360.

Facer, K. (2021). Futures in Education: Towards an Ethical Practice. Paper Commissioned for the UNESCO Futures of Education Report. UNESCO Futures of Education Background Papers, 29.

Facer, K., & Selwyn, N. (2021). Digital Technology and the Futures of Education – Towards 'Non-stupid' Optimism (Issue April).

Filho, W. L., Shiel, C., Paço, A., Mifsud, M., Avila, L. V., Brandli, L. L., Molthan-Hill, P., Pace, P., Azeiteiro, U. M., Vargas, V. R., & Caeiro, S. (2019). Sustainable Development Goals and Sustainability Teaching at Universities: Falling behind or Getting Ahead of the Pack? *Journal of Cleaner Production*, 232, 285–294. https://doi.org/10.1016/j.jclepro.2019.05.309.

GNI. (2022). The Global Network Initiative Welcomes the "Declaration for the Future of the Internet." https://globalnetworkinitiative.org/declaration-future-internet/.

Gungordu, N., Yalcin-Celik, A., & Kilic, Z. (2017). Students' Misconceptions about the Ozone Layer and the Effect of Internet-Based Media on It. *International Electronic Journal of Environmental Education*, 7(1), 1–16.

Habib, A. (2016). Goals and Means: Reimagining the South African University and Critically Analysing the Struggle for Its Realisation. *Transformation: Critical Perspectives on Southern Africa*, 90(1), 111–132.

Heath, G., & O'Donoghue, R. (2021). Supporting Student Teachers to Teach Catchment and River Management in Geography. In I. Schudel, Z. Songqwaru, S. Tshiningayamwe, & H. Lotz-Sisitka (Eds.), *Teaching and Learning for Change. Education and Sustainability in South Africa* (pp. 80–91), African Minds, Cape Town.

Hoover, E., & Harder, M. K. (2014). What Lies Beneath the Surface? The Hidden Complexities of Organizational Change for Sustainability in Higher Education. *Journal of Clean Production*, 175(e188). https://doi.org/10.1016/j.jclepro.2014.01.081.

Horsthemke, K. (2018). African Philosophy and Education. In *The Palgrave Handbook of African Philosophy*. https://doi.org/10.1057/978-1-137-59291-0.

Howard, N., & Cambria, E. (2013). Intention Awareness: Improving Upon Situation Awareness in Human-Centric Environments. *Human-Centric Computing and Information Sciences*, 3(1), 1–17. https://doi.org/10.1186/2192-1962-3-9.

Kamwendo, G. H. (2016). Unpacking Africanisation of Higher Education Curricula: Towards a Framework. In V. Msila & M. T. Gumbo (Eds.), *Africanising the Curriculum: Indigenous Perspectives and Theories* (pp. 17–33). African Sun Media.

Kutywayo, A., Chersich, M., Naidoo, N.P., Scorgie, F., Bottoman, L., Mullick, S. (2022). Climate change knowledge, concerns and experiences in secondary school learners in South Africa. *Jàmbá: Journal of Disaster Risk Studies* 14, 1–7.

Kymäläinen, T., Plomp, J., Heikkilä, P., & Ailisto, H. (2014). Intention Awareness: A Vision Declaration and Illustrating Scenarios. Intelligent Environments (IE), In I. International Conference (Ed.), ieeexplore.ieee.org (pp. 214–217).

Lange, L. (2017). 20 Years of Higher Education Curriculum Policy in South Africa. *Journal of Education (University of KwaZulu-Natal)*, 68, 31–57.

Leichenko, R., Gram-Hanssen, I., & O'Brien, K. (2022). Teaching the "How" of Transformation. *Sustainability Science*, 17(2), 573–584. https://doi.org/10.1007/s11625-021-00964-5.

Lotz-Sisitka, H., & Lupele, J. (2017). ESD, Learning and Quality Education in Africa: Learning Today for Tomorrow. In H. Lotz-Sisitka, O. Shumba, J. Lupele, & D. Wilmot (Eds.), *Schooling for Sustainable Development in Africa* (pp. 3–24). Springer.

Lotz-Sisitka, H., Mandikonza, C., Misser, S., & Thomas, K. (2021). Making Sense of Climate Change in a National Curriculum. In I. Schudel, Z. Songqwaru, H. Lotz-Sisitka, & S. Tshiningayamwe (Eds.), *Teaching and Learning for Change. Education and Sustainability in South Africa* (pp. 92–111). African Minds.

Lotz-Sisitka, H., Rosenberg, E., & Ramsarup, P. (2021). Environment and Sustainability Education Research as Policy Engagement: (Re-) invigorating 'Politics as Potential' in South Africa in South Africa. *Environmental Education Research*, 27(4), 525–553. https://doi.org/10.1080/13504622.2020.1759511.

Mashigo, M. (2018). *Intruders*. Picador Africa.

Mazrui, A. (2001). Ideology and African Political culture. In T. Kiros (Ed.), *Exploration in African Political Thought: Identity, Community Ethics* (pp. 97–131). Routledge.

Mazrui, A & Mutunga, W. (2004). *Debating the African Condition: Race, Gender, and Culture Conflict* Vol 1. Africa Research and Publications.

Mezirow, J. (2003). Transformative Learning as Discourse. *Journal of Transformative Education*, 1, 58–63.

Mezirow, J. (2000). Learning to Think like an Adult. Core Concepts of Transformation Theory. In J. Mezirow, & Associates (Eds.), *Learning as Transformation. Critical Perspectives on a Theory in Progress* (pp. 3–33). San Francisco, CA: Scientific Research Publishing.

Ndlovu-Gatsheni, S., & Ndlovu, P. P. (2021). The Invention of Blackness on a World Scale. In *Decolonising the Human. Reflections from Africa on Difference and Oppression* (p. 251). WITS University Press. https://library.oapen.org/bitstream/handle/20.500.12657/46908/9781776146789_WEB.pdf?sequence=1.

Nhamo, G. (2021). Localisation of SDGs in Higher Education: Unisa's Whole Institution, All Goals, and Entire Sector Approach. *Southern African Journal of Environmental Education*, 37(1), 63–85. https://doi.org/10.4314/sajee.v37i1.5.

O'Brien, K., & Sygna, L. (2013). Responding to Climate Change: The Three Spheres of Transformation. In: Proceedings of Transformation in a Changing Climate International Conference, Oslo, 19–21 June, 16–23.

O'Neill, B. C., Carter, T. R., Ebi, K., Harrison, P. A., Kemp-Benedict, E., Kok, K., Kriegler, E., Preston, B. L., Riahi, K., Sillmann, J., van Ruijven, B. J., van Vuuren, D., Carlisle, D., Conde, C., Fuglestvedt, J., Green, C., Hasegawa, T., Leininger, J., Monteith, S., & Pichs-Madruga, R. (2020). Achievements and Needs for the Climate Change Scenario Framework. *Nature Climate Change*, 10(12), 1074–1084. https://doi.org/10.1038/s41558-020-00952-0.

Okeke, C. (2012). A Neglected Impediment to True Africanisation of African Higher Education Curricula: Same Agenda, Differential Fee Regimes. *Journal of Higher Education in Africa*, 8(2), 39–52.

Rieley, J. B. (1997). Scenario Planning in Higher Education. In: Walking the Tightrope: The Balance between Innovation and Leadership. Proceedings of the Annual International Conference of the Chair Academy (6th, Reno, NV, February 12-15, 1997), p 11.

RSA DEA. (2019). Department of Environmental Affairs, Acts and Regulations. https://www.dffe.gov.za/parliamentary_updates.

Sardesai, S., Stute, M., & Kamphues, J. (2021). A Methodology for Future Scenario Planning. In A. López-Paredes (Ed.), *Lecture Notes in Management and Industrial Engineering* (pp. 35–57). Springer. https://www.springer.com/series/11786.

Serrano, R., Rodrigues, L. H., Lacerda, D. P., & Parboni, P. B. (2018). Systems Thinking and Scenario Planning: Application in the Clothing Sector. *Systemic Practice and Action Research*, 31, 509–537.

Shava, G. N., & Heystek, J. (2021). Managing Teaching and Learning: Integrating Instructional and Transformational Leadership in South African Schools. *International Journal of Educational Management*. https://doi.org/10.1108/IJEM-11-2020-0533.

Sinnot, G. (2020). The How of Scenario Planning: Systems Thinking in Action. Active Partnership, Engaging Communities, Transforming Lives. https://www.activepartnerships.org/news/how-scenario-planning-systems-thinking-action.

Ulmer, N., & Wydra, K. (2020). Sustainability in African Higher Education Institutions (HEIs): Shifting the Focus from Researching the Gaps to Existing Activities. *International Journal of Sustainability in Higher Education*, 21(1), 18–33. https://doi.org/10.1108/IJSHE-03-2019-0106.

UNDP. (2019). Sustainable Development Goals | UNDP in South Africa. Undp Sdg. https://sustainabledevelopment.un.org/content/documents/24474SA_VNR_Presentation__HLPF_17_July_2019._copy.pdf.

UNESCO. (2017). Changing Minds, Not the Climate: The Role of Education. UNESCO, Paris.

UNESCO. (2021). A Transformative Agenda: Outcomes of the CONFINTEA VII Regional Preparatory Conferences. 15.

Waghid, F. (2018). Action Research and Educational Technology: Cultivating Disruptive Learning. *South African Journal of Higher Education*, 32(4), 1–11. https://doi.org/10.20853/32-4-3097.

15 Summary
Creating systems innovation platforms for transformative pathways in circular economy

Nafiisa Sobratee-Fajurally, Luxon Nhamo, and Tafadzwanashe Mabhaudhi

15.1 TOWARDS AN EVOLUTIONARY SYSTEMS INNOVATION FRAMEWORK FOR CIRCULAR ECONOMY (CE) TRANSITIONING

The chapters in this book cover multi-dimensional aspects of resource use efficiency framed around the circular economy (CE) and how its intertwined domains enable transformative approaches. Various epistemic standpoints are discussed, ranging from systems optimisation versus systems application challenges in sludge waste management in Sweden, gendered and intersectionality considerations, the efficacy of user interface in WASH initiatives from a socio-technical perspective, integrated strategies for climate change adaptation and pandemic preparedness, and CE implications at multiple governance levels with examples from Sub-Saharan Africa. This leads to the insight that whereas the various resource strategies grouped under the CE's banner are not new individually, the concept offers a new framing of these strategies by drawing attention to their capacity to prolong resource use and sustain ecosystems as well as to the interrelationships between these strategies. This chapter aims at synthesising the learning outcomes from each chapter. More specifically, it seeks to demonstrate how tension arising between the dichotomy of short-term efficiency versus large-scale transition or mandated sector-specific achievements versus long-term systemic resilience outcomes can be accommodated if we can shift our perspective from seeing these as dichotomies to one where the evolutionary principle of complex systems is envisioned (Siegenfeld & Bar-Yam, 2020). Since transformative approaches to CE are complex, tensions due to competing strategies are inevitable. One of the ways to view the transformative potential of CE is to understand the nature and scale of change that any intervention seeks to address. Considering the topics covered in this work, six leverage points framed around transformative CE are discussed.

DOI: 10.1201/9781003327615-15

These are organised as (i) realising that the current equilibrium is outdated and skewed towards linearity such that impacts are additive but non-systemic, (ii) creating a new culture for enabling transformative patterns by connecting actors who share a new set of values, (iii) enabling Community of Practice (CoP) that share a common identity, (iv) connecting resources in novel ways by repurposing existing capacities, that is, extending the ontology of the CE, (v) institutionalising and supporting new networked configurations till normalisation, and (vi) impacts become systemic through synergies, trade-offs and comprise are negotiated and new patterns co-exist and are visible. The theme from each chapter is plotted against these leverage points, as shown in Figure 15.1.

The starting point is the recognition that the current dominant linear approach to steer large-scale transition is no more relevant to address the global grand challenges as discussed in Chapters 1 and 7 (Figure 15.1). This warrants the creation of new and contextualised patterns for new life cycles, which should essentially be rooted in connecting people who share a new set of values to drive the transformative pathways. Chapter 4 considers CE concerning the sustainability of food systems. The authors posit that successful transformative change involves understanding the type of societal transitions required. Hence, the cross-scale and cross-domain identification of relevant intervention strategies become a pre-requisite to determining the eligibility criteria for impact evaluation and the identification of leverage points. That is to say that each socio-technical transition must account for relevance and context-specificity.

Interestingly, Chapters 8 and 9 illustrate the latter point whereby the authors demonstrate the role of gender mainstreaming and the relevance of other intersectional inequalities in understanding norms for common access in agricultural transformation and sustainable management of water resources. A comprehensive understanding of patterns of culture and how these impact the transformative trajectories in socio-ecological interactions are exemplified in Chapter 10. Patterns of culture are essential in creating sufficient momentum to establish a consolidated CoP, the aim of which is to foster a path-dependent trajectory to reconfigure resource use efficiency towards sustainable targets. The case study on sustainable sewage sludge management in Sweden (Chapter 2) covers how a voluntary certification system, if instituted by CoPs, could counteract the undesirable impact of technological lock-in. In the early stages of the transition to CE, there might be a lag in demonstrating system optimisation of a new process that is still in its infancy, which, in effect, cannot garner sufficient support compared to other established resource transformation processes.

Thus, the strategic enactment for transitioning to CE is essential (Chapter 3), as demonstrated through the International Water Association framework whereby joined opportunities to explore interrelated water, energy and material pathways are created to derive maximum benefits from resource recovery processes. Another example of connecting resources in new ways by repurposing existing capacities is using remote-sensing data to compare the spatio-temporal variation in atmospheric pollution in South Africa before and during the COVID-19 lockdowns. Along the same line, Chapter 6 evaluates the current resource recovery technologies capable of enabling CE transitioning. It highlights the crucial role

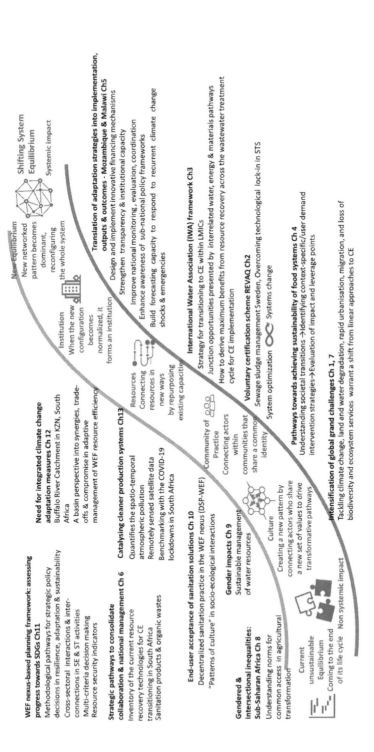

FIGURE 15.1 Understanding the contributed chapters through the scaling innovation framework. A key question in any systems innovation initiative is how to scale to a level where the whole system is influenced. Large-scale transition in CE is full of many small initiatives, but the essence is in realising where meaningful change is most needed to generate beneficial cascading effects and where trade-offs can be negotiated. Many of our current approaches use linear thinking and outdated assumptions, which is unsustainable in a world where radical CE transformation has the potential for inclusive and exponential change. There is scope to think about scaling systems to leverage the power of connectivity/feedback loops and network effects and remaining agile. Such an approach uses the evolutionary principle of systems transition to demonstrate how to get from just changing parts where the impact is not systemic to re-patterning whole systems with systemic and meaningful impacts. CE: circular economy, LMIC: low- and middle-income countries, SE: socio-ecological, STS: socio-technical system, ST: socio-technical.

of strategic pathways to consolidate multi-stakeholder collaboration and national management. This entails shaping new norms to accommodate the CE transition and its institutional settings.

When institutions are fully functional, this creates 'windows of opportunities' (Herrfahrdt-Pähle et al., 2020), a term used in the transition systems literature, for new networked patterns to become dominant and replace unsustainable trajectories with the desired CE configurations. Three chapters provide deep but disparate insights into how systemic impacts can be derived at this level. Chapter 5 covers several contexts in Sub-Saharan Africa whereby adaptation strategies could be implemented to address bottlenecks in achieving the desirable translational outputs and outcomes. Chapter 11 contributes towards the methodological pathways to steer strategic policy decisions in resilience, adaptation and sustainability. Such knowledge bases do not belong to a single ontological domain and, hence, require transdisciplinary approaches to be dynamic to accommodate the complex cross-sectoral interrelationships in socio-ecological and socio-technical activities. The authors demonstrate the importance of identifying resource security indicators supported through multi-criteria decision-making in the water-energy-food (WEF) nexus-based planning framework to assess progress towards the Sustainable Development Goals (SDGs). Lastly, the three important notions of efficient resource management in the real world, namely synergies, trade-offs, and compromise, were considered essential to driving systemic impact (Chapter 12).

15.2   OVERCOMING THE CONSTRAINTS OF THE 'SHIFTING THE BURDEN ARCHETYPE'

Chapters 1 and 7 have explicitly covered aspects whereby existing approaches to drive socio-technical and socio-ecological transitions are limited due to the obsolete dominance of linear configurations. In essence, the evolutionary pathway towards a systems-stable CE requires systemic and analytical thinking. In Figure 15.2, a "Shifting the burden" system archetype illustrates how some chapters in the book relate to these aspects and the limitations that arise for transformative CE. Systems archetypes are patterns of behaviour of a system arranged in causality loops, within which the notion that every action creates a reaction demonstrates the feedback mechanism occurrence. The idea of leverage points arises when the nature of these feedbacks is evaluated against the ripple effect generated across the interrelated loops (Wolstenholme & Wolstenholme, 2003). In the literature developed in this synthesis, mention has been made of the extrinsic systemic effects such as (i) synergies, (ii) trade-offs, and (iii) competing interests generating complex interactions across nexus domains which are beyond the scope of ontological analytical frameworks such as process design, process indicators or end-product quality parameters. Such issues extend the boundary critique of CE processes and technologies and emphasise the need to scale innovation systems that accommodate an embedded notion of an evolutionary principle. Figure 15.2 shows that fundamental processes of systems innovations should accompany symptomatic and/or short-term solutions in CE transition to improve coherence in governance mechanisms. This is supported by having a wide evidence base.

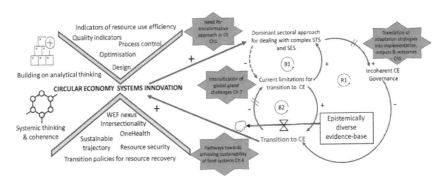

FIGURE 15.2 The evolutionary pathway towards a systems-stable CE requires both systemic and analytical thinking. A "shifting the burden" system archetype illustrates how some chapters cover these aspects and are highlighted in the causal loop diagram ![dot]. The linear transformative approach's shortcomings are consolidated within the systems innovation framework. The Balancing loop B1: the current emphasis on tackling the limitations in implementing CE indicates the dominance of sectoral approaches. However, because CE constitutes complex socio-technical and socio-ecological systems, there is an accumulation of unintended consequences, which means that issues are not addressed fundamentally but at a 'symptomatic' level. Balancing loop B2: the current limitations for the successful transition to CE necessitate an evolutionary transitions approach where diverse evidence bases are applied. However, there is a perceived delay in its implementation. Reinforcing loop R1: The accumulation of siloed governance mechanisms increases incoherence, negatively impacting the transition to CE. Adopting a systems innovation approach would counteract the unintended consequences of siloed approaches. Within SI, both analytical and systemic approaches co-exist to improve the transition mechanism. //: systems delay, ⊠: action required for system change, ⌣: system boundary, Dotted arrow: the sectoral approach can address ontological (reductionist) issues, but these are not fundamental solutions to implement a sustainable transition to CE.

15.3   RECOMMENDATIONS FOR RESEARCH-DRIVEN IMPACTS FOR CE INITIATIVES

Since we have argued that scaling CE from technological artefacts to successful transformative transition ought to follow the evolutionary principle of complex systems, we posit that evidence-based adaptive co-management is the 'engine of growth' for establishing the transition to CE. The question which then follows pertains to what type of knowledge base is required to exploit this evolutionary potential such that the contribution of the leverage points is recognised in the CE systems innovation framework.

Figure 15.3 shows the areas where transdisciplinary research can be pursued to improve the efficiency of CE initiatives. In essence, adaptive co-management needs to be established to assess the effectiveness of the value chain from performance to outcomes. Subsequently, the transition is perceived in terms of socio-ecological resilience, socio-economic viability, and achievement of global sustainable targets.

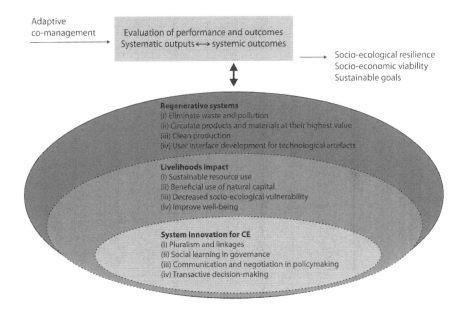

FIGURE 15.3 Knowledge areas as an engine of growth capable of harnessing translational research and practice in CE.

Within the 'engine of growth', the first outer layer is about creating the operational space for regenerative systems in multiple sectors, such as waste elimination and pollution, recycling processes aiming to recirculate materials at their highest value and user interface development for technological artefacts. An interrogation at a deeper level of analysis should be able to translate how sub-systems behave over time in terms of impacts on livelihoods in terms of sustainable resource utilisation, beneficial use of natural capital, decreased socio-ecological vulnerability and enhanced well-being. The core part refers to the system innovation's evolutionary principal catalyst. This is the system's structure and its elements, namely pluralism and linkages, social learning in governance, and negotiation in policy-making. It can be viewed as a recursive 'memory' within all other sub-systems by applying transactive rationality at all stages to promote inclusivity in the transformative outcomes.

REFERENCES

Herrfahrdt-Pähle, E., Schlüter, M., Olsson, P., Folke, C., Gelcich, S., & Pahl-Wostl, C. (2020). Sustainability Transformations: Socio-Political Shocks as Opportunities for Governance Transitions. *Global Environmental Change*, 63, 102097. https://doi.org/10.1016/J. GLOENVCHA.2020.102097.

Siegenfeld, A. F., & Bar-Yam, Y. (2020). An Introduction to Complex Systems Science and Its Applications. *Complexity*, 2020, 1–16. https://doi.org/10.1155/2020/6105872.

Wolstenholme, E. F., & Wolstenholme, E. (2003). Towards the Definition and Use of a Core Set of Archetypal Structures in System Dynamics. *System Dynamics Review*, 19(1), 7–26. https://doi.org/10.1002/SDR.259.

Index

Printed and bound by CPI Group (UK) Ltd, Croydon, CR0 4YY

17/10/2024

01775656-0013